Drinking Water Disinfection By-products

Sughosh Madhav • Mohd Aamir Mazhar •
Sirajuddin Ahmed • Pramod Kumar •
Pradeep Kumar Mishra
Editors

Drinking Water Disinfection By-products

Sources, Fate and Remediation

Springer

Editors
Sughosh Madhav
Department of Civil Engineering
Jamia Millia Islamia
New Delhi, India

Mohd Aamir Mazhar
Department of Civil Engineering
Jamia Millia Islamia
New Delhi, India

Sirajuddin Ahmed
Department of Environmental Science
Jamia Millia Islamia
New Delhi, India

Pramod Kumar
Department of Chemistry
University of Delhi
New Delhi, India

Pradeep Kumar Mishra
Department of Chemical Engineering
Indian Institute of Technology (BHU)
Varanasi, India

ISBN 978-3-031-49046-0 ISBN 978-3-031-49047-7 (eBook)
https://doi.org/10.1007/978-3-031-49047-7

This Springer imprint is published by the registered company Springer Nature Switzerland AG
The registered company address is: Gewerbestrasse 11, 6330 Cham, Switzerland

Paper in this product is recyclable

Preface

Water purification means purifying water to make it fit for drinking purposes. Water treatment improves the quality of water by passing it through various processes. After treatment, a disinfectant (for example, chlorine, chloramines, ozonation) may be added to kill any remaining microorganisms and protect the water from pathogens when it is piped to the end users. But during the disinfection process, many disinfectant by-products (DBPs) are formed unintentionally in the drinking water when chlorine or other disinfectants react with natural organic matter (NOM) present in water. These include trihalomethanes (THMs), haloacetic acids (HAAs), and other emerging by-products. Today, the occurrence of DBPs is a significant issue in the drinking water community since some DBPs are considered potentially carcinogenic, and toxicological studies suggest that some DBPs may cause adverse reproductive and developmental outcomes. Thus, there is a dreadful need to identify and quantify the DBPs formed in the drinking water so that suitable treatment strategies can be chalked out to save public health from the consequences related to these DBPs.

In this book, different disinfection processes (Chlorination, ozonation, UV radiation, and thermal treatment) and the formation of DBPs are well explained. The combined or integrated disinfection processes are also be part of this book. This book is focused on different types of disinfection by-products, including halogenated, non-halogenated, organic and inorganic DBPs in drinking water. The health issues related to DBPs are also discussed in this book. The analytical measurement techniques drinking water are also part of the book. Recent trends for controlling the formation of DBPs are also included in this book. The remedial measure of DBPs from drinking water, including nanotechnology, will also be another focused theme of this current book. Case studies provided a clearer view and practical aspects of the significance and management of DBPs. Thus the current book contains fundamental knowledge of DBPs, their precursors, significance and remediation measures.

The chapters in the book address the impending problems with a certain degree of solutions to manage disinfection by-products in the drinking water. This book contains both practical and theoretical latest and broad aspects of drinking water disinfection. An emphasis has been made on the recent research on the environmental impact of disinfectant by-products. This book will be helpful for undergraduate and university students, teachers, researchers, and scientists especially those working in areas of water treatment related studies.

We tried a humble attempt to reflect upon the various aspects of disinfection by-products, hoping that it would be a significant addition to the already available literature. The contributors to the book having different backgrounds provide a holistic approach to the topic imbibing diverse practices and perspectives. We express our sincere gratitude to all the contributors and publishers for producing a remarkable and meaningful edited volume on an important issue.

New Delhi, India Sughosh Madhav
Mohd Aamir Mazhar
Sirajuddin Ahmed
Pramod Kumar
Varanasi, India Pradeep Kumar Mishra

Contents

Water Disinfection: Background, Current Trends, Challenges, and Gaps 1
Sadiqa Abbas, Jyoti Chawla, Shagufta Jabin, and Rajeev Kumar

Drinking Water Chlorination and Disinfection by-Products: Formation, History, and Regulations 21
Mohd. Aamir Mazhar, Sughosh Madhav, Sirajuddin Ahmed, Pramod Kumar, and Christian Springer

Recent Advancements in Chlorine Applications for Water Quality Control 35
Abdul Gani, Mohit Singh, Shray Pathak, and Athar Hussain

A Comparative Study of Different Disinfection Processes, and Management Practices to Control the Formation of Disinfection by-Products (DBPs) 59
Namrata Malik and Rahul Kumar

Various Disinfection Processes and Formation of Disinfection by-Products in Drinking Water 85
Rafael Audino Zambelli and Ítalo Waldimiro Lima de França

Disinfection by-Products (DBPs) and their Toxicological Risk on Human Wellbeing: A Public Health Concern 109
Dharmeswar Barhoi, Sweety Nath Barbhuiya, Arpana Rabha, and Sarbani Giri

Route of Exposure, Mechanism of Action, and Toxicological and Human Health Impacts of Disinfection By-Products 135
Astha Pathak and Avinash Pratap Gupta

Detection and Measurement of Disinfection By-Products in Drinking Water 161
Vivek Dalvi, Afzal Ansari, Shweta Kalia, Rubia Gaur, Beni Lew, and Abid Ali Khan

Recent Trends in Controlling the Disinfection By-Products Before their Formation in Drinking Water: A Review 177
Mehreen Shah, Mohd. Aamir Mazhar, Sirajuddin Ahmed, Beni Lew, and Nadeem Khalil

Removal of Disinfection By-Products by Physico-Chemical Treatment Methods 193
Deeksha Ranjan

Removal of Disinfection By-Products from Drinking Water: Influencing Factors and Physico-Chemical Treatment Processes 219
Surya Singh and Sarvesh Rai

Availability and Minimization of Nitrogenous Disinfectant By-Products in Drinking Water 239
Richa Saxena, Himanshu Gupta, and Soniya Dhiman

GIS-Based Trends Analysis of THMs Compounds in Indian Drinking Water Supplies 263
Jaydev Kumar Mahato, Shivam Saw, B. D. Yadav, and Sunil Kumar Gupta

Recent Development in Nanotechnology for the Removal of Disinfection By-Products 273
Bhawana Raj, Rachna, and Umesh Kumar Singh

Water Disinfection: Background, Current Trends, Challenges, and Gaps

Sadiqa Abbas, Jyoti Chawla, Shagufta Jabin, and Rajeev Kumar

Abstract World's population is growing rapidly and it will touch 10 million by the end of 2050. Availability of clean and safe water for all is a global challenge. Disinfection of water is the most important step in water treatment to ensure the inactivation of pathogens from water. There are many techniques/materials available for disinfection of water. However, there are also many challenges associated with successful implementation of methods of disinfection including cost, type of pathogens, and other health implications caused due to formation of disinfection byproducts during the disinfection process. In this chapter, the background of water disinfection techniques as well as recent trends of the area has been discussed in detail with an objective to suggest safer strategies for water disinfection. Disinfection by-products often lead to health implications due to their long-lasting and toxic behavior. There are many other factors including pH, residence period, temperature, natural organic matter, and other water characteristics that significantly affect the efficiency of disinfectants and fate of disinfection by-products. Some by-products are carcinogenic, reproductive toxicants as well as mutagenic in nature. However, with advancements in disinfection techniques formation of toxic by-products can be avoided during water treatment. These methods have been discussed in the chapter along with the challenges and knowledge gaps to provide a basic outlook for future research in the development of safer disinfection alternatives.

Keywords Disinfection · Impurities · Pathogens · Water contamination · Water quality

S. Abbas (✉)
Department of Civil Engineering, School of Engineering and Technology, Manav Rachna International Institute of Research and Studies, Faridabad, India
e-mail: sadiqa.fet@mriu.edu.in

J. Chawla · S. Jabin · R. Kumar
Department of Applied Sciences (Chemistry), School of Engineering and Technology, Manav Rachna International Institute of Research and Studies, Faridabad, India

S. Madhav et al. (eds.), *Drinking Water Disinfection By-products*,
https://doi.org/10.1007/978-3-031-49047-7_1

1 Introduction

The well-being of mankind mainly depends upon ecological and environmental factors. Ecological and environmental factors depend upon the quality of water available for various purposes in life. Despite the great effort, human being in the present era continues to face challenges with respect to water pollution. In recent decades, water contamination has become a universal concern as the call for better quality water has increased many times due to industrial pollution and overpopulation (Calderon 2000; Richardson and Postigo 2012; Saini et al. 2019). The kinds of contaminants present in water are a decisive factor in the selection of treatment schemes. Pathogens present in water are in suspended form. Pathogens as well as life-threatening chains get to pollute the groundwater system through leaching. Drinking water is mainly contaminated by suspension, pathogens, and physical changes such as temperature and turbidity.

Water contaminants can be physical that may change the color/taste/odor/temperature, or increase turbidity, biological may include various pathogens and chemicals including toxic heavy metals/chlorides/nitrates/fluorides/dyes (Richardson et al. 2007). Water treatment has become essential since the last decade due to water quality degradation (Jabin et al. 2021; Kapoor et al. 2015; Crider et al. 2023; Jabin et al. 2023). Apart from remediation of chemical pollutants, biological pollutants are also of great concern due to health implications mostly in developing countries (Plewa and Wagner 2011; Costet et al. 2011). Fecal materials are a significant provider of biological pollutants in water which causes diseases in living organisms. These diseases are caused due to the presence of pathogens such as, *Shigella*, *Ascaris lumbricoides*, *Vibrio cholera*, *Cryptosporidium parvum*, *Salmonella typhi*, *Clostridium perfringens*, *Schistosoma*, *Escherichia coli*, and Intestinal enterococci (Costet et al. 2011) in water containing fecal materials. Microbial organisms present in water should be removed through disinfection before their final intake so that severe diseases can be avoided (King et al. 2000; Bove et al. 2007; Cahoon 2019).

Disinfection of drinking water is generally done by chemical process or ultraviolet radiation or by combination of both (Rook 1974). Chlorine/bromine-based chemicals and ozone (O_3) are generally exploited for the disinfection of water. Among all, chlorine is mostly preferred (Hu et al. 2018; Hrudey 2009; Lindmark et al. 2022) because of its better efficiency and low cost (Hong et al. 2013; Yang et al. 2014). However, low dosages of chlorine cannot kill a few bacterial species such as *Cryptosporidium* and *Giardia* (Park et al. 2009; Shinde et al. 2021). Few chemicals released or remain unused during disinfection methods are called disinfection by-products (DBPs) (Kumar et al. 2023). They are mostly carcinogenic in nature. But their chemical behavior mostly depends upon the kinds of chemicals used during disinfection (Richardson and Postigo 2012). DBPs are generally present due to a reaction between free chlorine with the organic matter present in water (Richardson and Postigo 2012; Mazhar et al. 2020). In addition, it was observed that DBPs are present in water for longer periods as chlorinated compounds (Muellner et al. 2007). Here, significant methods of reducing biological pollutants are discussed along with their limitations of producing by-products in treated water. The rough estimation numbers of DBPs are found to be around the range of 600–700 in aqueous medium (Richardson and Postigo 2012) but toxic effects of only 80 DBPs could be reported to date in the literatures (Muellner et al. 2007).

Further, pH, residence period, and temperature also significantly affect the performance of disinfectants (Ghernaout, Elboughdiri et al. 2019; Kristiana et al. 2017; Ghernaout, Alghamdi et al. 2019). Chemicals used as disinfectants combine with natural organic matter in treated water and form DBPs, which form haloacetic acids (HAAs), trihalomethanes (THMs), and iodo and fluoro compounds. Among these DBPs, HAAs and THMs are the most toxic in nature. As per literature, few DBPs are carcinogenic, reproductive toxicants, and mutagenic in nature and cause serious human health hazards (Kristiana et al. 2017). They are fatal when inhaled through air and when it is present in water, it goes inside the human body through dermal contact. In potable water, level of DBPs present in treated water depends on the dosage of chemicals added in raw water, physicochemical features of water, the temperature conditions, treatment plant running cases, and the contact period of the water in the distribution system (Marais et al. 2018; Ghernaout and Ibn-Elkhattab, 2019). DBPs are mostly formed due to chlorination (Ghernaout 2014). However, other disinfectants such as chloramines and ozone do not form so toxic byproducts. Further, Ozone-based disinfectants produce bromate which is injurious to human health. It has been reported that chloramine decreases the generation of DBPs and other chemicals such as 2-methyl isoborneol which causes toxicity in treated water (Ghernaout 2017a; Kellali and Ghernaout 2019; Eslami et al. 2019). However, chloramine has the capacity to develop biofilms because it permeates biofilms more efficiently as compared to other disinfectants. That is why treatment technology like coagulation, advanced oxidation process, ion exchange, and membrane processes are far better and can remove dissolved impurities more efficiently. Generally, all microbes are present in water in the form of dissolved impurities (Ghernaout 2017b; Carter et al. 2019). Chaukura et al. did a comprehensive study to analyze the impact of organic matter on the production of disinfection by-products.

In another study, haloaldehyde, iodo-THMs, and haloacetonitriles have also been found in significant amounts in water after treatment (Bond et al. 2012). Aldehydes, haloacetamides, cyanides, and halomethane come under the latest category of disinfection by-products (Carter et al. 2019). Limited understanding of such DBP is there till date (Ghernaout and Elboughdiri 2020). Even anthropogenic pollutants such as iodide and bromide compounds also interact with chloramine, chlorine, ozone, and chlorine dioxide to generate DBPs (Ghernaout and Naceur 2011). Organic pollutants contaminants also enter water through different manufacturing industries and wastewaters. These pollutants combine with disinfectants employed in water treatment and form DBPs (Ghernaout et al. 2011; Ghernaout 2017c). Organic contaminants that generate DBPs through interaction with disinfectants come from discarded personal care products/pharmaceuticals, pesticides, ethoxylate surfactants, antibacterial agents, etc. (Richardson et al. 2007). The iodide and bromide ions act as the inorganic precursors of DBPs.

In this chapter, the historical background of water disinfection has been discussed followed by analysis of disinfectants. Assessment of disinfection by-products has been done to propose safer strategies for water disinfection. Challenges and knowledge gaps in the area have also been highlighted to provide a basic outlook for future research in the area.

2 Historical Background of Water Disinfection

Disinfection of water is an important exercise that has been followed since ages for the removal of pathogens from drinking water. It is an important step for water treatment as it prevents serious waterborne diseases that may outbreak because of presence of harmful microbes in drinking water. Disinfection has been proposed by different methods including exposing the water to sunlight, boiling, using silver, copper, chlorine, peroxide, bromine, ozone, and ultraviolet radiation. Each method has some advantages or limitations. Various water disinfection options have been depicted in Fig. 1. Boiling is considered the most ancient approach used for water disinfection. However, wine was also used as a disinfectant by Romans. However, the efficiency of different disinfectants/methods was not studied before the discovery of microscope and identification of bacteria/ microorganisms in the seventeenth century. It further took two more centuries to recognize the activities of harmful microorganisms. This is when disinfectants were formally introduced and their effectiveness has been studied by different researchers. In 1902, calcium hypochlorite Ca $(ClO)_2$ and ferric chloride $(FeCl_3)$ were used for water disinfection in Belgium. In 1906, ozone (O_3) was first utilized in France for disinfection. In 1904 and 1908 chlorine was used in New Jersey and London respectively as a disinfectant. Chlorine (a yellow-green gas) dissolves in water to form HOCl hypochlorous acid that gives hypochlorite ion OCl^-. Inactivation of microorganisms depends on the time (t) of exposure of microorganism to specific disinfectants which was explained using Chick's Law. It is clear from the equation given below that it is a first-order reaction:

$$dn / dt = kN$$

dn/dt refers to the time for inactivation of N number of microorganisms, k constant depends on nature of disinfectant, microorganism as well as water characteristics including pH, temperature, other impurities, etc.), and number of active microorganisms after treatment. Disinfection efficiency can be evaluated through changes in concentrations of microorganisms over time. Ultrasound rays with a frequency 20–40 kHz ultrasound have also been proposed for water disinfection but require high energy, although it can be applied with other techniques like chlorination to improve the performance. Carbon- and metal-based nanomaterials have also been proposed for disinfection and some of the common examples are carbon nanotubes (CNTs), fullerol, Silver/Copper/Zinc/Nickel/Cobalt/Zinc oxide nanoparticles

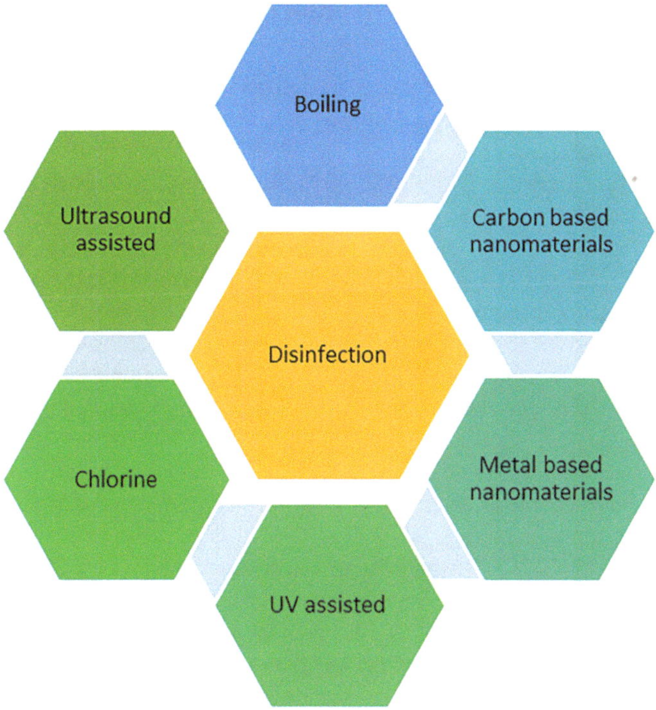

Fig. 1 Methods/materials for disinfection

(Song et al. 2020). Arias and Yang (2009) explored the antimicrobial activities of different types of CNTs with various functional groups. Kang et al. 2007, studied the antibacterial activity of single-walled carbon nanotube to study the antimicrobial activity (on bacterial cells) using *E. coli* as a target organism. Liu et al. 2007a observed that modified copper-coated carbon nanotube has good bactericidal agents for *E. coli* and *Staphylococcus aureus* bacteria. Liu et al. 2007b also observed that silver-coated carbon nanotubes have a better bactericidal rate in comparison to silver-coated pyrolytic carbon. Akhil et al. 2016 confirmed the antibacterial properties of zinc oxide nanoparticles. Disinfectants like chlorine, ozone, and chloramine can effectively control harmful microorganisms but the formation of disinfection by-products is still under question due to the toxic behavior of by-products formed due to certain transformations (Krasner et al. 2006). In some cases, the dose of disinfectant used is very high and that leads to high concentration of by-products. In today's scenario though a number of options are available for disinfection of water but still there is a challenge to find the safer and economical approach for disinfection of water.

3 Advancements in Water Disinfection

Drinking water purification procedures come in a variety of forms, from relatively simple (boiling water) to highly advanced (membrane filtration systems), and from widely utilized (chlorination) techniques to experimental size methods (sonication) (Cahoon 2019), adsorption (Kaur et al. 2022; Nayyar et al. 2022). Treatment facilities disinfect water using primary and secondary methods. In order to keep microorganisms from regrowing as water travels through the distribution system, secondary disinfection keeps disinfectants residual in the finished drinking water. Primary disinfection accomplishes the desired level of microbial death or deactivation (Mala-Jetmarova et al. 2017; Chaudhari et al. 2017). Chlorine was previously the preferred disinfectant, but more recently, water has been purified using chemicals like ozone, chlorine dioxide, and chloramines (Bhatt and Paneria 2017). The efficacious secondary disinfectants are Chlorine and Chloramines, on addition to water Chlorine produces nascent oxygen, which eliminates bacteria in a quick, inexpensive, and reliable manner. When chlorine gas is dissolved in water, it quickly forms hypochlorous acid (HOCl), which then dissociates into hypochlorite ion (OCl) (Datturi et al. 2015; Onyango et al. 2010; Bhatt and Paneria 2017).

People are susceptible to contracting infectious diseases if their drinking water is not disinfected (Ngwenya et al. 2012). Selection of an appropriate disinfection method is crucial not only from the perspective of removal of microbial population but also from the techno-economic and sustainability aspects. Developed and developing nations have been using a variety of methods for water disinfection. About 69% of the treatment facilities employ chlorine gas, 15% make use of sodium hypochlorite, and 14% do so with calcium hypochlorite. One of the most widely utilized disinfectants in major water treatment facilities is chlorination, but occasionally, chlorine-based disinfectants might not be enough to keep the water's microbiological purity in the distribution system. As a result, chloramines are also employed as a secondary disinfectant. (Genthe and Kfir 1996). The usage of O_3 in the water treatment process is continuously increasing according to a survey on the subject (Rajagopaul et al. 2008). This is true despite the high operational expenses. Owing to its brief half-life, chlorine was almost universally favored over ozone as the last disinfectant. According to Momba et al. (1998), the South African potable water industry has seen a rise in popularity for the disinfection technique known as ultraviolet (UV) irradiation. The capacity of UV disinfection to deactivate *Cryptosporidium* oocysts (typically resistant to conventional chemical disinfectants) is its most significant advantage (Bukhari et al. 1999).

USEPA (2007) has reported that chlorine dioxide, chloramines, ozone, and UV radiation are the most widely utilized disinfectants/oxidants in the United States. An increase of 9% in usage of chlorine as a disinfectant has been reported for the duration between 1998 and 2007. The usage of chlorine has been linked to the occurrence of unfavorable DBPs in the distribution system. Maximum Contamination Levels for total trihalomethanes (TTHMs) is 80 mg/L and haloacetic acids is 60 mg/L as per USEPA guidelines (2006a).

It has been reported that sodium hypochlorite (NaOCl) is the most common disinfectant used in 78% of water treatment plants in Canada, while chlorine gas was used by 19%, calcium hypochlorite by 1.4%, and alternative disinfectants were used by less than 0.5% of facilities (Health Canada, 2009). Chlorine is preferred in Australia because of its good performance against a variety of infectious microbes and low cost. It can also safeguard water inside of the pipe system (Hunterwater, 2009). Chloramines, ozone, and UV radiation are other frequently employed chemical disinfection methods in Australia.

In Europe, Chlorine is the primary disinfectant that is most frequently used, followed by ozone and chlorine dioxide (Smeets et al. 2006). UK is one of the only European nations that employ chloramines to prevent the creation of disinfection by-products and to reduce residual infection in the distribution network. Chloramines are also used occasionally for disinfection in Finland, Spain, and Sweden. As a major oxidant and disinfectant, France primarily uses ozone, whereas Italy and Germany either employ ozone (O_3) or chlorine dioxide (ClO_2) (Lenntech 2011; European Commission 2008).

TiO_2 is a popular photocatalyst for water disinfection and decontamination. In a process known as heterogeneous photocatalysis, highly reactive species are produced when the system is exposed to UV light. These species are capable of transforming contaminants. It is typically used as a powder suspended in contaminated water. The use of particle TiO_2, however, necessitates the addition of further purification steps, such as separation and catalyst recycling. By fixing TiO_2 to the proper supports, these stages can be avoided (Litter et al. 2020). Low cost, low toxicity, and good chemical and biological stability are all characteristics of TiO_2 photocatalyst. However, only UV light can activate TiO_2, and only around 5% of solar radiation is UV light (Zhang et al. 2003). It also has a significant band gap, 3.02 eV for the rutile form and 3.18 eV for the anatase form. Additionally, there are significant problems with charge carrier recombination and delayed migration (Sharma et al. 2020; Dutta et al. 2019).

Bismuth-based photocatalysts have replaced photocatalysts like ZnO, TiO_2, Fe_2O_3, and CdS that have a high band gap and absorb in the UV area as a more effective method of using solar energy (Anku et al. 2018). This is because most Bi-based photocatalysts demonstrated visible light activation because of their comparably narrower band gaps and solar light contains around 45% of visible light. Bi-based photocatalysts are often inexpensive, simple to make, non-toxic, and extremely stable (Meng and Zhang 2016). Additionally, photocatalysts have limitations such as charge carrier recombination, smaller surface area, and limited light absorption efficiency (Pankaj et al. 2019).

Many different methods, including chlorination, ozonation, UV radiation, and others, are used to disinfect water, however, the majority of them are either ineffective or result in additional hazardous compounds. Of late, photocatalysis has emerged as one of the most effective methods of water disinfection. This process utilizes solar light for destruction of pollutants or microorganisms present (Kumar et al. 2021). When sufficient frequency light ($h_{v>=}E_g$; Eg = band gap) strikes a semiconductor photocatalyst in this process, electrons excite from the photocatalyst's valence band (VB) to the conduction band (CB), leaving holes behind. As a result of their interactions with H_2O and O_2, these photogenerated electrons and holes produce a variety of reactive species, such as O_2 OH, and H_2O_2. The contaminants in the water are degraded by these highly reactive species (Kumar et al. 2020).

Numerous nanomaterials, such as nano-Ag, nano-ZnO, nano-TiO_2, nano-Ce_2O_4, and fullerenes, show potent antibacterial activities (Li et al. 2008 and Klaine et al. 2008). These nanomaterials render microorganisms inactive by releasing harmful metal ions (such as Ag^+ and Zn^{2+}), impairing the integrity of cell membranes upon direct contact (such as CNTs, nC60, and nano-Ce_2O_4), or producing reactive oxygen species (ROS) with a lower propensity to form DBPs (such as nano-TiO_2, fullerene, and aminofullerene).

These nanomaterials render microorganisms inert by releasing harmful metal ions (such as Ag and Zn_2), impairing the integrity of cell membranes upon direct contact (such as CNTs, nC60, and nano-Ce_2O_4), or producing (Lee et al. 2011).

NASA researchers developed acoustic nanotube technology to clean water for render it useful in the international space station. This invention can also be applied to cleanse potable, pharmaceutical-grade water on earth. In this method, water is guided through tiny nanotubes using acoustics rather than pressure. An acoustically powered molecular screen combined with carbon nanotubes makes up this technology. Only water molecules can flow through these carbon nanotubes, which inhibit bigger molecules and pollutants. Instead of eliminating impurities from water, this technique drives water away from them. In addition, compared to other filtration methods, it uses less power. Additionally, the filter system does not need to be flushed throughout this process.

Cavitation is a rapid physical process that results from a sudden reduction in pressure. Cavities are created when the liquid medium is disturbed at one or more locations, and the flow structure has a significant impact on how they are shaped. When the unstable vapor structures get close to an area of high pressure, they frequently suddenly collapse. The collapse may result in strong shear flows, jets, high local temperatures, shock waves, quick depressurization, and supersonic flow (Shamsborhan et al. 2010; Kosel et al. 2017). Acoustic cavitation and hydrodynamic cavitation are the two main kinds of cavitation. The difference between the two is caused by a process that causes the local pressure to fall; otherwise, the mechanisms that control the hydrodynamic and the acoustic bubbles are almost the same. In Acoustic cavitation, acoustic wave propagation reaches the necessary low pressures to disturb the liquid and create cavitation whereas the liquid's current speed in hydrodynamic cavitation will result in a lowered local pressure below the liquid's temperature of saturation, which will cause cavitation to occur (Zupanc et al. 2019).

Few research have examined the effects of cavitation on aquatic viruses, therefore the specific process by which cavitation inactivates viruses is still not entirely understood (Filipić et al. 2022). Heat and high pressure were known to inactivate viruses, and OH was associated with damage to viral capsid proteins (Chen et al. 2021). Su et al. (2010) hypothesized that cavitation-induced damage to the exterior protein capsid or the recognition sites located on the capsid's exterior could be the cause of virus inactivation. According to Kosel et al. (2017), virus inactivation may be caused by the modification of the viral capsid or genome by OH generated during cavitation mixed with mechanical impacts.

Hydrodynamic cavitation costs more than chlorination and ozonation even though it is a less expensive technique of disinfection than Acoustic cavitation (Holkar et al. 2019). Additionally, it needs a constant energy source, and the amount of water that can be treated using this method is limited (Dular et al. 2016). Hydrodynamic cavitation disinfection technology is now only being used in laboratories due to high operational costs (Pichel et al. 2019); large-scale applications have not yet been adopted (Sun et al. 2020).

Another method employed for disinfection is the electrochemical method. The capacity of disinfectants to oxidize in the electrode layer or electrode bulk is the main tenet of the electrochemical disinfection approach. Direct anode oxidization and indirect product oxidation are the two categories into which this disinfection approach, which is characterized by the formation of intermediates, is divided. Direct anode oxidation relies on the transfer of electrons between the electrode and the target material in order to inactivate it without the use of toxic substances or other organic compounds. Contrarily, concentrated saline solution is required for anode oxidation, and as a result, the accumulated molecules and ions (such as H_2O and Cl) at the electrode are oxidized to produce chlorine- and oxygen-active substances (such as oscillation of [O(3P)], H_2O_2, and O_3). Intermediate products are vital in moving the electrons from the target chemical to the electrode's surface in order to oxidize and remove it (Panizza and Cerisola 2005; Santana et al. 2005; Yang et al. 2018; Jung et al. 2020). The main cause of the inactivation of microorganisms when electrochemical disinfection is employed is damage to the intracellular enzyme system of the microbes (Long et al. 2015).

4 Challenges and Opportunities in Water Disinfection

Eliminating the pathogens or microorganisms that cause waterborne diseases is the main aim of disinfection of public water supplies. Treatments that significantly lower the overall number of living pathogens or microorganisms in the water can stop the spread of diseases like cholera, typhoid, salmonellosis, shigellosis, and paratyphoid fevers (Schoenen 2002; Gandhi and Prakash 2023). The most popular technique for disinfection of water sources is chlorination throughout the world. It has been so effective that not having any waterborne disease epidemics is now practically expected. According to the National Academy of Sciences' 1977 publication Drinking Water and Health, "chlorination is the standard method for disinfection of water (Nielsen et al. 2022). Chlorine has literally saved lives in terms of drinking water disinfection. Nevertheless knowing that chlorination can lead to the creation of disinfection byproducts trihalomethanes, chlorite, bromated, and other halogenated hydrocarbons has motivated a review of the present disinfection methodology to identify alternate agents or procedures (Sun et al. 2023; Chen et al.

2022; Dong et al. 2023; Ebsa and Dibaba 2022; Jiang et al. 2022). These by-products are mainly formed by the reaction of chlorine with organic compounds present in water. By-product creation is often a greater concern for those who are using surface water, rivers, streams and lake for daily purpose. As per the World Health Organisation, "under any circumstances, disinfection performance should not be compromised in order to meet disinfection by-products guidelines, to decrease amount of the substance. There are also difficulties in balancing several disinfection by-products (Peng et al. 2023; Wang et al. 2023). In a distribution system, disinfection by-products act differently. Some disinfection by-products may be present in low concentrations in areas where water has only recently entered the distribution system. Others may be present in low amounts in regions where water has been distributed for a longer period of time. Disinfection by-product concentrations vary with temperature, pH, and seasons in water quality (Qadafi et al. 2023; Chang et al. 2020; Ibrahim and Abu-Shanab 2013).

There are lots of challenges and opportunities in water disinfection techniques. A variety of materials are available (natural or synthetic) which can be applied as a disinfectant for drinking water. Some techniques are also available which do not have any by-products after disinfection. Monitoring systems and sensors are also easily available for monitoring microorganisms and other contaminants in water resources. These monitoring systems provide day-to-day data with high accuracy which helps to assess the disinfection method properly. However, water disinfection is a very challenging technique and lots of variables should be considered at the time of disinfection of water. The technique or materials used to disinfect water sources for human consumption should be perfectively evaluated and assessed easily. The effectiveness of the material and methods against waterborne microorganisms is the most important variable to disinfect the water source. Also, the post-treatment role of residual materials, the accuracy of the method, availability and abundance of materials, and cost of the materials play crucial roles in adopting the technology and materials for disinfection of water sources. When deciding which methods or materials to be used for disinfection of drinking water, it is very important to consider the potential role of undesirable by-products to be eliminated after disinfection (Jin et al. 2022; Ramírez et al. 2023). By-products due to disinfection of water play a key role in selecting the material and method for disinfection of water. Sometimes by-products are in high amounts and may cause various side effects in human beings by drinking this type of water. So every water industry should be very carefully assessed such data before disinfection of water and distributing drinking water. In the below section, what type of challenges and opportunities associated with disinfection of drinking water are discussed in detail.

Monitoring of emerging contamination and microorganisms in drinking water is one of the main parameters and challenging tasks to protect human health and other living things. Monitoring attempts to shield consumers from illness brought on by dangerous organisms including bacteria and viruses (Liao et al. 2023; Favere et al. 2021).

Disinfection of water is done mainly by using chlorine, chloramines, chlorine dioxide, ozone, and ultraviolet light. All these disinfectants are widely used for microorganisms (Ishaq et al. 2018). Chlorination is widely used in the United Kingdom because it is very successful in inactivating many hazardous viruses and bacteria, it is safe for people at the doses used, and it is reasonably inexpensive. It also protects the water from pathogens and other impurities after it has left the treatment works. Chlorine is sometimes coupled with ammonia to generate monochloramine, which is the recommended residual for drinking water exiting a treatment plant (Rice and Gomez-Taylor 1986). In chlorination method pathogenic bacteria and poliomyelitis viruses will be removed by maintaining a 0.5-mg/L residual of free chlorine for a half an hour contact time at pH < 8; nevertheless, there is a risk of creating taste and trihalomethanes in the mains system in the presence of total organic carbon (Gray 2014).

Chloramines are almost never used for their bactericidal effect. However, it has high persistent residual effect at high temperatures because it is more stable than free chlorine at higher temperatures. Thus, in those countries where a high degree of residual disinfectant is permitted at the consumer tap, chloramines are being used more following ozone or chlorine disinfection (Deng et al. 2023). Maintaining a rate of 0.2 mg/L for a quarter of an hour provides an effective degree of protection from pathogens by using chlorine dioxide. However, it is not encouraged, and in some countries, such as France, the use of excessive amounts of ClO_2 for disinfection may be prohibited. The ClO_2 present in water oxidizes organic matter and releases the ClO_2^- ion, which is toxic and provides unpleasant odor in water (Barbeau et al. 2005).

Maintaining a rate of 0.4 mg/L for 4 min is recommended and it provides an effective degree of protection from pathogens by using ozone. Time will be different for different types of pathogens. It is important to make sure that this form of treatment does not produce undesirable oxidation by-products. Water must be free from soluble Mn(II) prior to ozonation for ozone disinfection. If this is not the case, the water will turn pink which turns light-dark brown due to precipitates of $MnO_{2.}$ Thus it is not recommended that ozone be used as a final stage disinfectant. The water must be filtered using granular activated carbon to reduce biologically degradable dissolved organic carbon and reduce the possibility of regrowth in distribution, followed by a "belt-and-braces" chlorination (Morrison et al. 2022).

To consider the excellent transmission levels of UV light in water a dose range of 20–40 mWs/cm^2 can be advised, and the usage of medium-pressure lamps will become mandatory. UV disinfections are the only disinfectants that do not produce harmful byproducts and successfully inactivate all microorganisms, including protozoan cysts. However, the following should be taken care of before applying this method for disinfection. Unlike chemical oxidants, it is not possible to check the effectiveness of the delivered dose by measuring the residual; therefore, the equipment must be equipped with UV intensity sensors to ensure that the true radiation emitted by each lamp can be continuously checked. No residual effect of UV disinfection is observed in distribution network. Therefore, it should be supplemented with another disinfectant that has a residual effect in distribution network (Zhao et al. 2022; Deng et al. 2023; Ngwenya et al. 2012; Li and Mitch 2018; Dieter 2014).

Disinfection of drinking water is a very sensitive process. The following key points should be considered at the time of disinfection of water:

1. The amount of emerging contaminants and pathogens in water varies substantially. Water in the deep well is relatively free from waterborne microorganisms. However, surface water is contaminated with emerging contaminants and pathogens due to mixing of sewage and other polluted water. A risk assessment is required to determine the type of disinfection needed based on the type of waterborne pathogens.
2. The presence of pathogens in the source water should be properly evaluated in terms of risk assessment. Risk assessment studies help in identifying the emerging contaminants and pathogen sources within the catchment and counting the quantity of *Escherichia coli* bacteria and coliform.
3. Surface water must be appropriately cleaned and disinfected. Prior to the addition of a disinfectant, this may necessitate a number of processes, i.e., coagulation, flocculation, sedimentation, and filtration. These processes must be continuously monitored to ensure that the required amount of microorganisms and other contaminants are removed.
4. Every water business develops a site-specific disinfection strategy that specifies the treatment conditions, as well as their operation and control. All procedures must be validated to ensure that they can achieve the requisite degree of disinfection and that adequate measures are in place to validate disinfection effectiveness.
5. Turbidity can also suggest the presence of particles in water, which can obstruct disinfection operations by protecting pathogens. Before the final step of chlorination, the turbidity of the water is measured and should be less than 1 Nephelometric turbidity unit (NTU). In the United Kingdom, most of the water treatment plants reach substantially lower turbidity (0.1 NTU) being regularly targeted and accomplished.
6. Chlorination must be regularly monitored. This includes ensuring that the appropriate dose is used to achieve the desired amount of pathogen inactivation after accounting for the impact of flow rate, pH, and temperature.
7. In practice, there is a trade-off between keeping chlorine residual effective enough to preserve the water and delivering water that is aesthetically acceptable in terms of taste and aroma. People may sense a small taste or odor of chlorine, albeit their ability to detect chlorine varies and individual tastes are subjective. To remove the chlorine taste, place a jug of water in the refrigerator for a few hours.
8. *Cryptosporidium* cannot be destroyed by chlorine at the amounts employed for disinfection of water. Physical removal, i.e., filtration along with application of ultraviolet disinfection, can provide excellent *Cryptosporidium* management.
9. If there is a problem with the disinfection process or any other sign that the water supply is contaminated or unsafe to drink, public health advice may be required. Typically, such decisions are made in consultation with local health authorities. Customers may be advised to boil water before consuming or brushing their teeth as a result of this. Boiling is an efficient method of destroying dangerous microorganisms such as bacteria, viruses, and *Cryptosporidium*.

Competing objectives must be met by drinking water disinfection activities. They must guard or protect against germs, viruses, and microbes while lowering disinfection by-products to fulfill EPA or WHO guidelines. It is a difficult undertaking that demands close and continual attention in all respects.

5 Conclusion

The major problem in the water disinfection process is the formation of hazardous disinfectant by-products (DBPs). Disinfection using UV disinfections however do not produce harmful by-products and successfully inactivate all microorganisms, including protozoan cysts. However, unlike chemical oxidants, it is not possible to check the effectiveness of the delivered dose by measuring the residual; therefore, the equipment must be equipped with UV intensity sensors for proper quality analysis. UV disinfection should be supplemented with another disinfectant that has a residual effect in distribution network. Effective and safe disinfection of water can be achieved by further monitoring the concentration of disinfection by-products. Sometimes strong Oxidizing agents are used for better performance but such agents also have more probability to react with matter present in water. Another important factor is the correct dose of disinfectant that can be chosen with the help of appropriate sensing and monitoring tools to choose the correct dose for water to be treated as per the water quality characteristics. Nanomaterials have also been proposed but their risk assessment is in the budding stage. Nanomaterials can be modified to get safer options keeping in mind the toxicity of proposed nanomaterials. Moreover, integrated technologies can be preferred for better and safer results. Optimization tools may further help to optimize the dose of disinfectant as per water quality to get the desired outcome.

References

Akhil K, Jayakumar J, Gayathri G, Khan SS (2016) Effect of various capping agents on photocatalytic, antibacterial and antibiofilm activities of ZnO nanoparticles. J Photochem Photobiol B Biol 160:32–42

Anku WW, Oppong, SO and Govender, PP, 2018. Bismuth-based nanoparticles as photocatalytic materials. Bismuth-advanced applications and defects Characterization

Arias LR, Yang L (2009) Inactivation of bacterial pathogens by carbon nanotubes in suspensions. Langmuir 25(5):3003–3012

Barbeau B, Desjardins R, Mysore C, Prévost M (2005) Impacts of water quality on chlorine and chlorine dioxide efficacy in natural waters. Water Res 39(10):2024–2033

Bhatt BV, Paneria DB (2017) Modernization in water distribution system. Sarvajanik College of Engineering and Technology, India, p 296

Bond T, Goslan EH, Parsons SA, Jefferson B (2012) A critical review of trihalomethane and haloacetic acid formation from natural organic matter surrogates. Environ Technol Rev 1(1):93–113

Bove GE, Rogerson PA, Vena JE (2007) Case-control study of the effects of trihalomethanes on urinary bladder cancer risk. Arch Environ Occup Health 62(1):39–47
Bukhari Z, Hargy TM, Bolton JR, Dussert B, Clancy JL (1999) Medium-pressure UV for oocyst inactivation. J Am Water Works Ass 91(3):86–94
Cahoon LB (2019) Water purification: treatment of microbial contamination. In: Advances in water purification techniques. Elsevier, pp 385–395
Calderon RL (2000) The epidemiology of chemical contaminants of drinking water. Food Chem Toxicol 38:S13–S20
Carter RA, Liew DS, West N, Heitz A, Joll CA (2019) Simultaneous analysis of haloacetonitriles, haloacetamides and halonitromethanes in chlorinated waters by gas chromatography-mass spectrometry. Chemosphere 220:314–323
Chang CJ, Huang CP, Chen CY, Wang GS (2020) Assessing the potential effect of extreme weather on water quality and disinfection by-product formation using laboratory simulation. Water Res 170:115296
Chaudhari AG, Joshi AK, Bhosale NS, Dalavi NK, Khode PS (2017) Review study: experimental investigation by WaterGEMS software for redesign of water distribution system of Bhavani Mata ESR. Guru Gobind Singh College of Engineering and Research Center, pp 604–608
Chen H, Wang J, Zhao X, Wang Y, Huang Z, Gong T, Xian Q (2022) Occurrence of dissolved black carbon in source water and disinfection byproducts formation during chlorination. J Hazard Mater 435:129054
Chen L, Deng Y, Dong S, Wang H, Li P, Zhang H, Chu W (2021) The occurrence and control of waterborne viruses in drinking water treatment: a review. Chemosphere 281:130728
Costet N, Villanueva CM, Jaakkola JJK, Kogevinas M, Cantor KP, King WD, Lynch CF, Nieuwenhuijsen MJ, Cordier S (2011) Water disinfection by-products and bladder cancer: is there a European specificity? A pooled and meta-analysis of European case–control studies. Occup Environ Med 68(5):379–385
Crider YS, Tsuchiya M, Mukundwa M, Ray I, Pickering AJ (2023) Adoption of point-of-use chlorination for household drinking water treatment: a systematic review. Environ Health Perspect 131(1):016001
Datturi S, Van Steenbergen F, Van Beusekom M, Kebede S, Ababa A (2015) Comparing defluoridation and safe sourcing for fluorosis mitigation in the Ethiopian central rift valley. Fluoride 48(4):293–314
Deng L, Wang T, Shen J, Tan C, Hu J, Singh RP (2023) Comparison of UV/chloramine disinfection of methylamine water in the absence and presence of bromide: Halonitromethanes formation, toxicity alteration, and reaction mechanisms. J Environ Chem Eng 11(3):109749
Dieter HH (2014) Drinking-water criteria (safety, quality, and perception)
Dong F, Zhu J, Li J, Fu C, He G, Lin Q, Li C, Song S (2023) The occurrence, formation and transformation of disinfection byproducts in the water distribution system: a review. Sci Total Environ 867:161497
Dular M, Griessler-Bulc T, Gutierrez-Aguirre I, Heath E, Kosjek T, Klemenčič AK, Oder M, Petkovšek M, Rački N, Ravnikar M, Šarc A (2016) Use of hydrodynamic cavitation in (waste) water treatment. Ultrason Sonochem 29:577–588
Dutta V, Sharma S, Raizada P, Hosseini-Bandegharaei A, Gupta VK, Singh P (2019) Review on augmentation in photocatalytic activity of CoFe2O4 via heterojunction formation for photocatalysis of organic pollutants in water. J Saudi Chem Soc 23(8):1119–1136
Ebsa DG, Dibaba WT (2022) Assessment of drinking water treatment and disinfection by-products Dessalegn Geleta Ebsa a,*, Wakjira Takala Dibaba b. S. Afr J Chem Eng 41(1):85–92
Eslami H, Ehrampoush MH, Esmaeili A, Salmani MH, Ebrahimi AA, Ghaneian MT, Falahzadeh H, Fard RF (2019) Enhanced coagulation process by Fe-Mn bimetal nano-oxides in combination with inorganic polymer coagulants for improving as (V) removal from contaminated water. J Clean Prod 208:384–392
European Commission (2008) The quality of drinking water in the European Union. Synthesis report on the quality of drinking water in the member states of the European Union in the period

1999–2001 Directive 80/778/EEC, 14 April 2008, https://circabc.europa.eu/sd/a/58220131-ecc4-49f9-9ad1-bbb6e9e79578/report1999-2001.pdf. Accessed 26 Mar 2023
Favere J, Waegenaar F, Boon N, De Gusseme B (2021) Online microbial monitoring of drinking water: how do different techniques respond to contaminations in practice? Water Res 202:117387
Filipić A, Lukežič T, Bačnik K, Ravnikar M, Ješelnik M, Košir T, Petkovšek M, Zupanc M, Dular M, Aguirre IG (2022) Hydrodynamic cavitation efficiently inactivates potato virus Y in water. Ultrason Sonochem 82:105898
Gandhi J, Prakash H (2023) Photo-disinfection processes for bacterial inactivation and underlying principles for water constituents' impact: a review. Chem Eng J Adv 14:100482
Genthe B, Kfir R (1996) Studies on microbiological drinking water quality guidelines. Water Research Commission
Ghernaout D (2014) The hydrophilic/hydrophobic ratio vs. dissolved organics removal by coagulation–a review. J King Saud Univ-Sci 26(3):169–180
Ghernaout D (2017a) Water treatment chlorination: an updated mechanistic insight review. Chem Res J 2:125–138
Ghernaout D (2017b) Microorganisms' electrochemical disinfection phenomena. EC Microbiology 9:160–169
Ghernaout D (2017c) Water reuse (WR): the ultimate and vital solution for water supply issues. Int J Sustain Dev Res 3(4):36
Ghernaout D, Alghamdi A, Ghernaout B (2019) Microorganisms' killing: chemical disinfection vs. electrodisinfection. Appl Eng 3:13–19
Ghernaout D, Elboughdiri N (2020) Antibiotics resistance in water mediums: background, facts, and trends. Appl Eng 4:1–6. https://doi.org/10.4236/oalib.1106003
Ghernaout D, Elboughdiri N, Al Arni S (2019) Water reuse (WR): dares, restrictions, and trends. Appl Eng 3:159–170
Ghernaout D, Ibn-Elkhattab RO (2019) Drinking water reuse: one-step closer to overpassing the "Yuck Factor". Open Access Libr J 6(11):1
Ghernaout D, Naceur MW, Aouabed A (2011) On the dependence of chlorine by-products generated species formation of the electrode material and applied charge during electrochemical water treatment. Desalination 270:9–22. https://doi.org/10.1016/j.desal.2011.01.010
Gray NF (2014) Free and combined chlorine. In: Microbiology of waterborne diseases. Academic Press, pp 571–590
Health Canada (2009) Guidelines for Canadian drinking water quality guideline technical document chlorine. Water, Air and Climate Change Bureau, Healthy Environments and Consumer Safety Branch, Health Canada, Ottawa, Ontario (Catalogue No. H128-1/09-588E). https://sswm.info/sites/default/files/reference_attachments/HEALTH%20CANADA%202009.%20Guidelines%20for%20Canadian%20drinking%20water%20quality.pdf. Accessed 25 Apr 2023
Holkar CR, Jadhav AJ, Pinjari DV, Pandit AB (2019) Cavitationally driven transformations: a technique of process intensification. Ind Eng Chem Res 58(15):5797–5819
Hong H, Xiong Y, Ruan M, Liao F, Lin H, Liang Y (2013) Factors affecting THMs, HAAs and HNMs formation of Jin Lan reservoir water exposed to chlorine and monochloramine. Sci Total Environ 444:196–204
Hrudey SE (2009) Chlorination disinfection by-products, public health risk tradeoffs and me. Water Res 43(8):2057–2092
Hu J, Chu W, Sui M, Xu B, Gao N, Ding S (2018) Comparison of drinking water treatment processes combinations for the minimization of subsequent disinfection by-products formation during chlorination and chloramination. Chem Eng J 335:352–361
Hunterwater (2009) Chlorination of water supplies. Factsheet. http://www.hunterwater.com.au/Resources/Documents/Fact-Sheets/Water-Quality/Chlorination-of-Water-Supplies.pdf. Accessed 20 Mar 2023
Ibrahim HZ, Abu-Shanab MA (2013) Monitoring of some disinfection by-products in drinking water treatment plants of El-Beheira governorate, Egypt. Appl Water Sci 3:733–740

Ishaq MS, Afsheen Z, Khan A Khan A, (2018). Disinfection methods. Photocatalysts-applications and attributes, pp 3–19

Jabin S, Gupta P, Sharma M (2021) Polyelectrolytes as a material of value in water treatment: a review. Asian J Water Environ Pollut 18(3):109–115

Jabin S, Kapoor JK, Jadoun S, Chandna N, Chauhan NPS (2023) Synthesis and characterization of polyamine-based polyelectrolytes for wastewater treatment in the sugar industry. J Mol Struct 1275:134573

Jiang H, Kaw HY, Zhu L, Wang W (2022) Halonaphthoquinones: a group of emerging disinfection byproducts of high toxicity in drinking water. Water Res 217:118421

Jin Y, Chen Z, Chen X, Huang P, Chen X, Ding R, Liu J, Chen R (2022) The drinking water disinfection performances and mechanisms of UVA-LEDs promoted by electrolysis. J Hazard Mater 435:129099

Jung E, Shin H, Hooch Antink W, Sung YE, Hyeon T (2020) Recent advances in electrochemical oxygen reduction to H2O2: catalyst and cell design. ACS Energy Lett 5(6):1881–1892

Kang S, Pinault M, Pfefferle LD, Elimelech M (2007) Single-walled carbon nanotubes exhibit strong antimicrobial activity. Langmuir 23(17):8670–8673

Kapoor JK, Jabin S, Bhatia HS (2015) Optimization of coagulation-flocculation process for food industry waste water treatment using polyelectrolytes with inorganic coagulants. J Indian Chem Soc 92(11):1697–1703

Kaur I, Kaur N, Singh BP, Kumar R, Chawla J (2022) 4-Formylphenyl boronic acid grafted amino MCM-41 for efficient adsorption of Cu (II) ions in aqueous medium: isotherm, kinetic and optimization studies. Toxin Rev 41(2):551–563

Kellali Y, Ghernaout D (2019) Physicochemical and algal study of three dams (Algeria) and removal of microalgae by enhanced coagulation. Appl Eng 3(1):56

King WD, Marrett LD, Woolcott CG (2000) Case-control study of colon and rectal cancers and chlorination by-products in treated water. Cancer Epidemiol Biomark Prev 9(8):813–818

Klaine SJ, Alvarez PJ, Batley GE, Fernandes TF, Handy RD, Lyon DY, Mahendra S, McLaughlin MJ, Lead JR (2008) Nanomaterials in the environment: behavior, fate, bioavailability, and effects. Environ Toxicol Chem: Int J 27(9):1825–1851

Kosel J, Gutiérrez-Aguirre I, Rački N, Dreo T, Ravnikar M, Dular M (2017) Efficient inactivation of MS-2 virus in water by hydrodynamic cavitation. Water Res 124:465–471

Krasner SW, Weinberg HS, Richardson SD, Pastor SJ, Chinn R, Sclimenti MJ, Onstad GD, Thruston AD (2006) Occurrence of a new generation of disinfection byproducts. Environ Sci Technol 40(23):7175–7185

Kristiana I, Liew D, Henderson RK, Joll CA, Linge KL (2017) Formation and control of nitrogenous DBPs from Western Australian source waters: investigating the impacts of high nitrogen and bromide concentrations. J Environ Sci 58:102–115

Kumar M, Shekhar S, Kumar R, Kumar P, Govarthanan M, Chaminda T (2023) Drinking water treatment and associated toxic byproducts: concurrence and urgence. Environ Pollut:121009

Kumar R, Raizada P, Verma N, Hosseini-Bandegharaei A, Thakur VK, Van Le Q, Nguyen VH, Selvasembian R, Singh P (2021) Recent advances on water disinfection using bismuth based modified photocatalysts: strategies and challenges. J Clean Prod 297:126617

Kumar R, Sudhaik A, Raizada P, Hosseini-Bandegharaei A, Thakur VK, Saini A, Saini V, Singh P (2020) An overview on bismuth molybdate based photocatalytic systems: controlled morphology and enhancement strategies for photocatalytic water purification. J Environ Chem Eng 8(5):104291

Lee J, Hong S, Mackeyev Y, Lee C, Chung E, Wilson LJ, Kim JH, Alvarez PJ (2011) Photosensitized oxidation of emerging organic pollutants by tetrakis C60 aminofullerene-derivatized silica under visible light irradiation. Environ Sci Technol 45(24):10598–10604. https://watermanaustralia.com/acoustic-nanotube-technology/. Accessed 26 Apr 2023

Lenntech BV (2011) Water disinfection application standards (for EU). http://www.lenntech.com/processes/disinfection/regulation-eu/eu-water-disinfection-regulation.htm. Accessed 21 Mar 2023

Li Q, Mahendra S, Lyon DY, Brunet L, Liga MV, Li D, Alvarez PJ (2008) Antimicrobial nanomaterials for water disinfection and microbial control: potential applications and implications. Water Res 42(18):4591–4602
Li XF, Mitch WA (2018) Drinking water disinfection byproducts (DBPs) and human health effects: multidisciplinary challenges and opportunities. Environ Sci Technol 52(4):1681–1689
Liao C, Tian L, Wang Z, Zhu X, Han Y, Li T, Wang X (2023) Toxicity warning and online monitoring of disinfection by-products in water by electroautotrophic biocathode sensors. Biosens Bioelectron 219:114799
Lindmark M, Cherukumilli K, Crider YS, Marcenac P, Lozier M, Voth-Gaeddert L, Lantagne DS, Mihelcic JR, Zhang QM, Just C, Pickering AJ (2022) Passive in-line chlorination for drinking water disinfection: a critical review. Environ Sci Technol 56(13):9164–9181
Litter MI, Vera ML, Traid HD (2020) TiO2 coatings prepared by sol-gel and electrochemical methodologies. In: Sol-gel derived optical and photonic materials. Woodhead Publishing, pp 39–74
Liu T, Tang H, Zhao J, Li D, Li R, Sun X (2007a) A study on the bactericidal properties of Cu-coated carbon nanotubes. Front Mater Sci China 1:147–150
Liu T, Tang HQ, Cai XM, Zhao J, Li DJ, Li R, Sun XL (2007b) A study on bactericidal properties of Ag coated carbon nanotubes. Nucl Instrum Methods Phys Res, Sect B 264(2):282–286
Long Y, Ni J, Wang Z (2015) Subcellular mechanism of Escherichia coli inactivation during electrochemical disinfection with boron-doped diamond anode: a comparative study of three electrolytes. Water Res 84:198–206
Mala-Jetmarova H, Sultanova N, Savic D (2017) Lost in optimisation of water distribution systems? A literature review of system operation. Environ Model Softw 93:209–254
Marais SS, Ncube EJ, Msagati TAM, Mamba BB, Nkambule TT (2018) Comparison of natural organic matter removal by ultrafiltration, granular activated carbon filtration and full scale conventional water treatment. J Environ Chem Eng 6(5):6282–6289
Mazhar MA, Khan NA, Ahmed S, Khan AH, Hussain A, Changani F, Yousefi M, Ahmadi S, Vambol V (2020) Chlorination disinfection by-products in municipal drinking water – a review. J Clean Prod 273:123159
Meng X, Zhang Z (2016) Bismuth-based photocatalytic semiconductors: introduction, challenges and possible approaches. J Mol Catal A Chem 423:533–549
Momba MN, Cloete TE, Venter SN, Kfir R (1998) Evaluation of the impact of disinfection processes on the formation of biofilms in potable surface water distribution systems. Water Sci Technol 38(8–9):283–289
Morrison CM, Hogard S, Pearce R, Gerrity D, von Gunten U, Wert EC (2022) Ozone disinfection of waterborne pathogens and their surrogates: a critical review. Water Res 214:118206
Muellner MG, Wagner ED, McCalla K, Richardson SD, Woo YT, Plewa MJ (2007) Haloacetonitriles vs. regulated haloacetic acids: are nitrogen-containing DBPs more toxic? Environ Sci Technol 41(2):645–651
National Research Council (1977) Drinking water and health: Volume 1. National Research Council
Nayyar M, Kumar R, Chawla J (2022) Development and characterization of N-substituted derivative of 2-sulfanylacetamide on Phyllanthus emblica seed coat as novel adsorbent for remediation of As (III) from water. Water Pract Technol 17(8):1742–1757
Ngwenya N, Ncube EJ, Parsons J (2012) Recent advances in drinking water disinfection: successes and challenges. Rev Environ Contam Toxicol:111–170
Nielsen AM, Garcia LAT, Silva KJS, Sabogal-Paz LP, Hincapié MM, Montoya LJ, Galeano L, Galdos-Balzategui A, Reygadas F, Herrera C, Golden S (2022) Chlorination for low-cost household water disinfection–a critical review and status in three Latin American countries. Int J Hyg Environ Health 244:114004
Onyango MS, Masukume M, Ochieng A, Otieno F (2010) Functionalised natural zeolite and its potential for treating drinking water containing excess amount of nitrate. Water SA 36(5):655–662
Panizza M, Cerisola GJEA (2005) Application of diamond electrodes to electrochemical processes. Electrochim Acta 51(2):191–199

Pankaj R, Anita S, Singh VP, Gupta VK, Hosseini-Bandegharaei A, Rajesh K, Pardeep S (2019) Solar light assisted degradation of oxytetracycline from water using Bi2O3/Fe3O4 supported graphitic carbon nitride photocatalyst. Desalin Water Treat 148:338–350

Park SH, Piyachaturawat P, Taylor AE, Huang CH (2009) Potential N-nitrosodimethylamine (NDMA) formation from amine-based water treatment polymers in the reactions with chlorine-based oxidants and nitrosifying agents. Water Sci Technol Water Supply 9(3):279–288

Peng F, Lu Y, Dong X, Wang Y, Li H, Yang Z (2023) Advances and research needs for disinfection byproducts control strategies in swimming pools. J Hazard Mater:131533

Pichel N, Vivar M, Fuentes M (2019) The problem of drinking water access: a review of disinfection technologies with an emphasis on solar treatment methods. Chemosphere 218:1014–1030

Plewa MJ, Wagner ED (2011) Drinking water disinfection by-products: comparative mammalian cell cytotoxicity and genotoxicity. In: Nriagu JO (ed) Encyclopedia of environmental health, vol 1. Elsevier, Burlington, pp 806–812

Qadafi M, Rosmalina RT, Pitoi MM, Wulan DR (2023) Chlorination disinfection by-products in Southeast Asia: a review on potential precursor, formation, toxicity assessment, and removal technologies. Chemosphere:137817

Rajagopaul R, Mbongwa NW, Nadan C (2008) Guidelines for the selection and effective use of ozone in water treatment. Water Research Commission

Ramírez Á, Muņoz-Morales M, de la Morena A, Sánchez N, Peņuela L, Sánchez A, Llanos J (2023) Screening of technologies for limiting the occurrence of disinfection by-products in urban water systems. J Water Process Eng 53:103660

Rice RG, Gomez-Taylor M (1986) Occurrence of by-products of strong oxidants reacting with drinking water contaminants—scope of the problem. Environ Health Perspect 69:31–44

Richardson SD, Plewa MJ, Wagner ED, Schoeny R, DeMarini DM (2007) Occurrence, genotoxicity, and carcinogenicity of regulated and emerging disinfection by-products in drinking water: a review and roadmap for research. Mutat Res Rev Mutat Res 636(1–3):178–242

Richardson SD, Postigo C (2012) Drinking water disinfection by-products. *Emerging organic contaminants and human health*, pp.93–137

Rook JJ (1974) Formation of haloforms during chlorination of natural waters. J Soc Water Treat Exam 23:234–243. Find this article online

Saini S, Kumar R, Chawla J, Kaur I (2019) Adsorption of bivalent lead ions from an aqueous phase system: equilibrium, thermodynamic, kinetics, and optimization studies. Water Environ Res 91(12):1692–1704

Santana MH, De Faria LA, Boodts JF (2005) Electrochemical characterisation and oxygen evolution at a heavily boron doped diamond electrode. Electrochim Acta 50(10):2017–2027

Schoenen D (2002) Role of disinfection in suppressing the spread of pathogens with drinking water: possibilities and limitations. Water Res 36(15):3874–3888

Shamsborhan H, Coutier-Delgosha O, Caignaert G, Abdel Nour F (2010) Experimental determination of the speed of sound in cavitating flows. Exp Fluids 49(6):1359–1373

Sharma K, Raizada P, Hosseini-Bandegharaei A, Thakur P, Kumar R, Thakur VK, Nguyen VH, Pardeep S (2020) Fabrication of efficient CuO/graphitic carbon nitride based heterogeneous photo-Fenton like catalyst for degradation of 2, 4 dimethyl phenol. Process Saf Environ Prot 142:63–75

Shinde SR, Apte Dr SD, Khare Dr KC Otter Mr P (2021) Chlorination as drinking water disinfection technique and disinfection by products: a Scientometric analysis

Smeets P, Rietveld L, Hijnen W, Medema G Stenström TA (2006) Efficacy of water treatment processes-report of the microrisk pro2ect. European Union pro2ect on microbiological risk assessment. http://www.microrisk.com

Song B, Gong J, Tang W, Zeng G, Chen M, Xu P, Shen M, Ye S, Feng H, Zhou C, Yang Y (2020) Influence of multi-walled carbon nanotubes on the microbial biomass, enzyme activity, and bacterial community structure in 2, 4-dichlorophenol-contaminated sediment. Sci Total Environ 713:136645

Su X, Zivanovic S, D'Souza DH (2010) Inactivation of human enteric virus surrogates by high-intensity ultrasound. Foodborne Pathog Dis 7(9):1055–1061

Sun G, Kaw HY, Zhou M, Guo P, Zhu L, Wang W (2023) Chlorinated nucleotides and analogs as potential disinfection byproducts in drinking water. J Hazard Mater 452:131242

Sun X, Liu J, Ji L, Wang G, Zhao S, Yoon JY, Chen S (2020) A review on hydrodynamic cavitation disinfection: the current state of knowledge. Sci Total Environ 737:139606

USEPA: The effectiveness of disinfectant residuals (2007). https://www.epa.gov/sites/default/files/2021-05/documents/effectiveness_of_disinfectant_residuals_final_-_3-7-07.pdf. Accessed 25 Apr 2023

Wang X, Qian Y, Chen Y, Liu F, An D, Yang G, Dai R (2023) Application of fluorescence spectra and molecular weight analysis in the identification of algal organic matter-based disinfection by-product precursors. Sci Total Environ 882:163589

Yang S, Verdaguer-Casadevall A, Arnarson L, Silvioli L, Colic V, Frydendal R, Rossmeisl J, Chorkendorff I, Stephens IE (2018) Toward the decentralized electrochemical production of H2O2: a focus on the catalysis. ACS Catal 8(5):4064–4081

Yang Y, Komaki Y, Kimura SY, Hu HY, Wagner ED, Mariñas BJ, Plewa MJ (2014) Toxic impact of bromide and iodide on drinking water disinfected with chlorine or chloramines. Environ Sci Technol 48(20):12362–12369

Zhang L, Kanki T, Sano N, Toyoda A (2003) Development of TiO2 photocatalyst reaction for water purification. Sep Purif Technol 31(1):105–110

Zhao J, Peng J, Yin R, Fan M, Yang X, Shang C (2022) Multi-angle comparison of UV/chlorine, UV/monochloramine, and UV/chlorine dioxide processes for water treatment and reuse. Water Res 217:118414

Zupanc M, Pandur Ž, Perdih TS, Stopar D, Petkovšek M, Dular M (2019) Effects of cavitation on different microorganisms: the current understanding of the mechanisms taking place behind the phenomenon. A review and proposals for further research. Ultrason Sonochem 57:147–165

Drinking Water Chlorination and Disinfection by-Products: Formation, History, and Regulations

Mohd. Aamir Mazhar, Sughosh Madhav, Sirajuddin Ahmed, Pramod Kumar, and Christian Springer

Abstract Water is a vital element which supplies essential nutrients to the plants and animals on Earth and without water no life will exist on Earth. The raw water is conventionally treated by processes such as coagulation, flocculation, sedimentation, and filtration, followed by disinfection before supplying the treated water to the consumers. Disease-causing pathogens enter the water supply network through leakages thus contaminating water which may cause illness. During disinfection process disinfectants like chlorine and chloramine are used to kill the disease-causing microorganisms. When chlorine combines with the natural organic matter (NOM) in raw water, disinfection by-products (DBPs) are created. This led researchers to discover more DBPs formed in drinking water and to find their harmful health effects. According to a USEPA assessment, the four THMs are likely human carcinogens for chloroform, bromodichloromethane, bromoform, and dibromochloromethane, respectively. A few animals' toxicity studies have reported a strong relationship between the DBPs and carcinogenic diseases. THMs were first regulated by the United States Environmental Protection Agency (USEPA) at 100 μg/L (ppb) in drinking water in 1979. In 1998, the Stage 1 Disinfection By-Products Rule limited THMs to 80 μg/L and for the first time regulated haloaceticacids (HAAs), chlorate, chlorite, and bromate. Since then, many countries have regulated the DBPs. There are several ways to manage DBP concentrations, including eliminating the precursor NOM before adding chlorine, removing DBPs once they have formed, or utilising other disinfectants.

M. A. Mazhar (✉) · S. Madhav
Department of Civil Engineering, Jamia Millia Islamia (A Central University), New Delhi, India

S. Ahmed
Department of Environmental Science, Jamia Millia Islamia, New Delhi, India

P. Kumar
Department of Chemistry, Sri Aurobindo College, University of Delhi, New Delhi, India

C. Springer (✉)
Fachhochschule Erfurt, Erfurt, Germany
e-mail: christian.springer@fh-erfurt.de

S. Madhav et al. (eds.), *Drinking Water Disinfection By-products*,
https://doi.org/10.1007/978-3-031-49047-7_2

Keywords Chlorination · Disinfection by-products · DBPs formation · DBPs regulations

1 Introduction

Water is a vital substance that plays a very important role in the human body. It transports essential nutrients to the plants and without water no life will exist on earth. The presence of water in a human body is around 70%, without eating food a human can survive a couple of weeks but without water one can only survive for a few days. Only clean water should be used for drinking and other close to body purposes because the contaminated water can cause sickness. Disease-causing pathogens enter the water supply system through leakages and the water can get contaminated. With an increase in the world's populace, there is an increase in the demand for clean drinking water (Kanan and Karanfil 2011). The drinking water must be dealt with and tested to guarantee its security, notwithstanding its point of source. With considerable number of headways made conceivable by science and innovation, the treatment and conveyance of clean drinking water are the finest progressions of the twentieth century.

From the watershed to the appropriate supply network, there are numerous procedures and testing that happen before drinking water arrives at the taps (Liu et al. 2019). Depending on the size of the network, the type of water resources (groundwater or surface water) and thus the water quality, the types of treatment applied by a given municipal water supply differ (Ashley et al. 2020; Cai et al. 2019; Taylor et al. 2019; Zhang et al. 2019). Before drinking water treatment came into existence around the globe, a large number of individuals died of cholera, typhoid fever, diarrhoea, and hepatitis A (WHO 2018). In the greater part of the nations in the world, protective barriers (pressure piping systems, water protection areas, water treatment by filtration e.g.) and particularly disinfection (chlorination, UV disinfection) of drinking water have served to get rid of these infections (Calderon 2000; D Nikolaou 2005).

Water-borne diseases are prevented by chlorine disinfection, but unidentified disinfection by-products (DBPs) are formed when chlorine reacts with natural organic matter (NOM) present in the drinking water. The International Agency for Research on Cancer (IARC) does not classify chlorinated water as carcinogenic. However, it does classify two DBPs—chloroform and bromochloromethane—as Group 2B carcinogens (meaning they are "possibly carcinogenic to humans") (Zeng et al. 2014). DBPs formed have a chronic effect in long term whereas the pathogens present in water have an acute effect on human beings. Only a few classes of DBPs like Trihalomethanes (THMs) and Haloacetic Acids (HAAs) are being regulated out of the discovered DBPs. For the first time in 1979, the United States Environmental Protection Agency (USEPA) regulated THMs in drinking water at 100 μg/L (ppb). Since then, many countries have regulated the disinfection by-products, for instance, in Germany the limit value for drinking water is 0.05 mg/L in the distribution network and 0.01 mg/L at the outlet of the waterworks (Roeske 2007).

However, in India DBPs were not regulated until 2012. In 2012, the second revision of India Standard Drinking Water Specification (IS:10500, 2012) regulated the four THMs for the first time. DBPs' regular ingestion in modest amounts may have a negative impact on human health, with a large portion of the emphasis in the past being on their cancer-causing nature (Mills et al. 1998). Increased risk of cancer has been regularly shown by the epidemiological studies.

There are several ways to manage DBP concentrations, including eliminating the precursor NOM before adding chlorine, removing DBPs once they have formed, or utilising other disinfectants (Singer 1994). For example, chlorine dioxide is an option, in which virtually no halogenation occurs. Reverse osmosis, coagulation, ion exchange, co-precipitation, herbal filtration, catalytic reduction, adsorption, and electrodialysis are just a few of the processes that can be used to remove DBPs. A relationship between pathogens and DBPs is shown in Fig. 1. The WHO stated that the death risk from pathogens is 100–1000 times greater than the cancer risk from DBPs and the risk of illness from pathogens is 10,000–1 million times greater than the cancer risk from DBPs (Amy et al. 2004).

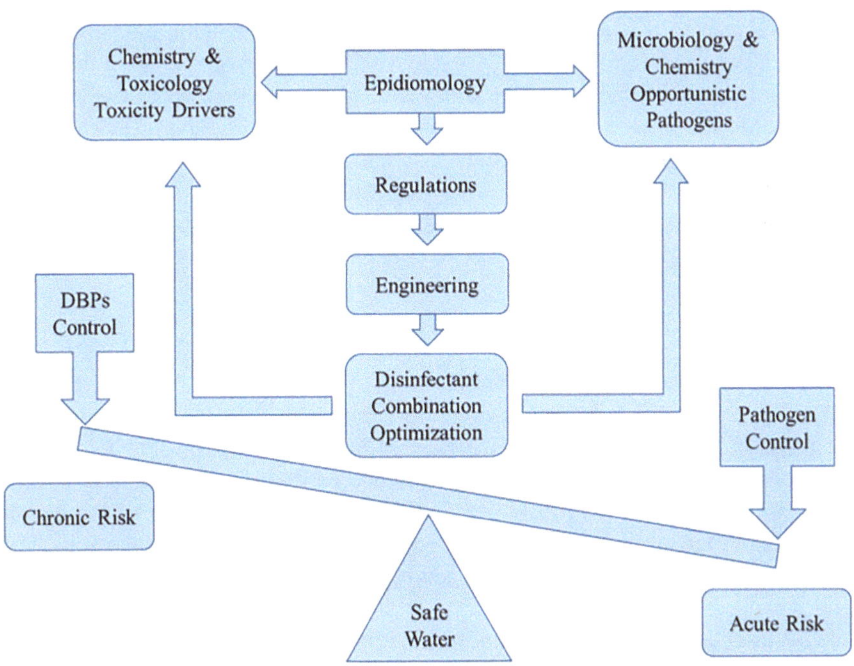

Fig. 1 Relationship between pathogens and DBPs for safe drinking water

1.1 Chlorine: The Disinfectant of Choice

Chlorine was first used in the early 1900s as a disinfectant in drinking water supplies and is the most widely recognised technique for disinfection which decimates the pathogenic life forms around the globe (Li et al. 2019; Wu et al. 2019). Chorine is one of the most efficient and cost-effective disinfectants for eliminating microorganisms and, additionally, ensuring the lingering fixation for assured microorganism development in the water supply network. Waterborne illnesses like cholera, typhoid fever, and dysentery substantially decreased because of chlorine treatment, insuring human health. The residual Chlorine present in the water supplied prevents microbial regrowth in the water distribution network until the consumers tap. Chlorine may be used in several forms such as gas (elemental chlorine), liquid (sodium hypochlorite solution), and solid (calcium hypochlorite). Two chemical species are formed together known as "free chlorine" when chlorine reacts with water, the chemical reaction takes place as follows:

$$Cl_2 + H_2O \rightleftharpoons HOCl + H^+ + Cl^- \quad (1)$$

$$HOCl \rightleftharpoons H^+ + OCl^- \quad (2)$$

These two compounds behave in an unexpected way as nature of hypochlorous acid (HOCl) and hypochlorite ion (OCl^-) is electrically neutral and electrically negative, respectively. As a disinfectant and oxidant, HOCl is stronger and more reactive than the OCl^-. We can determine the HOCL-to-OCl^- ratio based on the pH of the water. Hypochlorous acid is more reactive at higher acidity, i.e. lower pH values, whereas hypochlorite ion is more reactive at higher pH. Therefore, pH plays a key role in the speed and viability of chlorine disinfection against pathogens in water treatment. The end users are at incredible risk of getting waterborne diseases if chlorination or other purification processes are not done in water treatment. Chlorine-based processes were being used around the world for disinfecting the drinking water systems by the late nineteenth century.

1.1.1 Advantages of Using Chlorine

Disinfection by chlorine is one of the greatest innovations. At present disinfection by chlorine is the most economical when compared to the other disinfection techniques available. It is the most potent and effective method against killing a wide range of microorganisms. Since the characteristics of the raw water change its flexible dosing empowers more prominent control over disinfection. While disinfecting, chlorine also removes harmful odours. Certain organic and inorganic compounds are also effectively oxidised by chlorine. Chlorine also removes some nitrogenous compounds, ammonia, and hydrogen sulphide which have unpleasant taste and obstruct disinfection. It also removes iron and manganese. In the water distribution network, until the consumers tap the residual chlorine present in the water supplied prevents the microbial regrowth.

1.1.2 Disadvantages of Using Chlorine

The aquatic life is at threat because at low concentrations, the residual chlorine is toxic, and the water should be dechlorinated. The storage, transport, and handling are to be taken care of as all forms of chlorine are profoundly corrosive and harmful. More hazardous DBPs like THMs and HAA form when NOM present in the water gets oxidised, which can be harmful to humans (Rook 1974). In treated water, its use increases chloride content, and the level of total dissolved solids also increases. At a lower dose of chlorine, some species of microorganisms were unaffected. The impacts of releasing dechlorinated mixes in the environment are unidentified. Sometimes due to higher concentrations of chlorine demanding groups of microorganisms residual chlorine is ineffective and requires more doses for sufficient disinfection.

1.2 Microbiological Drinking Water Quality and Human Health

In today's time, the safety of municipal drinking water can be taken for granted, but earlier human beings were at risk without filtration and chlorination process of the drinking water. A century ago, thousands of people died annually due to the spread of infectious diseases like typhoid, cholera, hepatitis A, and dysentery (WHO 2018). These diseases owing to the existence of microorganisms in drinking water got viably reduced after disinfection methods were employed in the treatment system. Water is related to numerous different diseases in many ways. By hygienic utilisation of water, it controls some of the diseases positively, but for others, it may act as a vector where diseases are transmitted through water or where water is required to complete the life cycle by the agent vectors. The different types of water-borne diseases are outlined in Table 1.

In the year 2017, about 6.8 billion individuals (90% of the world's population) got the necessary drinking water as compared to 5 billion individuals (82% of the world's population) in 2000. With this trend, it is estimated that in the year 2030, around 96% of the world's population will be covered. Drinking water services approximately increased from 95% to 97% in the urban area, 69% to 81% in rural areas from 2000 to 2017 (UNICEF and WHO 2019). It is estimated by the WHO (WHO 2018) that due to water-borne diseases, 3.4 million individuals die every year, mostly children (Hinckley et al. 2005). In the United States (US), most incidence is related to improper disinfection of contaminated groundwater. Moreover, due to leakages of water main and other crisis circumstances, contamination occurs

Table 1 Water-borne diseases

Disease group	Name of disease	Route leaving host	Infection route	References
Waterborne	Cholera	Faeces	Oral	(Chowdhury 2013; Pan et al. 2014; Zhang et al. 2010)
	Typhoid	Faeces/ urine	Oral	
	Infectious hepatitis	Faeces	Oral	
	Giardiasis	Faeces	Oral	
	Dracunculiasis	Cutaneous	Percutaneous	
	Amoebiasis	Faeces	Oral	
Due to lack of hygiene	Bacillary dysentery	Faeces	Oral	(Chowdhury 2013; Liu et al. 2017; Zhang et al. 2011)
	Pinworm (Enterobius)	Faeces	Oral	
	Enteroviral diarrhoea	Faeces	Oral	
	Scabies	Cutaneous	Cutaneous	
	Paratyphoid fever	Faeces	Oral	
	Skin sepsis	Cutaneous	Cutaneous	
	Trachoma	Cutaneous	Cutaneous	
	Amoebiasis	Faeces	Oral	
	Conjunctivitis	Cutaneous	Cutaneous	
	Lice and typhus	Bite	Bite	
Due to scarce sanitation	Trichuriasis	Faecal	Oral	
	Ascariasis	Faecal	Oral	
	Hookworm (Ancylostoma/ Necator)	Faecal	Oral/ percutaneous	(Liu et al. 2011; Thokchom et al. 2020; Zeng et al. 2014)
Due to parasites living completely in water	Schistosomiasis	Urine/ Faeces	Percutaneous	(Liu et al. 2011; Thokchom et al. 2020)
Due to vectors living partly in water	Dracunculiasis	Cutaneous	Percutaneous	(Chowdhury 2013)

in the supply networks. Worldwide in many countries, human beings and health services are under extreme load due to these water-borne diseases.

In drinking water bacteria, viruses, and parasitic protozoa are the three standard classes of the microorganisms present in it. In both surface water and groundwater, bacteria and viruses are present, and parasitic protozoa mainly appear in surface water. The microorganisms get killed or inactivated due to disinfection of water such that they will not contaminate human hosts. Due to chlorine disinfection, the bacteria and viruses are well under control, but more refined control measures are required to restrict parasitic protozoa. Therefore, in areas where disinfection is done, bacterial or viral infections are exceptionally low as compared with parasitic protozoan infections.

1.3 Disinfection by-products: A Challenge

By doing chlorination in drinking water there is an increase in life expectancy and a dramatic turn down in water-borne diseases around the world. Due to chlorination humans assume that the water which they are drinking is free from pathogens. THMs are mostly produced as DBPs, which include chloroform (CF), dibromochloromethane (DBCM), bromodichloromethane (BDCM), and bromoform, as a result of the reaction between the NOM precursors, such as humic and fulvic acid, and chlorine (BF) (Mazhar et al. 2020). Table 2 lists some of the DBPs created by various common disinfectants (Craun et al. 2004). Chloroform is typically the most dominant THM, and brominated THMs are created when bromide is oxidised with precursors. Some THMs have been identified as potentially carcinogenic to humans. Contrary to chlorination, chloramination yields low THM concentrations while also creating additional DBPs, such as cyanogen chloride (Duong et al. 2003). By oxidising bromide, a brominate precursor, ozone can also produce hypobromous acid, a brominated THM.

The present evidence about the carcinogenic nature of DBPs in humans is not sufficient, but the presence of these chemicals in more quantity is undesirable. Various economical methods are available these days to reduce their formation. Through the International Programme on Chemical Safety (IPCS), WHO along with the United Nations Environment Program, the International Labour Organization warns that the effect of DBPs is low at which they occur when compared to dangers related to inadequate disinfection. Hence, we should understand the importance of controlling these DBPs formed due to disinfection. Understanding

Table 2 DBPs of different disinfectants

Common disinfectants	Organohalogen by-products	Inorganic by-products	Nonhalogenated by-products
Chlorine/ hypochlorous acid	THMs, HAAs, HANs, chloral hydrate (CH), chloropicrin, chlorophenols, N-chloramines, halofuranones, bromohydrins	Chlorate (mostly from hypochlorite use)	Aldehydes, cyanoalkanoic acids, alkanoic acids, benzene, carboxylic acids
Chlorine dioxide		Chlorite, chlorate	Unknown
Chloramine	HANs, cyanogen chloride, organic chloramines, chloramino acids, chloral hydrate, haloketones (HK)	Nitrate, nitrite, chlorate, hydrazine	Aldehydes, ketones
Ozone	Bromoform, MBA, DBA, DBAC, cyanogen bromide	Chlorate, iodate, bromate, hydrogen peroxide, hypobromous acid, epoxides, ozonates	Aldehydes, ketoacids, ketones, carboxylic acids

the gravity of the problem the USEPA has additionally constrained DBPs present in the drinking water and most of the municipal drinking water systems are fulfilling these new guidelines by removing the NOM before disinfection.

1.4 Disinfection By-Products Chemistry

Natural humic acid and fulvic acid present in raw water are the antecedents of these halogenated compounds. THMs, HAAs, haloacetonitriles, chloral hydrate, and chlorophenols are the most generally discovered DBPs (Bellar et al. 1974; Rook 1974). Chlorinated furanone MX, halopicrins, cyanogens halides, haloketones, and haloaldehydes are some other found DBPs. It is found that a greater number of DBPs formed are known today, and many more are yet to be discovered but their toxicity is unknown. The following equation generalises the creation of DBPs:

$$\mathrm{NOM} + \mathrm{HOCL} + \mathrm{Br}^{-} \rightarrow \mathrm{DBPs} \qquad (3)$$

1.5 Disinfection By-Products History and Regulation

To reduce the number of bacteria, chlorine was first added to an urban water system in Chicago, Illinois, in 1908. Chlorinating drinking water became the most popular method of municipal water disinfection swiftly throughout the United States, which reduced the mortality and morbidity caused by water-borne diseases (Cantor 1997). An American scientist named Beller in 1971 found that there was no chloroform in the Ohio River but after treatment chloroform appeared in the treated water (Bellar et al. 1974). This indicated that the DBPs are formed in drinking water after chlorination. In 1974, Rooks discovered the formation of the first DBPs, i.e. chloroform and other trihalomethanes (THMs) during chlorination of Rotterdam water supply (Rook 1974). Furthermore, this led researchers to discover more DBPs formed in drinking water and to find their undesirable health effects. The DBPs discovered till now are shown in Fig. 2.

In 1979, USEPA first regulated these THMs concerning their carcinogenic nature to human beings. Since then, an abundance of research has improved knowledge about DBPs formation, their health risk and their control. Now it is possible to understand that potential DBPs are formed from all chemical disinfectants and are thoroughly studied for their toxicity. To reduce the level of risk to human health, the USEPA proposed in 1986 that the maximum contaminant level (MCL) for total THMs be set at 80 μg/L under Stage I rules, and the MCL is anticipated to decrease to 40 μg/L under Stage II (Pontius 1999). Additionally, HAAs were further limited to 60 μg/L for the combined total of five species in treated waters. In

Fig. 2 Historical review of DBPs

contrast, the European Union and the United Kingdom set a 100-μg/L limit for THMs. While the WHO considered the allowable limit of THMs for CF to be 300 μg/L, BDCM to be 60 μg/L, DBCM to be 100 μg/L, and BF to be 100 μg/L, separately. In addition, the Indian government recently set the THMs limit at 200 μg/L for CF, 60 μg/L for BDCM, 100 μg/L for DBCM, and 100 μg/L for BF in 2012. Table 3 displays the DBPs allowable limits established by several organisations for treated water.

According to the USEPA, the four THMs are carcinogenic to people, and CF, BDCM, and BF have been classified as type B2 probable human carcinogens and DBCM as type C possible human carcinogens (Gates et al. 1999). A few investigations on the toxicity of animals have found a direct link between DBPs and cancer-causing illnesses. The National Cancer Institute in the United States stated in 1976 that rodents develop cancer as a result of ingesting water containing chloroform. Later investigations found that chlorination of drinking water increased the incidence of bladder, rectum, colon, and unfavourable effects on fertility (Cantor et al. 1999). Animals exposed to THMs and HAAs showed negative health impacts, according to laboratory studies. THM exposure results in the development of tumours in the livers and kidneys of these animals, as well as pregnancy loss, decreased sperm motility, and diminished fetotoxicity. HAA exposure also resulted in liver tumours, as well as unfavourable health consequences on rats that were just born and pregnant. Though sufficient evidence of a cancer-causing nature in experimental animals exists, there is a dearth of such evidence in humans.

Table 3 DBPs limits around the globe

Disinfectant	DBPs		Cancer group	WHO	US EPA	United Kingdom	European Union	China	India
Chlorine or chloramine	**Trihalo methanes**	Chloroform	2B	0.3	0.08	–	–	0.06	0.2
		Bromodichloromethane	2B	0.06	0.08	–	–	0.06	0.06
		Dibromochloromethane	3	0.1	0.08	–	–	0.1	0.1
		Bromoform	3	0.1	0.08	–	–	0.1	0.1
		Total THM		*	0.08	0.1	0.1	*	–
	Halo acetic acids	Monochloroacetic acid		0.02	0.06	–	–	–	–
		Dichloroacetic acid	2B	0.05	0.06	–	–	0.05	–
		Trichloroacetic acid	2B	0.2	0.06	–	–	0.1	–
		Dibromoacetic acid	2B	–	–	–	–	–	–
		Bromochloroacetic acid	2B	–	–	–	–	–	–
		Total HAA		–	0.06	–	0.06	–	–
	Chlora mines	Monochloramines	3	3	4	–	–	3	–
	NDMA	N-Nitrosodimethylamine	2A	0.0001	–				
	Haloacetonitrile	Dichloroacetonitrile	3	0.02	–	–	–	–	–
		Trichloroacetonitrile	3	–	–	–	–	–	–
		Dibromoacetonitrile	2B	0.07	–	–	–	–	–
		Bromochloroacetonitrile	3	–	–	–	–	–	–
	Cyanide	Cyanogen chloride as cyanide		–	–	–	–	0.07	–
		Cyanide		–	0.2	0.05	0.05	0.05	–
		Nitrite	2A	3	1	0.5	0.5	1	
		Nitrate	2A	50	10	50	50	10	
	Chloral hydrate	Trichloroacetaldehyde	2A	–	–	–	–	0.01	–

Chlorine dioxide	**Cholorophenols**	2-Cholorophenol		–	–	–	–	–	–
		2,4-dicholorophenol		–	–	–	–	–	–
		2,4,6-tricholorophenol	2B	0.2	–	–	–	0.2	–
		2,3,4,6-tetracholorophenol		–	–	–	–	–	–
		Pentachlorophenol	1	0.009	0.001	–	–	0.009	–
	Oxyhalides	Chlorite		0.7	1	–	0.25	0.7	–
		Chlorate		0.7	–	–	0.25	0.7	–
Ozone		Bromate	2B	0.01	0.01	0.01	0.01	0.01	–
		Formaldehyde	1	–	–	–	–	0.9	–
References			(IARC, 2019)	(WHO, 2017)	(USEPA, 2018)	(UKWQR, 2016)	(EUDWR, 2019)	(CDWQ, 2006)	(IS 10500, 2012)

Note: *The sum of ratio of the concentration of each THMs species to its respective guideline values should not exceed 1

Cancer Group: 1 (Carcinogenic to humans), 2A (Probably carcinogenic to humans), 2B (Possibly carcinogenic to humans), 3 (Not classifiable as to its carcinogenicity to humans)

2 Conclusion

During disinfection when chlorine reacts with the organic matter disinfection by-products are formed in water. For the first time in 1979, the USEPA regulated THMs in drinking water and in India it was regulated in 2012 by the Bureau of Indian Standards. USEPA reported that these disinfection by-products are carcinogenic in nature and can be probable carcinogens to humans. Disinfection by-products are not regulated in many countries and new by-products species are being discovered around the world. Disinfection by-products can be controlled by various methods like filtration, coagulation, or by using alternate disinfectants. A constant effort is required for reducing DBPs formed by protecting the source water from pollution, for this more stringent rule to be made by the government for protecting the rivers. The DBPs precursors, i.e. the organic matter present in raw water can be removed to some extent by coagulation, membranes, granular activated carbon, ozone-biofiltration etc. or by a combination of these. To control the DBPs formation pre-chlorination should be done to reduce post chlorination or for primary and secondary disinfection different combinations of disinfectants can be used. Health risk associated with chlorination is unknown therefore more research is needed to determine the health risk associated with DBPs. Thus, there is a dreadful need to identify the DBPs present in the drinking water and to quantify DBPs in the effluent of the existing water treatment plants, so that suitable treatment strategies can be chalked out to save the health risk from the consequences of the presence of these DBPs due to inadequately disinfecting treated water.

References

Amy G, Craun G, Craun GF, Siddiqui M (2004) EHC 216: disinfectants and disinfectant by-products. Environ Health Criteria 216

Ashley DL, Smith MM, Silva LK, Yoo YM, De Jesús VR, Blount BC (2020) Factors associated with exposure to Trihalomethanes, NHANES 2001-2012. Environ Sci Technol 54:1066–1074. https://doi.org/10.1021/acs.est.9b05745

Bellar TA, Lichtenberg JJ, Kroner RC (1974) Occurrence of Organohalides in chlorinated drinking waters. J Am Water Works Assoc 66:703–706. https://doi.org/10.1002/j.1551-8833.1974.tb02129.x

Cai L, Li L, Yu S, Guo J, Kuppers S, Dong L (2019) Formation of odorous by-products during chlorination of major amino acids in East Taihu Lake: impacts of UV, UV/PS and UV/H2O2 pre-treatments. Water Res 162:427–436. https://doi.org/10.1016/j.watres.2019.07.010

Calderon RL (2000) The epidemiology of chemical contaminants of drinking water. Food Chem Toxicol 38:13–20. https://doi.org/10.1016/s0278-6915(99)00133-7

Cantor KP (1997) Drinking water and cancer. Cancer Causes Control 8:292–308. https://doi.org/10.1023/A:1018444902486

Cantor KP, Lynch CF, Hildesheim ME, Dosemeci M, Lubin J, Alavanja M, Craun G (1999) Drinking water source and chlorination byproducts in Iowa. III. Risk of brain cancer. Am J Epidemiol 150:552–560. https://doi.org/10.1093/oxfordjournals.aje.a010052

CDWQ (2006) National Standard of the people ' s Republic of China standards for drinking water quality. Drink Water Qual GB 5749:1–13

Chowdhury S (2013) Exposure assessment for trihalomethanes in municipal drinking water and risk reduction strategy. Sci Total Environ 463–464:922–930. https://doi.org/10.1016/j.scitotenv.2013.06.104

Craun GF, Craun RA, Gunther F, Pegram M, Siddiqui M (2004) Environmental health criteria 216 disinfectants and disinfectant by-products. World Health Organisation

D Nikolaou SKGA (2005) Survey of disinfection by-products in drinking water in Athens, Greece. Desalination 176:13–24. https://doi.org/10.1016/j.desal.2004.10.029

Duong HA, Berg M, Hoang MH, Pham HV, Gallard H, Giger W, Von Gunten U (2003) Trihalomethane formation by chlorination of ammonium- and bromide-containing groundwater in water supplies of Hanoi, Vietnam. Water Res 37:3242–3252. https://doi.org/10.1016/S0043-1354(03)00138-6

EUDWD (2019) Drinking water directive. Counc EU 2019:1–124

Gates D, Kranser S, Rice R, Mark J, Blake B, Maggie M (1999) Alternative disinfectants and oxidants. USEPA, 1999. EPA Guid. Man., Table 3–3, pp. 3–10

Hinckley AF, Bachand AM, Reif JS (2005) Late pregnancy exposures to disinfection by-products and growth-related birth outcomes. Environ Health Perspect 113:1808–1813. https://doi.org/10.1289/ehp.8282

IARC (2019) International Agency for Research on Cancer. Agents classified by the IARC monographs. WHO Report 1–123:1–37

IS:10500 (2012) Drinking water — specification (second revision) IS 10500 (2012): drinking water. Bur India Stand 25:1–18

Kanan A, Karanfil T (2011) Formation of disinfection by-products in indoor swimming pool water: the contribution from filling water natural organic matter and swimmer body fluids. Water Res 45:926–932. https://doi.org/10.1016/j.watres.2010.09.031

Li L, Yang S, Yu S, Zhang Y (2019) Variation and removal of 2-MIB in full-scale treatment plants with source water from Lake Tai, China. Water Res 162:180–189. https://doi.org/10.1016/j.watres.2019.06.066

Liu C, Ersan MS, Plewa MJ, Amy G, Karanfil T (2019) Formation of iodinated trihalomethanes and noniodinated disinfection byproducts during chloramination of algal organic matter extracted from Microcystis aeruginosa. Water Res 162:115–126. https://doi.org/10.1016/j.watres.2019.06.053

Liu S, Li Z, Dong H, Goodman BA, Qiang Z (2017) Formation of iodo-trihalomethanes, iodo-acetic acids, and iodo-acetamides during chloramination of iodide-containing waters: factors influencing formation and reaction pathways. J Hazard Mater 321:28–36. https://doi.org/10.1016/j.jhazmat.2016.08.071

Liu S, Zhu Z, Fan C, Qiu Y, Zhao J (2011) Seasonal variation effects on the formation of trihalomethane during chlorination of water from Yangtze River and associated cancer risk assessment. J Environ Sci 23:1503–1511. https://doi.org/10.1016/S1001-0742(10)60573-6

Mazhar MA, Khan NA, Ahmed S, Khan AH, Hussain A, Changani F, Yousefi M, Ahmadi S, Vambol V (2020) Chlorination disinfection by-products in municipal drinking water – a review. J Clean Prod 273:123159. https://doi.org/10.1016/j.jclepro.2020.123159

Mills CJ, Bull RJ, Cantor KP, Reif J, Hrudey SE, Huston P (1998) Workshop report. Health risks of drinking water chlorination by-products: report of an expert working group. Chronic Dis Can 19:91–102

Pan S, An W, Li H, Su M, Zhang J, Yang M (2014) Cancer risk assessment on trihalomethanes and haloacetic acids in drinking water of China using disability-adjusted life years. J Hazard Mater 280:288–294. https://doi.org/10.1016/j.jhazmat.2014.07.080

Pontius FW (1999) Complying with future water regulations. J Am Water Works Assoc 91:46–58. https://doi.org/10.1002/j.1551-8833.1999.tb08599.x

Roeske W (2007) Trinkwasserdesinfektion (drinking water disinfection). Oldenbourg Industrieverlag GmbH, München, pp 37–38

Rook JJ (1974) Formation of haloforms during chlorination of natural waters. Water Treat Exam 23:234–243
Singer PC (1994) Control of disinfection by-products in drinking water. J Environ Eng 120:727–744. https://doi.org/10.1061/(ASCE)0733-9372(1994)120:4(727)
Taylor A, Zrinyi N, Mezyk SP, Gleason JM, MacKinnon L, Przepiora A, Pham ALT (2019) In-situ chemical oxidation of chlorendic acid by persulfate: elucidation of the roles of adsorption and oxidation on chlorendic acid removal. Water Res 162:78–86. https://doi.org/10.1016/j.watres.2019.06.061
Thokchom B, Radhapyari K, Dutta S (2020) Occurrence of trihalomethanes in drinking water of Indian states: a critical review, disinfection byproducts in drinking water. LTD. https://doi.org/10.1016/B978-0-08-102977-0.00004-4
UNICEF, WHO (2019) Progress on household drinking water, sanitation and hygiene 2000–2017: special focus on inequalities, launch version July 12 Main report Progress on drinking water, sanitation and hygiene
UKWQR, 2016. WATER, ENGLAND AND WALES The Water Supply (Water Quality) Regulations 2016. Water, Engl Wales 614, 52
USEPA (2018) Edition of the drinking water standards and health advisories tables. Water Supply 25:1205–1241
WHO, 2017. Guidelines for Drinking-water Quality, Fourth Edition Incorporating the First Addendum, Fourth Inc. ed.
WHO (2018) Typhoid vaccine: WHO position paper – March 2018. Wkly Epidemiol Rec 13:153–172
Wu QY, Zhou YT, Li W, Zhang X, Du Y, Hu HY (2019) Underestimated risk from ozonation of wastewater containing bromide: both organic byproducts and bromate contributed to the toxicity increase. Water Res 162:43–52. https://doi.org/10.1016/j.watres.2019.06.054
Zeng Q, Chen Y-Z, Xu L, Chen H-X, Luo Y, Li M, Yue J, Liu A-L, Li Y-F, Lu W-Q (2014a) Evaluation of exposure to trihalomethanes in tap water and semen quality: a prospective study in Wuhan, China. Reprod Toxicol 46:56–63. https://doi.org/10.1016/j.reprotox.2014.03.005
Zeng Q, Zhou B, Cao WC, Wang YX, You L, Huang YH, Yang P, Liu AL, Lu WQ (2014b) Predictors of urinary trichloroacetic acid and baseline blood trihalomethanes concentrations among men in China. Sci Total Environ 493:806–811. https://doi.org/10.1016/j.scitotenv.2014.06.067
Zhang J, Yu J, An W, Liu J, Wang Y, Chen Y, Tai J, Yang M (2011) Characterization of disinfection byproduct formation potential in 13 source waters in China. J Environ Sci 23:183–188. https://doi.org/10.1016/S1001-0742(10)60440-8
Zhang SH, Miao DY, Liu AL, Zhang L, Wei W, Xie H, Lu WQ (2010) Assessment of the cytotoxicity and genotoxicity of haloacetic acids using microplate-based cytotoxicity test and CHO/HGPRT gene mutation assay. Mutat Res – Genet Toxicol Environ Mutagen 703:174–179. https://doi.org/10.1016/j.mrgentox.2010.08.014
Zhang X, He J, Lei Y, Qiu Z, Cheng S, Yang X (2019) Combining solar irradiation with chlorination enhances the photochemical decomposition of microcystin-LR. Water Res 159:324–332. https://doi.org/10.1016/j.watres.2019.05.030

Recent Advancements in Chlorine Applications for Water Quality Control

Abdul Gani, Mohit Singh, Shray Pathak, and Athar Hussain

Abstract Chlorine has been widely used for water disinfection since the early twentieth century, and recent advancements in its applications have led to more efficient and effective water treatment. The main focus of these advancements has been on reducing the formation of disinfection by-products (DBPs), It may be hazardous to human health. One approach to reducing DBP formation is the use of chlorine dioxide (ClO_2), which has a higher oxidation potential than chlorine and can therefore more effectively oxidize organic matter in water. Another approach is the use of monochloramine, a less reactive and less volatile form of chlorine, that can be more stable in water and reduce DBP formation. Other advancements in chlorine applications include the use of automated monitoring and control systems to optimize chlorine dosing and reduce over- or under-dosing, and alternate methods of disinfection, include using ozone and ultraviolet (UV) light, which can be used in conjunction with chlorine to further reduce DBP formation. Overall, these advancements in chlorine applications have contributed to safer and more effective water treatment, with reduced risks of DBP exposure for consumers. This chapter discusses the recent advancements in chlorination and disinfection by-products, their types, guidelines for disinfection by-products, and methods for removing disinfection by-products.

Keywords Disinfection · Chlorination · Water quality · Pathogen · Disinfection by-products

A. Gani · M. Singh · S. Pathak · A. Hussain (✉)
Department of Civil Engineering, Netaji Subhas University of Technology, New Delhi, India
e-mail: athar.hussain@nsut.ac.in

S. Madhav et al. (eds.), *Drinking Water Disinfection By-products*,
https://doi.org/10.1007/978-3-031-49047-7_3

1 Introduction

Many of the organisms in the raw water are removed or inactive as the water travels through a treatment facility. Nevertheless, disinfection of processed water is the final step in a water treatment facility. The crucial phase in water purification for public supply is this one, as it eliminates the most harmful pathogens from the water that cause waterborne diseases (Sayers 2012).

1.1 Disinfection

Potable water disinfection is a global procedure for removing microbiological pollutants and is regarded as one of the main municipal health breakthroughs of the twentieth period (Boone et al. 2012). Since the usage of disinfectants, there has been a significant decrease in aquatic contagious illnesses such as polio, typhoid fever, hepatitis, and cholera, which constituted a public health hazard for several years. The precise description of 'disinfection' is under threat, and the term is frequently misused. It is not the same as sterilization, which includes deactivating all living things; instead, it involves eliminating dangerous pathogenic organisms that are the source of diseases (Sayers 2012). Disinfection is often the most crucial procedure in terms of providing clean water. Disinfection has two components: the disinfection of the water to remove any germs that have been allowed to survive different stages of treatment, and the application of a vestigial disinfectant to ensure that water departing the treatment mechanism stays harmless as it travels through the distribution network to the spot of use.

The significance of the process can be well understood by considering the fact that in case of emergencies or wartime when the treatment of water cannot be possible by using processes such as sedimentation and filtration, the only process of water disinfection is generally carried out. The disinfectant not only removes bacteria from the treated water but also confirms the entire safety of water in the distribution system as it reaches the end users. Therefore, a chemical that can provide a residual sterilizing effect for a longer period with protection against recontamination may be recommended. However, the disinfectant should be economical and can be easily measurable by simple test procedures. It should destroy 100% harmful pathogens in a short duration and make it safe for use. It should be readily available and should not require high skills for its applications. The water does not become poisonous and offensive against the consumer after treatment. The concentration of such disinfectant can be calculated easily and expeditiously.

Obviously, the disinfectant should indeed be killing the microorganisms of concern but remaining non-toxic to mankind. Throughout the treatment facility, a variety of the mentioned methods may be used for disinfection:

a) Ozonation
b) Dosing of chlorine dioxide
c) UV disinfection
d) Chlorination

1.2 Chlorination

The most commonly employed type of disinfection is chlorination. Chlorination refers to the method of using chlorine to kill microorganisms in the water. For communal potable supply systems, the method is widely used in all developing countries. Ever since the early 1900s, chlorine has been employed as a disinfectant. It is the most popular primary and secondary disinfection used across the world, as well as the most popular and cost-effective approach. Because it eliminates targeted species by oxidizing cell membranes, chlorine is the most extensively used disinfection for municipal effluent. It is an effective oxidant and disinfectant, as well as a chlorinating agent (Wang et al. 2006a). If the water is devoid of organic molecules and ammonia, which require chlorine, it offers a steady chlorine residual for the consumable water transportation network. Chlorine oxidizes and disinfects pathogens and pollutants in the effluent, septage, sludge, swimming pool water, and cooling tower water, in addition to drinking water disinfection. Because chlorine may create trihalomethanes (THMs) and other halogenated (TOX) and nonhalogenated organic substances, it should be used with caution, especially when THM and TOX intermediates are available (Wang et al. 2006a; Wang 2006). A certain number of the non-halogenated oxidation products generated by fulvic and humic acid chlorination are comparable to all those generated by other oxidants/disinfectants notably ozone, potassium permanganate, and chlorine dioxide. For the chlorination process, chlorine can be delivered in a number of forms, including gaseous, solid, and aqueous solutions. The three types of chlorine are available:

a. Liquid (liquid Cl_2)
b. Solid (bleaching powder)
c. Gaseous form (chlorine gas) (Fava and Thomas 1978)

Chlorine gas is utilized most commonly, particularly by large agencies, because it is considerably more economical cost than others. Calcium hypochlorite, also known as $Ca(OCl)_2$, is the solid substance of chlorine. Sodium hypochlorite solution, also known as NaOCl, is the viscous liquid of chlorine.

The major advantage of chlorine is that it can be manufactured easily by using electrolytes of common salts (NaCl). It is also a potent oxidant, which can penetrate the organism's cell wall and its contents. The chlorine dosage can be precisely regulated and can be easily detected by using a basic ortho-toluidine test. It does not form harmful constituents upon inorganic reaction in water. It also leaves the required residue in the water to protect the water from recontamination. The major disadvantage of using chlorine is that it is extremely corrosive when used as chlorine gas for water disinfection and lethal to inhalation. Hence, it must be carefully placed at a distance in a sealed jar. Also, the chlorine when added to water containing phenolic compounds reacts with them and results in cancer-causing substances.

1.3 Residual Chlorine and Chlorine Demand

A sufficient chlorine quantity will effectively disinfect presently controlled bacteria. Since chlorine reacts with plenty of chemicals in water, the "chlorine demand" of these additional compounds to be satisfied before free chlorine in large quantities is available for disinfection (Wang et al. 2006a). When chlorine is introduced to water, some of it is utilized to eliminate bacteria in the water, while the rest is used to oxidize organic contaminants in the water. The chlorine that is still present after the dosage process is known as residual chlorine. The amount of chlorine still usable for disinfection when there is sufficient chlorine available is known as the 'available residual chlorine'. The chlorine residual might be either freely available, combined available, or a mix of the two.

When chlorine is added to water, whether in the shape of a molecule or a hypochlorite, it undergoes hydrolysis to produce chlorine, hypochlorous acid, and hypochlorite ions. The entire amount of hypochlorite ions and hypochlorous acid are described as free-accessible chlorine (APHA 2012; Nielsen et al. 2022). Both substances are germicides, but HOCl is much more efficient than OCl (Tvrdá and Benko 2020). Even at concentrations suitable for drinking water, free chlorine is less efficient against protozoa than a variety of viruses and bacteria linked to polluted drinking water (CDC 2014; Mohamed et al. 2015).

The term "combined accessible chlorine" refers to the total amount of dichloramine and monochloramine, as well as organic nitrogen chlorine and nitrogen trichloride-containing substances (Wang et al. 2006a). An estimated analytical testing approach is used to quantify this. As a result, its demand may be computed as the imbalance between the amount of chlorine combined per litre of the treated water and the residual chlorine after the procedure. Polluted waterways rise the demand for chlorine. Water pre-treatment using sedimentation and aeration can reduce chlorine consumption. For drinking water, a remainder chlorine of 0.2 mg/L is recommended for water distribution systems (Lu et al. 1999).

2 Chlorine as a Water Disinfectant

This section outlines the chlorination process, along with the physio-chemical variables influencing its effectiveness and suitability for certain sites. Later on, we will talk about the machinery, chemicals, and operational and maintenance issues that apply to the three different physical states of chlorine.

Once chlorine is introduced to water, chemical processes take place that inhibit the activity or kill undesired bacteria. In accordance with the following reaction, chlorine gas will rapidly produce hydrochloric and hypochlorous acids when it is introduced to water.

$$Cl_2 + H_2O \xrightarrow{pH>5} HOCl + H^+ + Cl^- \quad (1)$$

Also, the pH of the fluid affects how the hypochlorous acid reacts even further. It will activate more strongly at higher pH levels, as illustrated below:

$$HOCl \overset{pH>8}{\Leftrightarrow} H^{+} + OCl^{-} \tag{2}$$

The two-dominant species, hypochlorite ions (OCl^{-}) and hypochlorous acid (HOCl), are declared as readily accessible chlorine and accountable for water disinfection. Chlorine interacts with ammonia in water to produce monochloramines (NH_2Cl), dichloramines ($NHCl_2$), and trichloramine (NCl_3), the distribution of which is affected by the pH of the water. These reactions' balance points are pH dependent. The resultant hydroxides increase the pH level of the water, decreasing the concentration of hypochlorous acid and reducing disinfection efficiency (Wang 2006). When employing hypochlorite variants of chlorine, the capacity to alter and control pH is crucial. Exclusively OCl ions exist at pH values more than 10, whereas HOCl ions persist at pH values less than 7 and larger than 5, with no dissociation. At pH values below 5, however, chlorine does not interact and stays in the state of elemental chlorine. The most harmful version of chlorine is hypochlorous acid, which is around 80 times stronger than hypochlorite ions. As a result, the pH of the water is preserved marginally lower than 7 although during chlorination operation.

When sodium or calcium hypochlorite solutions are used, they dissociate in the procedure outlined below:

$$Ca(OCl)_2 \overset{yields}{\rightarrow} Ca^{2+} + 2OCl^{-} \tag{3}$$

$$NaOCl \overset{yields}{\Leftrightarrow} Na^{+} + OCl^{-} \tag{4}$$

However, Eq. 4's balance also holds true, and the calcium or sodium ions will also have an impact on the concentration of hydrogen ions. Therefore, calcium and sodium hypochlorite reacts in water in the same ways that chlorine does, but the occurrence of the metallic ions affects the hydroxide ion concentration, raising the pH value as opposed to chlorine gas, which decreases the pH value (Wang et al. 2006a). Figure 1 shows how pH affects the dissociation of hypochlorous acid (Hussain and Gupta 2022).

The interactions involving chlorine and ammonia are another key group of events in chlorination. There are significant for two reasons: first, they eliminate hypochlorous acid, and after that, the chemicals generated are disinfectants that are frequently utilized to produce persistent disinfection in distribution networks. Hypochlorous acid or chlorine combines with ammonium ion to exchange hydrogen atoms with chlorine one at a time (Sayers 2012).

$$NH_3 + HOCl \overset{yields}{\rightarrow} NH_2Cl + H_2O \tag{5}$$

$$NH_2Cl + HOCl \overset{yields}{\rightarrow} NHCl_2 + H_2O \tag{6}$$

Fig. 1 Impact of pH on hypochlorous acid dissociation

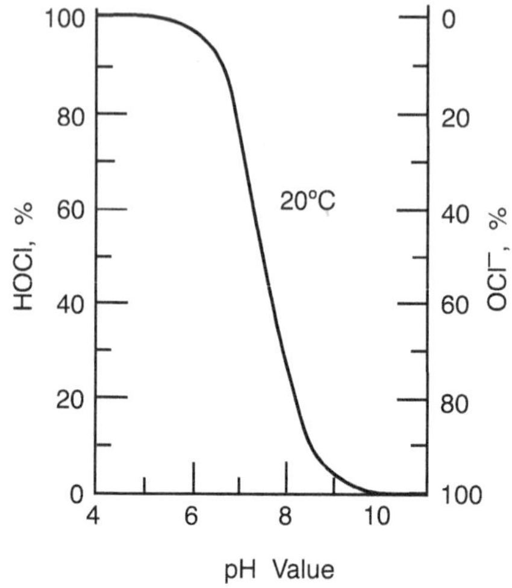

$$NHCl_2 + HOCl \xrightarrow{yields} NCl_3 + H_2O \quad (7)$$

These chemical reactions result in the creation of trichloramine, dichloramine, and monochloramine, as shown above (Binnie et al. 2017). The quantities of the three forms of chloramines are proportional to the ammonia concentration and pH in the water. At pH values more than 7.5, monochloramines (NH_2Cl) prevail, whereas dichloramines ($NHCl_2$) predominate at pH values within 5–6.5. Furthermore, tri-chloramines (NCl_3) are at their highest at pH values below 4.4. In water, these produced chloramines have disinfectant qualities. Consequently, combination chlorine, which is about 25 times less effective than ordinary chlorine, is produced when ammonia and chlorine in water mix to form chloramines. Even so, chlorination of ammonia-containing water results in the generation of nitrogen gas as shown:

$$2NHCl_2 + HOCl \xrightarrow{yields} N_2 + 3HCl + H_2O \quad (8)$$

2.1 Dosage of Chlorine

When chlorine is added to water it immediately reacts with inorganic impurities of the water such as S, Fe, Mn, and NO_2 that convert this chlorine into chlorides having no oxidizing power. Thereafter, the chlorine reacts with ammonia present in the water to form chloramines and at the same time the chlorine reacts with organic matter present in the water (Nielsen et al. 2022). The total amount of chlorine being utilized in all the above processes is known as chlorine demand of water. After all the above consumption of the total chlorine applied to water, the remaining chlorine is known as free chlorine. Free chlorine and combined chlorine cause the germicidal action on bacteria and pathogens present in water. This free available chlorine will kill the pathogens while the combined chlorine has a long-term germicidal effect to protect water for a long time. It is recommended that the free chlorine in water should be 0.2 mg/L after 10 min of its application.

As the concentration of free chlorine greater than 0.2 mg/L makes a bad taste in water and an even higher concentration is toxic to human beings. However, the free chlorine concentration of less than 0.2 mg/L can recontaminate the water after some time. Therefore, the optimum dosage of chlorine requirement in particular water before its distribution is required to be assessed and added accordingly. During the rainy season and epidemics high dose of chlorine should be added to water as compared to normal dosage. Also, the presence of different types of viruses may sometimes require a different dosage of chlorine in order to kill each type of virus individually. Poliomyelitis and hepatitis viruses require chlorine doses of 0.1 and 0.4 mg/L. However, the cysts of histolytic organisms that cause amoebic dysentery and tuberculosis organisms require a chlorine dose of 3 mg/L (Jaya Prakash et al. 2007). The coxsackievirus requires a chlorine dose of 21–138 mg/L. The contact time of 30 min at a pH value less than 8 may be considered after the application of chlorine in such types of water (Kalibbala 2007).

2.2 *Forms of Chlorine Applications*

Chlorine for the treatment of water for making it fit for drinking purposes, used in various forms and methods as per the purification requirement, contact period with microorganisms and cost-effective manner. This is done in various ways which are listed below:

A. **Plain chlorination:** It is a chlorine enrichment strategy employed when the water available on the surface requires no additional purification treatment. The lake and spring water are safe to drink after being chlorinated. Water with turbidity somewhere around 20.0–30.0 mg/L is readily cleansed, and harmful bacterial, organic waste, and colour found in the water are simply eliminated. To retain a residual chlorine content of 0.2 mg/L, the typical dosage is 0.8 mg/L/h at 15 N/cm^2 pressure.
B. **Pre-chlorination:** Pre-chlorination refers to the addition of chlorine prior to sedimentation and filtration (Chinn 2009). It is essential when the water is heavily contaminated and to reduce the taste, odour, colour, and algae development in water purification facilities. As the load on filters lowers, pre-chlorination enhances coagulation and therefore can minimize post-chlorination doses. The usual chlorine dosing is 5.0–10.0 mg/L, with a residual chlorine dosage of 0.1–0.5 mg/L. Following that, post-chlorination is required to guarantee that safe water is acquired during the treatment stage.

C. **Post-chlorination:** Post-chlorination is the standard method of applying chlorine after adequate water treatment. This is a procedure that is commonly used after the filtering phase until water is provided to the general population. After a 20-min contractual period, the residual chorine of roughly 0.1–0.2 mg/L assures that no additional contamination occurs in the water delivery system.
D. **Double chlorination:** The aforementioned two pre- and post-chlorination operations are referred to together as double chlorination. Post-chlorination is commonly employed in all water distribution networks.
E. **Break point chlorination:** It may be described as the amount of chlorine required to destroy the microorganisms and oxidize the organic matter and ammonia present in the water. The remaining chlorine after complete utilization acts as the residual chlorine as shown in Fig. 1. When chlorine is added to water it instantly reacts with ammonia present in it thus forming chloramines. As the free chlorine content increases it interacts with the chloramines and reduces the chlorine available. Some of the added chlorine is consumed in killing of microorganisms in water and thus the amount of chlorine added is less than the chlorine added as depicted through curve AB in Fig. 2 (Hussain and Gupta 2022). The fall down of curve BC indicates the chlorine consumption after its further addition in oxidizing the organic matter present in the water. The residual chlorine at point C is minimal and beyond this point any further addition of chlorine will act as free chlorine in the water to be treated. Therefore, the point C is called break point as any further addition of chlorine beyond this point will act as residual chlorine. The further addition of chlorine beyond this point is called break point chlorination. A residual chlorine dose of 0.2–0.3 mg/L is quite effective in controlling the potential recontamination of water in distribution systems, if sunlight is avoided.

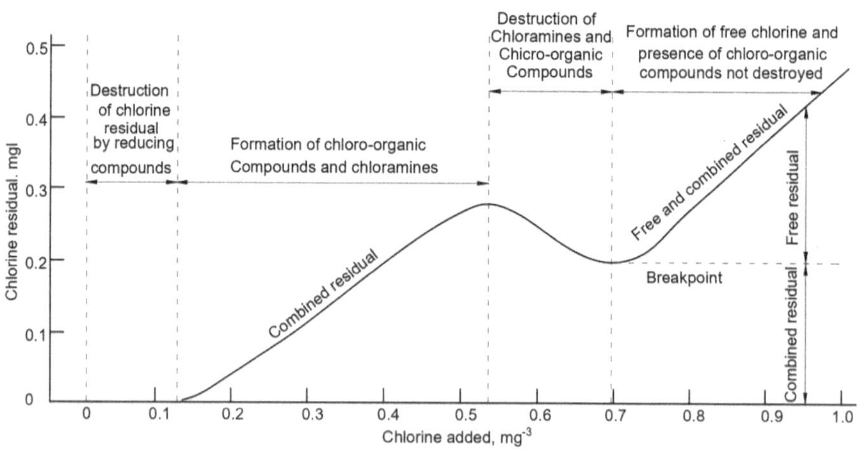

Fig. 2 Break point chlorination

F. **Super-chlorination:** Super-chlorination utilizes the most efficient method of chlorination. It is defined as the injection of a dosage significantly more than that required for appropriate bacterial filtration of water; in other sayings, enough chlorine is drugged to guarantee that the breakpoint is exceeded. This is the most often used form of disinfection since it is the safest. When there is an epidemic outbreak, the breakout water must be dechlorinated before being sent to the distributary network. A dosage of 10–15 mg/L is used, with a contact duration of 10–30 min.

G. **Marginal chlorination:** Dosing enough chlorine to achieve the required combined chlorine concentration is known as marginal chlorination. The disinfection becomes substantially less efficient when the chlorine is mixed. Marginal disinfection, according to some, should be used only on extremely high-quality waters, such as high-quality groundwaters or surface waters that have endured considerable treatment, including additional disinfection stages. Marginal chlorination can cause issues of taste and odour due to interactions between chlorine and the water being chlorinated, especially if organic debris is within the water. As a result, several providers favour super-chlorination.

H. **De-chlorination:** Chlorine residual can remain in the wastewater for several hours after disinfection. Because of the impact on aquatic life, numerous states are against the application of chlorination solely in pure receiving waters. To reduce the impact, chlorinated wastewater needs to be dechlorinated regularly. De-chlorination is the process of removing excess chlorine from water either entirely or in part. Excess chlorine in water corrosion occurs in the pipe edges and gives them a strong odour. Excess chlorine must consequently be removed prior to the actual distribution. Aeration, heating, and absorption are all physical procedures that may be used on charcoal. To remove chlorine from such fluids, chemical methods such as sulphur dioxide (SO_2) gas at 0.3–0.6 ppm for 10–15 min, sodium bi-sulphite ($NaHSO_3$), sodium thiosulphate ($Na_2S_2O_3$), sodium meta-bi-sulphate ($Na_2S_2O_5$), activated carbon, and ammonia (NH_4OH) are utilized. However, it is assumed that the residual chlorine dosage of 0.2–0.03 mg/L remains in the system.

I. **Re-chlorination:** In extended distribution lines, chlorine residuals might decline, leading to a dangerous water supply. Excess chlorine used to mitigate this will result in an unpleasant odour for consumers near the treatment site. In these circumstances, chlorine is injected again and re-chlorinated at intermediate sites, which are often utility reservoirs and booster pumping stations.

3 Kinetics of Disinfection

Chemical oxidants are thought to kill bacteria by perforating cell membranes and damaging enzymes. The disinfection process has several factors, all of which impact its effectiveness. The efficiency of chemical disinfectants is determined by various parameters, which are summarized here.

3.1 The Creatures That Will Be Eliminated

Most bacteria and viruses can be destroyed by chlorination, while some are more resistant than others. *E. coli* is thought to be more tolerant than most harmful bacteria and viruses, which is why it is utilized as a diagnostic organism. However, certain dangerous viruses are thought to be greater susceptible to chlorination than *E. coli*, therefore disinfection must include more than just eliminating *E. coli*. Chlorination and other disinfectants do not normally destroy protozoan oocysts, most prominently *Cryptosporidium parvum*, and must thus be eliminated or neutralized beforehand to chlorination.

3.2 The Disinfectant's Composition

Chlorine and hypochlorite can combine to generate a variety of chemical species in water, depending on the pH and the existence of other compounds. The efficacy of various disinfectants varies substantially.

3.3 The Disinfectant Concentration

The link between chlorine removal efficiency and diverse forms of free chlorine is pH dependent. Empirical evidence suggests that the chlorine dose be raised to accommodate for the higher pH levels. The level of chlorine with reference to contact time may be estimated theoretically using Eq. 9.

The concentration effect is given by:

$$C^n T_p = K \tag{9}$$

where,

C is referred to as disinfectant concentration in mg/L,

T_p is time required to accomplish a particular percentage kill in a minute,

n is the coefficient of dilution, and.

K is an experimentally derived constant.

It can be shown that if n < 1, contact period had a greater impact than concentration, but if n > 1, concentration has a bigger influence than contact period. The values of K and n are affected by a variety of parameters, including pH, temperature, the disinfectant employed, and the organism undergoing destruction. For chlorine, n is usually considered to be 1.

3.4 Contact Period

Theoretically, the rate of species death should follow Chick's law and it is described by the first order of reaction.

$$\frac{dN}{dt} = -kN \; or \; \ln\frac{N_t}{N_o} = -k.t \tag{10}$$

$$N_t = N_o\, e^{-k.t} \tag{11}$$

where,
N_o represents the number of microorganisms that survive initially,
N_t represents the number of microorganisms that survive at time t, and.
k is the killing rate constant for a certain organism/disinfectant pairing.

To put it another way, the kill rate is directly proportionate to the quantity of live organisms. Integration shows that at $k_t = 1$, 43.4% of the initial organisms will be killed. However, the effect of contact duration is more complicated. The frequency of kill rises with time for many species. It is believed to be connected to the time it takes for the disinfectant to penetrate and destroy organisms.

All pathogens must be killed by disinfection. In practise, the design criteria will be successful *E. coli* killing. The two critical parameters for this are the disinfectant dose (C) and contact duration (t). It is common practise to use contact period (C_t) values when designing disinfection systems. This is a simplified representation of a complicated connection, but it looks enough if a safety margin is added. The numerous chlorine compounds have very different levels of effectiveness. According to Hall and Hyde (1992), the contact period required for 99% removal of *E. coli* was assessed and they demonstrate two of the most important ideas about chlorine cleaning:

I. pH is significant. This is because hypochlorous acid, as opposed to the hypochlorite particle, is much more efficient. Therefore, the pH for chlorine disinfection should not be greater than 8.0. Above this, the dissociation of hypochlorous acid into hypochlorite and hydrogen ions increases.
II. Compared to mixed chlorine, free chlorine is much more efficient.

According to WHO (1993), successful chlorination calls for the following conditions: a pH of no higher than 8.0; a free chlorine residual of more than 0.5 mg/l; a contact time of at least 30 min. It is the same as a C_t of 15.

According to US guidelines (USA Surface Water Treatment Rule (SWTR) 1989), Giardia cysts must be removed or rendered inactive by a process that removes or reduces them by a three log, and viruses must be removed or reduced by a factor of four log. A C_t of 6 is needed for a four-log inactivation of viruses at 108 °**C** and a pH value of 6–8, and a C_t of 104 for a three-log inactivation of Giardia when disinfection alone is utilized without filtration or clarifying.

3.5 *Temperature*

Temperature rises cause an increase in chemical disinfection rates.

3.6 *Other Considerations*

Examples of these include the organism's concentration in the water and the existence of compounds that may safeguard organisms from the disinfectant. As a consequence, using low-turbidity water may be necessary since disinfecting turbid water may not be effective.

Disinfection is a complicated method that is challenging to simulate due to the large number of variables that influence its effectiveness. In practice, basic design requirements that provide reasonable assurance of successful disinfection are used; disinfection efficiency is validated utilizing biological indicator parameters, namely the absence of *E. coli* and coliform organisms in the treated water.

Any chlorine disinfection method can achieve the desired level of disinfection by varying the contact time and dose. The amount of chlorine used will vary depending on chlorine demand, effluent properties, and discharge requirements. Temperature, alkalinity, and nitrogen concentrations are some of the additional elements that contribute to optimal disinfection conditions. Prior to the full-scale implementation of a chlorine disinfection system, all important specifications should be pilot tested.

3.7 *Effectiveness of Chlorine*

The effectiveness of chlorine is affected by concentration, pH and contact duration, chlorine demand of the water, and the temperature (Black and Veatch Corporation 2010). When chlorine is combined with water, it produces hypochlorous acid (HOCl), which disintegrates into hydrogen ions (H^+) and hypochlorite (OCl^-) at pH values greater than 8, as the pKa is 7.53. HOCl is a powerful disinfectant because of its neutral charge, which agrees it to pass through the cell wall, change the structure of components in cells, and destroy pathogen activity. By a factor of 100, OCl^- is a weaker disinfectant. Thus, when HOCl is present, at pH 8.0, chlorination is most effective (WHO 2011).

The 'CT-factor' is an index of chlorine's efficiency in disabling pathogens that are determined empirically for all bacteria. The CT factor is calculated by multiplying the concentration of chlorine (in mg/L) by the contact period (in minutes) required to inactivate a certain pathogen. Chlorine may decrease disease-causing viruses by 99–99.99%, bacteria by 82–99.99%, and protozoa by 99–99.90% in the laboratory, however, it is ineffective against the helminth eggs such as *Ascaris lumbricoides* ova and protozoa *Cryptosporidium parvum*. The normal contact period at

25 °C is 30 min; for every 10 °C temperature reduction, the duration of contact should be increased by twofold. This advice is especially critical in cold weather circumstances. These pH and temperature effects have been confirmed in both laboratory and emergency conditions (Marois-Fiset et al. 2013; Elmaksoud et al. 2014; Ali et al. 2015).

When chlorine is introduced to water, it interacts with organic and inorganic elements (such as humic acid, fulvic acid, and metals) and loses its disinfectant properties (CDC 2008; Black and Veatch Corporation 2010). The quantity of chlorine consumed in these processes is known as the chlorine demand of water, and it is calculated experimentally. The leftover chlorine after the chlorine requirement has been fulfilled is referred to as total chlorine residual (TCR). TCR interacts with nitrogen-containing compounds to create combined chlorine, which functions as a mild disinfectant. The free chlorine residual (FCR) is the concentration of leftover free chlorine (in the form of HOCl and OCl^-, and at pH 2 chlorine gas). Breakpoint chlorination, or the time required to complete these processes, usually happens within half an hour following the addition of chlorine.

The occurrence of FCR in treated water is an indicator of potability, indicating that enough chlorine was applied to neutralize most bacteria and that water is protected against recontamination throughout storage, transit, and usage for a certain length of time (CDC 2008). The FCR content decreases throughout the distribution, transit, and storing of chlorinated water before consumption owing to exposure to ultraviolet (UV) radiation, heat, and added pollutants (Lantagne 2008).

3.8 Acquiescence with Chlorination

Chlorination has long been used as an efficient disinfectant. Before assessing if chlorine fits the demands of the municipality, advantages of chlorination must be accounted. The various merits of chlorination are listed below:

1. Chlorination is a well-known technique.
2. At the moment, chlorine disinfection is less expensive than ultraviolet rays or ozone disinfection (unless where de-chlorination is necessary and fire code standards must be satisfied).
3. The chlorine residual in wastewater effluent can extend disinfection even after the first treatment and can be evaluated to assess efficacy.
4. Chlorine disinfection is consistent and efficient against a broad range of harmful organisms.
5. Certain organic and inorganic substances can be oxidized by chlorine.
6. The dose of chlorine is adjustable.
7. During disinfection, chlorine can eradicate some offensive odours.

4 Challenges with Chlorination

The employment of chlorine for the treatment of water is a very old and proven method, but with the passage of time, various demerits of the use of chlorine have also highlighted in recent decades. These demerits are given below:

A. **Water Quality Source—Storage and Dosing:**
   Household water treatment systems must be strong enough to treat water by varying microbiological and physio-chemical impurities. When chlorine is introduced to water, it combines with metals, organic matter, and other substances, rendering it unsuitable for disinfection. Chlorine demand is defined as the disparity seen between quantity of added chlorine and the free residual chlorine after a specified contact duration. The chlorine demand of a particular water supply is consequently subject to temporal fluctuations since it is dependent on numerous characteristics such as turbidity, pH, temperature, and chemical characteristics of the water (Nielsen et al. 2022).
B. **Pathogens, Dose, and Contact Time:** The effectiveness of chlorination is determined by the dose and contact time. Most aquatic infections have been demonstrated to be 4-log inactivated at a dose of 0.5 mg/L and a contact period of 30–60 min. Furthermore, certain microorganisms are resistant to chlorination, which influences the needed chlorine level and contact duration for adequate inactivation. Pathogens, such as the protist *Cryptosporidium parvum*, are very resilient to ordinary chlorination because they produce resistant oocysts throughout their lifespan (Nielsen et al. 2022). If *Cryptosporidium* is a concern (primarily from surface sources), chlorination merely is insufficient; a filter or coagulation/flocculation phase is also advised. Ascaris, an additional chlorine-resistant pathogen with serious health consequences, can be eliminated from water using sedimentation, as well as filtration, or coagulation with flocculation.
C. **Disinfection by-products:** There is a danger of creating a diverse variety of disinfection by-products (DBPs) during chlorination, which can be hazardous to human health and cause worry. When chlorine combines with organic matter in water, DBPs are generated, and the kind and number of DBPs created are determined by parameters such as pH, temperature, water quality, and contact duration. Nonetheless, the concern of DBP creation and the associated severe health impacts may continue to be an impediment to high user compliance (Sandre et al. 2023). Water quality surveillance is widespread in centralized water distribution systems to prevent such issues, but the existence of DBP predecessors is unlikely to be routinely tested in self-supplied neighbourhoods, which might also portray an unnecessary risk towards this disinfection technique (Nielsen et al. 2022).
D. **Assessment of Domestic Chlorination:** When it comes to chlorination, numerous indications may be utilized to assess the effectiveness of an intervention. These consist of the reliability of chlorine products in the residence, free residual chlorine in the water, and better water quality as demonstrated by

microbiological indicators, according to the World Health Organization (2012), Free residual chlorine in water can be a sign of proper and constant usage, including water safety, and is thus important to monitor.

E. **Chlorination of Turbid Water:** Turbid water is difficult to cure since handling alternatives are complex and consumers might be opposed to the completed water's look, flavour, and actual or perceived THM danger. When the water supply is turbid, the following five techniques are available:

   I. Find a non-turbid alternative source.
   II. Employ an alternate treatment (such as filter systems/flocculation).
   III. Use a physical clarifying method, such as cloth filtration, having settled/transferring, or sand or clay filtering, prior to chlorination.
   IV. Use chemical coagulation/flocculation stage, such as moringa or alum, before chlorination.
   V. As is frequently advised, twice the chlorine dose.

5 Disinfection By-Products

When disinfectants like chlorine interact with naturally occurring organic substances in water, disinfection by-products (DBPs) are created. When taking a shower, bathing, doing the dishes, or swimming, DBPs can be discovered in the air. Trihalomethanes (THMs) are the most predominant kind of DBPs. These compounds are formed in drinking water because of several different variables, including pH, temperature, contact period, dosage, inorganic chemicals, and naturally occurring organic matter (Qadafi et al. 2023).

The emergence of chlorinated DBPs in drinking water, such as THMs, has highlighted the necessity to investigate alternative disinfectants and novel treatment approaches. The main by-products of chlorination are trihalomethanes, specifically chloroform, di-chloro-bromo-methane, di-bromo-chloro-methane, and bromoform. In addition to trihalomethanes, the by-products of chlorination and chloramination include halo-acetic acids and halo-aceto-nitriles. Additionally, cyanogen chloride and N-organo-chloramine are produced through chloramination (Álvarez et al. 2022). Over a wide pH range, ClO_2 is a potent water disinfectant. It is very effective in deactivating viruses and is most effective at killing bacteria. THMs are not formed, and the principal products are oxidized ones. One drawback is the potential for the formation of unwanted inorganic species, such as chlorate ions and chlorite. In comparison to chlorine disinfection, ClO_2 will yield fewer organic by-products.

During the chlorination process, organic material has also been reported to produce chlorinated acetic acid. According to Helte et al. (2022), the usual concentration of this chlorinated acetic acid in final drinking water supplies ranges from 30 to 160 g/L. Chloral hydrate, which has been found in drinking water in concentrations ranging from 0.01 to 100 g/L, is also a by-product of water chlorination. Chloramination is the most often used alternative to chlorine, and it results in a significant reduction in by-product formation compared to chlorine, but it requires a longer contact time because it is less effective than chlorine. Additionally, it frequently creates volatile compounds that give off tastes and odours.

The range of these by-products is thought to be influenced mostly by total organic carbon (TOC), humic acid content, and the number of inorganic compounds in the water supply. Since chlorination practices typically result in the production of organic disinfection by-products, the usage of chlorine dioxide is of special relevance. Additionally, it has been shown to be effective in inactivating pathogen species that are resistant to chlorine, such as *Cryptosporidium* and *Giardia*. Recently, it has been utilized to eliminate iron, manganese, and hydrogen sulphide as well as to manage taste and odour (USEPA 1999; Säve-Söderbergh et al. 2020).

The ability of disinfectants to inactivate or kill infections varies depending on the type and characteristics of the organisms and the process circumstances, such as pH, temperature, dose, and contact duration (Couto et al. 2021). Thermal fragmentation of longer halogenated classes results in the creation of small, chlorinated particles and the de-chlorination of organics, according to research on the effect of heating on the formation and behaviour of disinfection by-products (Wu 2001). Various types of disinfection by-products are given below:

5.1 Trihalomethanes

THMs at room temperature are flammable liquids, and both long- and short-term exposure to research animals to high dosages have been linked to a range of harmful consequences. In high-dose chronic investigations, the four most prevalent THMs—Chloroform—$CHCl_3$, Dibromochloromethane—$CHBr_2Cl$ (DBCM), Bromodichloromethane—$CHBrCl_3$ (BDCM), and bromoform—$CHBr_3$, have been proven to be hazardous to rats. $CHCl_3$ is the most common and thoroughly researched chemical in this class. The THM has a maximum contamination limit of 100 μg/L. THMs supplied by maize oil are much more hazardous than equal levels provided via aqueous emulsion. However, administration through drinking water showed no signs of harmfulness. Nevertheless, most of the research has been done with an oil vehicle (Stefán et al. 2019).

5.2 Bromoform ($CHBr_3$)

Bromoform dispersion and elimination were like those of chloroform. Bromoform removal by expired smell was more in the rat than in the mouse, but less in the rat than in the mouse. Although both reductive and oxidative channels were involved in oxidative metabolism, bromoform metabolism appears to be more prevalent. Bromoform, like DBCM, has a far larger potential for generating a mutagenic intermediate than BDCM.

5.3 *Bromodichloromethane ($CHBrCl_3$)*

Bromine substitution reduces volatility and improves the solubility of lipids (uptake into tissues), hence increasing biotransformation. BDCM seems to be the more powerful carcinogenic and rodent toxicant among the four trihalomethanes usually detected in drinking water. However, research on BDCM toxicity is limited. BDCM absorption seems to be quick and quite complete. The stomach, liver, and kidney had the greatest amounts. Absorption and elimination were faster with an aqueous vehicle than with an oil vehicle. BDCM, like chloroform, exhibits P-450-mediated oxidation and reduction metabolism, producing phosgene and dichloro-methyl radicals.

5.4 *Dibromochloromethane ($CHBr_2Cl$)*

DBCM has a lesser hazardous potency when contrasted with $CHCl_3$ and BDCM. DBCM's distribution and elimination pattern was remarkably comparable to that of BDCM, however, it is the least researched THM. Metabolism, it is assumed, follows the same biotransformation pathways as BDCM. The oxidative metabolism of BDCM should create bromo-chloro-carbonyl instead of phosgene, whereas reductive dehalogenation ought to generate a bromo-chloro-methyl radical.

5.5 *Chloroform ($CHCl_3$)*

Chloroform absorption is quick and thorough after dermal, oral, and inhalation routes. Chloroform seems to be broadly distributed in the body, with higher concentrations in the fat and liver and low concentrations in the brain, blood, muscle, kidney, and lung. Chloroform is quickly metabolized and passes through both reductive and oxidative biotransformations via cytochrome P-450. While oxidation metabolism generates phosgene, which is hazardous, reduction biotransformation generates dichloro-methyl radicals, which can combine with phospholipids to form adducts. The balance of oxidation and reduction pathways is affected by oxygen and Chloroform concentrations, strain, animal species, and metabolic location. However, oxidative metabolism takes precedence. Biotransformation of $CHCl_3$ occurs mostly in the kidney and liver. In human beings, inhalation and oral doses absorb 80% and 90% of the $CHCl_3$, respectively. Dermal exposure absorption was roughly similar to inhalation exposure. The oxidation of $CHCl_3$ is catalyzed by human cytochrome.

5.6 Halo-acetic Acid

After THMs, Halo-acetic acid (HAA) is among the next most common CBPs. As of now, chlorinated acetic acids have been better categorized as toxic compared to their brominated counterparts. Despite THMs, Halo-acetic acid is non-volatile and has poor cutaneous absorption (at low doses). Dichloroacetates (DCA) as well as trichloro-acetates (TCA) are found in much larger amounts than mono halo-acetates. TCA and DCA are cancerous metabolites of rodent carcinogens. The halo-acetic acid contamination limit is set at 60 μg/L.

5.7 Dichloroacetic Acid ($Cl_2CHCOOH$)

This chemical occurs as a salt in drinking water; despite this, most tests have been performed using free acid. As a consequence of the substantial pH artefacts that may be estimated when giving a strong acid, the application of the results of such research to evaluating human hazards will be dubious. DCA is rapidly absorbed from the intestine into the circulation. Once in the bloodstream, DCA is delivered to the muscles and liver, followed by the kidney, fat, and other tissues including the cerebral cortex and testes in lower amounts. DCA has a substantially greater systemic clearance. DCA metabolism is coordinated by a new GST, GST-zeta, present in the cytosolic portion. DCA indicates to cause autoinhibition of this enzyme. Despite significant species variations in DCA metabolism, autoinhibition appears to be universal, even in human beings.

5.8 Trichloroacetic Acid (Cl_3CCOOH)

TCA, like DCA, occurs nearly entirely in the salt form at the pH found in water for consumption. TCA is easily assimilated from the gastrointestinal tract in humans as well as animals, and its clearance from the blood is rather sluggish in comparison to other Halo-acetic acids. Around half of the governed dose was removed. There are significant variances in elimination by various varieties. TCA generates the same metabolites as DCA, whether it is transformed into DCA.

5.9 *Other Chlorination by-Products*

Brominated HAAs are generated in bromide-containing waters. There is very little information obtainable regarding the toxicological effects of these compounds. Toxicological data on haloketones and halo acetaldehydes in experimental animals and people are quite sparse. There were little effects on the central nervous system (CNS) and liver. Hepatocellular carcinomas in mice have been described, most likely as a result of a mutagenic influence. TCE, vinyl chloride, and dibromo chloropropane have all been identified as important metabolites by members of this class (Yang et al. 2012). This category contains significant chemicals such as trichloroacetaldehyde and chloral hydrate. Chloral hydrate is most recognized for its CNS depressive effects, with dosages ranging from 500 to 2000 mg producing depression of the CNS in humans. The liver can be harmed by it as well.

6 Rating Guidelines for Disinfection by-Products

A guideline rating is the concentration of a component that does not pose substantial harm to the consumer's health over the course of a lifetime of usage. Concerns about the potential detrimental health consequences of disinfection by-products have spurred significant research efforts in the United States and Europe in attempt to reduce the danger of cancer. Trihalomethanes in drinking water are now regulated by the World Health Organization (WHO), the United States Environmental Protection Agency (USEPA), and the European Union (EU). Under the Safe Drinking Water Act, the United States EPA established a regulation threshold of 100 μg/L for trihalomethanes in 1979. A threshold of 60 μg/L for the total of five halo-acetic acid and 10 μg/L for bromate has also been introduced. In addition to DBP control, coagulated and granulated activated carbon adsorption were mandated to eliminate DBP precursors. The regulatory position in Canada is analogous to that of Australia, and standards are set accordingly. Total THMs have a recommended value of 10 μg/L in Germany. The European Communities (EC) standards for drinking water quality of 100 μg/L for total THMs is presently being reviewed. To reassure consumers that drinking water is secure to consume, recommendations or regulations establishing maximum permitted quantities of chemicals in drinking water (Table 1) must be established.

Table 1 Chlorination by-products and drinking water guideline ratings (WHO 2011)

Chlorination by-products	WHO standard rating μg/L
Trihalomethanes (THMs)	
• Bromoform ($CHBr_3$)	100
• Bromodichloromethane ($CHBrCl_3$)	60
• Dibromochloromethane ($CHBr_2Cl$)	100
• Chloroform ($CHCl_3$)	200
Halo-acetic acid	
• Dichloroacetic acid ($Cl_2CHCOOH$)	50
• Trichloroacetic acid (Cl_3CCOOH)	100
Halo acetonitrile's	
• Dichloroacetonitrile (Cl_2CHCN)	90
Halo-carbonyl compounds	
• Chloral (CCl_3CHO, H_2O)	10

7 Toxicological Measures

Toxicological assessments can be used to measure exposure in numerous areas of environmental epidemiology. Toxico kinetics gives insight into the fate of environmental contaminants as well as the development of more sensitive laboratory methods with lower detection limits. Over the years, there has been significant progress in this area as epidemiologists have introduced more biomarkers of being exposed (Itoh and Matsuoka 1996). It is a complementary science to toxicology in which the facts pertaining to the effect of chemical dangers on animals, typically at considerably greater dosages and under much more controlled settings than occur in human situations. These advancements have provided environmental epidemiologists with a broader set of techniques for use in human investigations, which should boost the statistical strength of the investigations (Cantor et al. 1998). The nitro as well as carbonyl groups, were used to investigate the chromosomal abnormalities of disinfectant-treated water and its by-product. The carbonyl group contributes more to activity causing brevity. In the case of chlorinated water, organochlorine molecules, along with the carbonyl group, contribute to behaviour-inducing aberrations. According to epidemiological research, consuming chlorinated water may be linked to bladder, rectal, and colon cancer (Twerdok et al. 1997). The analysis of genetic mutations in mouse tumours may provide an improved comprehension of the molecular pathways driving chemical carcinogenesis, which is necessary for appropriate extrapolation of cancer models to humans. Small fish models are short-term models which are able to offer comparative toxicity details for each chemical family and chemical combination (Glaze 1993), formerly small fish models were used to investigate chlorination by-products. THMs have been linked to colon cancer in humans, and bromodichloromethane has been linked to a high rate of colon cancer in rodents (Itoh and Matsuoka 1996).

8 Methods for Removing Disinfection By-Products

Water pollution treatment methods include ion exchange, reverse osmosis, co-precipitation, coagulation, herbal filtration, catalytic reduction, adsorption, and electrodialysis. Chlorine-based remediation treatments retain their activity in water for so long as they are not exhausted by deactivation or competitive response. Water treatment designers and operators have just two options: restrict the creation of disinfection by-products through creative chlorination tactics or devise a procedure for elimination of organic and other chlorine vulnerable compounds. The majority of organic pollutants in drinking water are synthetic or natural. These substances include synthetic organic chemicals that cause taste and odour, insecticides, herbicides, colour, and trihalomethane precursors. Natural organic matter (NOM) is formed during the biological breakdown of organic compounds such as fatty acids, amino acids, steroids, phenols, sugars, urea, hydrocarbons, polymers, and porphyrins. Polysaccharides, lipids, Polypeptides, and humic compounds are among the polymers. Natural organic matter levels in drinking water range from 2 to 10 mg/L, while far higher amounts have been documented (Kim and Symons 1991). NOM or derived organic matter (DOM) dissolved and colloidal fractions are important precursors in these processes. Derived organic matter is a diverse combination of organic molecules derived from aquagenic and pedogenic activities. Previously, a variety of strategies for DOM isolation or fractionation were established, ranging from sorption using artificial resin to membrane separation.

The amount of elimination of disinfection by-products and their predecessors is relatively low when utilizing organic polymer coagulation, mechanical procedures such as deep bed filtering, settling, or pressure-controlled permeable membrane processes. As a result, the introduction of an adsorption procedure is shown to be the most appropriate and essential for its elimination. These organic compounds have been removed from water using powdered and granular activated carbon. The compounds formed with natural organic matter and chlorine are referred to as trihalomethane precursors, and the entire process is known as trihalomethane formation potential (THMFP), which may be evaluated directly or by measuring TOC.

Pre-chlorination is often used in treatment facilities to limit bacterial and algal development. In this situation, the trihalomethane production event occurs prior to the addition of PAC, and therefore the oxidation-reduction interaction that occurs between chlorine and activated carbon surface hindered the activated carbon's adsorption ability for phenolic chemicals (Hooper et al. 1996). As a result, the removal of trihalomethane precursors and trihalomethanes from drinking water by PAC varies. The proportion of natural organic matter that causes trihalomethane varies depending on the water sample. However, the data show that there is no relationship between the levels of trihalomethane precursor and total organic carbon in the various types of water.

Bacterial growth potential (BGP) and trihalomethane formation potential (THMFP) in relation to derived organic matter were investigated before and after alum treatment, and results showed that disinfection by-product of terrestrially derived DOM can be higher compared to the overall derived organic matter in natural water, and alum coagulation was not found to be sufficient to yield microbiologically dependable drinking water (USEPA 1999). The effectiveness of natural organic matter removal via adsorption in the column is much higher than that of traditional coagulation, especially at the start of a new absorbent surface. It has been demonstrated that activated carbon derived from apricot stones by pyrolysis may be utilized to remove trihalomethanes from chlorine-treated water.

9 Conclusion

Chlorination has a lot of benefits, including being very affordable, easy to use, and scalable. Due to the high risk of producing unfavourable taste, smell, and disinfection by-products when the source water contains a certain amount of organic matter, certain technological approaches are required to apply the correct chlorination dose, and the chlorine dose is not accurately estimated by experts. This is frequently ascribed to the end users' lack of motivation, highlighting the need for research on user behaviour changes in relation to water treatment, knowledge of water quality, and the dangers of consuming untreated water. Chlorine has long been used for water disinfection, but recent advancements have focused on reducing the formation of disinfection by-products, which can be harmful to human health. Some of these advancements include using chlorine dioxide (ClO_2) or monochloramine, which are less reactive forms of chlorine that can more effectively oxidize organic matter in water and reduce DBP formation. Other advancements include using automated monitoring and control systems to optimize chlorine dosing, as well as alternative disinfectants like UV light and ozone. These advancements have contributed to safer and more effective water treatment, but ongoing research is still needed to continue improving water treatment technologies and reducing the potential for harmful DBP formation.

References

Ali SI, Ali SS, Fesselet J-F (2015) Effectiveness of emergency water treatment practices in refugee camps in South Sudan. Bull World Health Organ 93(8):550–558. https://doi.org/10.2471/BLT.14.147645

Álvarez AR et al (2022) Chlorination by-products formation in a drinking water distribution system treated by ultrafiltration associated with pre-ozonation or coagulation/flocculation. J Water Process Eng 47:102779. https://doi.org/10.1016/j.jwpe.2022.102779

APHA (2012) Standard methods for the examination of water and wastewater, p 1496

Binnie C, Kimber M, Thomas H (2017) Disinfection. In: *Basic water treatment*. ICE Publishing, pp 193–210. https://doi.org/10.1680/bwtse.63341.193
Black & Veatch Corporation (2010) White's handbook of chlorination and alternative disinfectants.5th edn. John Wiley & Sons [Preprint], New York
Boone JD, Erickson BK, Huh WK (2012) New insights into cervical cancer screening. J Gynecol Oncol 23(4):282. https://doi.org/10.3802/jgo.2012.23.4.282
Cantor KP et al (1998) Drinking water source and chlorination byproducts. I. Risk of bladder cancer. *Epidemiology (Cambridge, Mass.)* 9(1):21–28
CDC (2008) Chlorine residual testing fact sheet, CDC SWS Project Chlorine Residual Testing
CDC (2014) Safe water system
Chinn TD (2009) Water treatment. In: *Environmental engineering*. 2009th edn. John Wiley & Sons, Inc., Hoboken, NJ, USA, pp 133–281. https://doi.org/10.1002/9780470432808.ch2
Couto M et al (2021) Health effects of exposure to chlorination by-products in swimming pools. Allergy 76(11):3257–3275. https://doi.org/10.1111/all.15014
Elmaksoud SA et al (2014) Use of household bleach for emergency disinfection of drinking water. J Environ Health 76(9):22–25
Fava JA, Thomas DL (1978) Use of chlorine to control OTEC biofouling. Ocean Eng 5(4):269–288. https://doi.org/10.1016/0029-8018(78)90004-5
Glaze WH (1993) Drinking water and health in year 2000
Hall T, Hyde R (1992) Water treatment processes and practice. WRc [Preprint]
Helte E, Säve-Söderbergh M, Ugge H, Fall K, Larsson SC, Åkesson A (2022) Chlorination by-products in drinking water and risk of bladder cancer–A population-based cohort study. Water Research 214:118202
Hooper SM et al (1996) Improving GAC performance by optimized coagulation. J Am Water Works Assoc 88(8):107–120
Hussain A, Gupta A (2022) Water supply and sanitation. I. K. International, New Delhi, India
Itoh S, Matsuoka Y (1996) Contributions of disinfection by-products to activity inducing chromosomal aberrations of drinking water. Water Res 30(6):1403–1410. https://doi.org/10.1016/0043-1354(96)00022-X
Jaya Prakash K et al (2007) Post-treatment of UASB reactor effluent by coagulation and flocculation process. Environ Prog 26(2):164–168. https://doi.org/10.1002/ep.10192
Kalibbala HM (2007) Application of indigenous materials in drinking water treatment
Kim P-S, Symons JM (1991) Using anion exchange resins to remove THM precursors. J Am Water Work Assoc 83(12):61–68
Lantagne DS (2008) Sodium hypochlorite dosage for household and emergency water treatment. J Am Water Work Assoc 100(8):106–119. https://doi.org/10.1002/j.1551-8833.2008.tb09704.x
Lu W, Kiéné L, Lévi Y (1999) Chlorine demand of biofilms in water distribution systems. Water Res 33(3):827–835. https://doi.org/10.1016/S0043-1354(98)00229-2
Marois-Fiset J-T et al (2013) Effects of temperature and pH on reduction of bacteria in a point-of-use drinking water treatment product for emergency relief. Appl Environ Microbiol 79(6):2107–2109. https://doi.org/10.1128/AEM.03696-12
Mohamed H et al (2015) Point-of-use chlorination of turbid water: results from a field study in Tanzania. J Water Health 13(2):544–552. https://doi.org/10.2166/wh.2014.001
Nielsen AM et al (2022) Chlorination for low-cost household water disinfection – a critical review and status in three Latin American countries. Int J Hyg Environ Health 244:114004. https://doi.org/10.1016/j.ijheh.2022.114004
Qadafi M et al (2023) Chlorination disinfection by-products in Southeast Asia: a review on potential precursor, formation, toxicity assessment, and removal technologies. Chemosphere 316:137817. https://doi.org/10.1016/j.chemosphere.2023.137817
Sandre F et al (2023) Occurrence and fate of an emerging drug pollutant and its by-products during conventional and advanced wastewater treatment: case study of furosemide. Chemosphere 322:138212. https://doi.org/10.1016/j.chemosphere.2023.138212

Säve-Söderbergh M et al (2020) Exposure to drinking water chlorination by-products and fetal growth and prematurity: a Nationwide register-based prospective study. Environ Health Perspect 128(5):057006. https://doi.org/10.1289/EHP6012

Sayers P (2012) Flood risk: planning, design and management of flood defence infrastructure. ICE Publishing

Stefán D et al (2019) Formation of chlorination by-products in drinking water treatment plants using breakpoint chlorination. Microchem J 149:104008. https://doi.org/10.1016/j.microc.2019.104008

Tvrdá E, Benko F (2020) Free radicals: what they are and what they do. In: Pathology. Elsevier, pp 3–13. https://doi.org/10.1016/B978-0-12-815972-9.00001-9

Twerdok LE et al (1997) The use of nontraditional assays in an integrated environmental assessment of contaminated ground water. Environ Toxicol Chem 16(9):1816–1820. https://doi.org/10.1002/etc.5620160908

USA Surface Water Treatment Rule (SWTR) (1989)

USEPA (1999) Alternative disinfection and oxidants guidance manual

Wang LK (2006) Water chlorination and Chloramination. In: Advanced physicochemical treatment processes. Humana Press, Totowa, NJ, pp 367–401. https://doi.org/10.1007/978-1-59745-029-4_11

Wang LK, Hung Y-T, Shammas NK (eds) (2006a) Advanced physicochemical treatment processes. Humana Press, Totowa, NJ. https://doi.org/10.1007/978-1-59745-029-4

WHO (1993) Guidelines for drinking water quality. World Health Organization, Geneva, pp 1–29

WHO (2011) Guidelines for drinking-water quality. World Health Organization

World Health Organization (2012) A toolkit for monitoring and evaluating household water treatment and safe storage programmes. World Health Organization

Wu W (2001) Effects of thermal treatment on halogenated disinfection by-products in drinking water. Water Res 35(15):3545–3550. https://doi.org/10.1016/S0043-1354(01)00080-X

Yang X et al (2012) Precursors and nitrogen origins of trichloronitromethane and dichloroacetonitrile during chlorination/chloramination. Chemosphere 88(1):25–32. https://doi.org/10.1016/j.chemosphere.2012.02.035

A Comparative Study of Different Disinfection Processes, and Management Practices to Control the Formation of Disinfection by-Products (DBPs)

Namrata Malik and Rahul Kumar

Abstract The disinfectant chemical that is most frequently utilized is chlorine. The most popular kinds are calcium or sodium hypochlorite and chlorine gas. Alternative disinfectants include gamma radiation, ozone, chlorine dioxide, chloramines, ultraviolet light, and others. Water flow rates, application and demand rates, water pH, equipment cost, chemical availability, and maintenance issues should all be carefully taken into account before choosing an appropriate methodology. When used improperly or in excess amount, several of these chemical disinfectants have an ability to respond with natural and inorganic forerunners to create DBPs that have negative health effects. This book chapter delves into the complex world of drinking water disinfection, exploring the different processes used to ensure the safety of our water supply. The chapter examines various key disinfection methods including chlorination, ozonation, thermal disinfection, and UV radiation, and compares their effectiveness and potential formation of DBPs. Through a comprehensive analysis of previous studies and new research, this chapter provides a detailed understanding of the advantages and disadvantages of each disinfection method, as well as their impact on water quality. The chapter also highlights the potential health risks associated with DBPs and provides recommendations for minimizing exposure to these harmful compounds.

Keywords Disinfection methods · Disinfection by-products (DBPs) · Water quality · Health risks · Waterborne pathogens

N. Malik · R. Kumar (✉)
Department of Life Sciences and Biotechnology, School of Science, Chhatrapati Shivaji Maharaj University, Panvel, Navi Mumbai, India

S. Madhav et al. (eds.), *Drinking Water Disinfection By-products*,
https://doi.org/10.1007/978-3-031-49047-7_4

1 Introduction

Disinfection by-products (DBPs) are a complicated collection of synthetic mixtures formed during water treatment when disinfectants react with the organic and inorganic matter in the water (Richardson et al. 2007; Li and Mitch 2018). DBPs are becoming a growing concern for water treatment and public health due to their potential health risks and potential effects on water safety and quality (Postigo and Richardson 2014; Benson and co. 2017).

DBPs can result from the use of disinfectants like chlorine, chloramines, or ozone to clean water (Ngwenya et al. 2013). Chlorine, the most widely used disinfectant, effectively kills bacteria and viruses in water. However, when chlorine interacts with water's natural organic matter (NOM), it can produce a number of DBPs, such as chlorite, haloacetic acids (HAAs), and trihalomethanes (THMs). According to Zazouli and Kalankesh (2017), the chemicals chloroform, bromodichloromethane, dibromochloromethane, and bromoform are all produced when chlorine and NOM react to produce a class of DBPs known as THMs. Chlorine reacts with humic and fulvic acids to produce HAAs (Bond et al. 2012; Parvez and co. 2019; Wang and co. 2023). Chlorite is a by-product of disinfecting with chlorine dioxide and can also be produced by using chlorine (Jefri et al. 2022).

The concentration of DBPs in drinking water is directed by the US Environmental Protection Agency (EPA), which sets the most extreme pollutant levels for THMs, HAAs, and chlorite. The MCL for THMs is 80 micrograms per liter (μg/L), while the most extreme pollutant level for HAAs is 60 μg/L (USEPA 2006; USEPA 2013). The MCL for chlorite is 1 milligram per liter (mg/L). These regulations are designed to protect public health and ensure safe drinking water (Benson et al. 2017).

DBPs can have potential health effects, and the risk of these effects relies upon the concentration, duration, and frequency of exposure, as well as individual susceptibility and other factors (Mishra et al. 2014). THMs and HAAs have been related to an expanded risk of bladder cancer, while chlorite has been related to an increased risk of anemia in infants (Weisman et al. 2022). Other potential health effects of DBPs incorporate the problems related to the reproductive system, developmental effects, and increased risk of asthma and other respiratory problems (Villanueva et al. 2015). The health risks associated with DBPs are complex and may differ relying on specific DBP, the concentration, and other factors (Anderson and Meade 2014).

To reduce the formation of DBPs, water treatment plants can utilize alternative disinfectants such as ozone or ultraviolet (UV) light, or reduce the concentration of natural organic matter in the source water (Tak and Vellanki 2018; Yang et al. 2021). Ozone is a powerful disinfectant that does not form THMs or HAAs, and can be effective at reducing DBPs (Chin and Berube 2005). UV light is also effective at killing bacteria and viruses and does not form DBPs (Mackenzie 2020). However, both ozone and UV treatments may be more costly or require additional treatment steps (Chin and Berube 2005). Another approach to reducing DBPs is to use treatment processes such as granular activated carbon (GAC) or ion exchange. GAC can be effective at removing NOM and reducing the formation of DBPs, while ion exchange can remove natural organic matter and other compounds that can react with disinfectants (Levchuk et al. 2018; Erdem et al. 2020). Disinfection by-products are a growing concern in water treatment and public health (Table 1). THMs, HAAs, and chlorite are the most common DBPs, and their concentration in drinking water is regulated by the EPA (Li and Mitch 2018; DeMarini 2020). To reduce the formation of DBPs, water treatment plants can use alternative disinfectants or treatment processes, but these methods may have additional costs or complexities (Ding et al. 2019; Leonard et al. 2022). It is important for water utilities and regulatory agencies to monitor and manage DBPs to ensure safe drinking water for the public.

The objectives of this book chapter are to: investigate the mechanisms by which DBPs are formed during disinfection processes; evaluate the health effects associated with exposure to DBPs; identify factors that influence the formation of DBPs in drinking water; and provide recommendations for monitoring and regulating DBPs in drinking water to ensure compliance with public health standards.

Table 1 DBPs present in disinfected waters

Disinfectant	Significant organohalogenic DBPs	Significant inorganic DBPs	Significant non-halogenic DBPs	References
Chlorine / hypochlorous acid	THM; HAA; Haloacetonitrils, Cl_2H_2O, Nitrochloroform, CPLs, N-chloramines, Halofuranones, Bromohydrins	Chlorates (mostly produced from the use of hypochlorite)	Aldehydes, Cyanoalkanoic acids, Alkanoic acids, benzene, R-COOH	Eryılmaz and Palabıyık (2013) Block and Rowan (2020)
Chlorine dioxide		Chlorite, chlorate	Unknown	Hua and Reckhow (2007) Jefri et al. (2022)
Chloramine	CK, organic chloramines, Chloramino acids, Haloketones, etc.	Nitrate, nitrite, chlorate, hydrazine	Aldehydes, ketones	Li and Mitch (2018)
Ozone	$CHBr_3$, MBA, DBA, DBAC, BrCN	Chlorate, iodate, bromate, H_2O_2, HBrO, epoxides, Ozonates	Aldehydes, Ketoacids, ketones, R-COOH	Mao et al. (2014) Pulicharla et al. (2020)

Table 2 Toxicological effects of DBPs

Class of DBPs	Compounds	Ratings	Effects
Trihalomethanes	Chloroform	B2	Cancer, liver, kidney, and reproductive effects
	Dibromochloromethane	C	Nervous system, liver, kidney, and reproductive effects
	Bromodichloromethane	B2	Cancer, liver, kidney, and reproductive effects
	Bromoform	B2	Cancer, nervous system, liver, and kidney effects
Haloacetic Acid	Monochloroacetic acid	–	Cancer and reproductive and developmental effects
	Dichloroacetic acid	B2	Cancer, reproductive, and developmental effects
	Trichloroacetic acid	C	Liver, kidney, spleen, and developmental effects
Haloacetonitrile	Trichloroacetonitrile	–	Cancer, mutagenic and clastogenic effects
Bromate	Bromide	–	Genotoxic carcinogen
	Bromate	B2	Cancer
Chlorite	Chlorite	D	Developmental and reproductive effects
	Chlorate	–	Irritation in the mouth, esophagus, or stomach, cancer or birth defects
Halophenol	2-Chlorophenol	–	Cancer and tumor promoter
Nitrosodimethylamine	–	–	Liver damage accompanied by internal bleeding, liver cancer and lung cancer, death of human babies

Source: Zazouli and Kalankesh (2017)

2 Methods of Disinfection

Disinfection can be defined as the killing, eradication, or deactivation of dangerous and harmful pathogenic microorganisms (Saqib Ishaq et al. 2019). The growth and reproduction of harmful microbes are stopped by their destruction or deactivation. Consuming contaminated water that is home to harmful microbes can make people sick. Sterilization kills all bacteria, both harmful and benign, while disinfection is a related process. Thus, disinfection is a better procedure to use.

2.1 Chlorine Gas

Chlorine gas is harmful and greenish-yellow in color. The high pressure makes the gas transform into fluid or liquid. Most commonly, chlorine gas is used to disinfect water. Chlorine is a very powerful way to get rid of practically all pathogenic germs

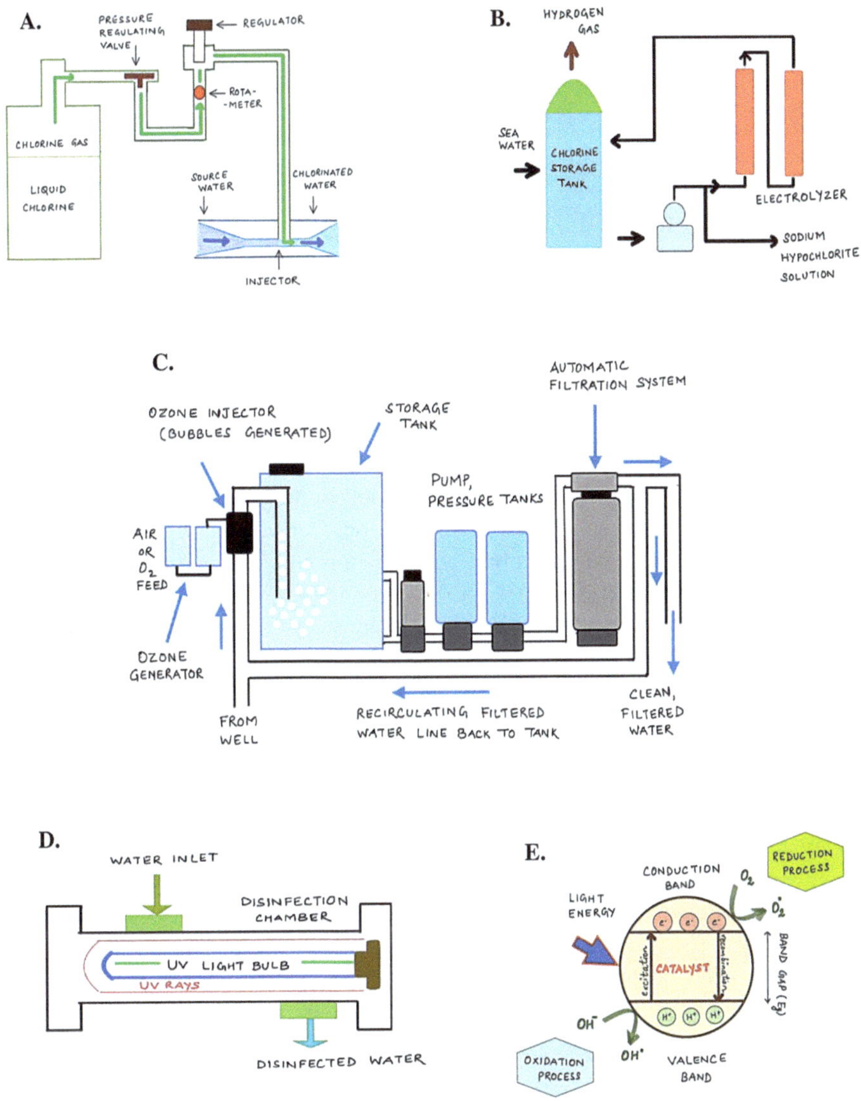

Fig. 1 Various disinfection methods: (**a**) Chlorination by gas method, (**b**) Chlorination by sodium hypochlorite solution, (**c**) Ozonation; (**d**) Disinfection by UV, and (**e**) Photocatalytic mechanism

in water (Saqib Ishaq et al. 2019). It has both primary and secondary disinfecting uses. Because the gas is so hazardous, it cannot be used in a residential system. Even at 0.1% air by volume concentrations, it is fatal (Khayyat 2000; Saqib Ishaq et al. 2019).

2.1.1 Process of Disinfection

The process of disinfecting drinking water using chlorine gas is a crucial procedure in ensuring that the water is safe for consumption (Wagenet et al. 1988; Lindsay 2011). This process involves several steps that are carefully monitored and regulated to ensure the quality of water (Fig. 1). The first step in this process is pre-chlorination. During this step, chlorine gas is forcefully added to the raw water to kill any bacteria or other microorganisms present in it. This step also helps in removing some organic matter present in the water. The addition of chlorine gas to the water creates a chemical reaction that kills most of the bacteria and viruses in the water (EPA 2022). The second step is flocculation. During this step, a coagulant is added to the water to create flocs or clumps of impurities. This makes it easier to remove them during the next step (EPA 2022).

The third step is sedimentation. In this step, the water is allowed to settle, and the flocs formed during the previous step settle to the bottom of the tank. This helps to remove any remaining impurities from the water (EPA 2022). The fourth step is filtration. To eliminate any leftover pollutants, the water is subsequently filtered through the effective filters. The filters help to remove any remaining impurities from the water (EPA 2022). The fifth step is disinfection. Chlorine gas is added to the filtered water to disinfect it. The amount of chlorine added is carefully controlled to ensure that it is effective in killing any remaining bacteria or viruses but not enough to be harmful to human health. This step ensures that the water is free from harmful pathogens that could cause illness in humans (EPA 2022). The final step is storage. The water is further stored in a tank until and unless it is made available to users. This step ensures that the water is available for use when needed (EPA 2022).

2.1.2 Disinfection by-Products

During the disinfection process, DBPs are formed when disinfectants like chlorine react with organic waste in the water (WHO 2011). Trihalomethanes, haloacetic acids, chlorite and chlorate, and bromate are a few of the DBPs that can form when chlorine gas is used to treat drinking water. Of these, trihalomethanes are known to be carcinogenic and can have severe health effects on human beings, while the others are typically less toxic (WHO 2011).

2.1.3 Advantages

Chlorination is a widely used method to destroy disease-causing microbes (CDC 2022). This method of disinfection is highly effective and can eliminate a wide range of harmful microorganisms, including those that are resistant to other disinfection methods. Chlorination is also an affordable method of disinfecting drinking water, as the cost of chlorine is quite low in comparison to other useful methods (CDC 2022).

One of the advantages of chlorination is that it is easy to apply and requires minimal equipment and training. This makes it a practical method of disinfection for both small and large water treatment systems. The ease of application is particularly important in developing nations, where access to clean water is often scarce, and advanced water treatment technologies may not be available. Chlorination provides an effective solution that can be implemented with minimal resources (EPA 2021).

Chlorine has a residual effect, meaning it remains in the water after disinfection to provide ongoing protection against recontamination (WHO 2017). This residual effect is particularly beneficial in large water distribution systems, where water may travel long distances before reaching the end user. Chlorine ensures that the water remains safe for consumption, even after extended periods in the distribution system (WHO 2017).

Chlorination is a well-established technology that has been used for over 100 years (Hua et al. 2020). As a result, there is a vast body of knowledge and expertise available on its use. This knowledge has been gained through extensive research, experimentation, and practical application, and it ensures that chlorination is a reliable and safe method of disinfection.

Perhaps the most significant advantage of chlorination is its link to considerable improvements in public health. Chlorination has been associated with a sharp decline in diseases like cholera, typhoid, and dysentery that are spread through contaminated water. These diseases are particularly prevalent in developing nations, where access to clean water is scarce.

2.1.4 Disadvantages

While chlorination is an efficient method of disinfecting drinking water, there are a number of drawbacks that must be considered. The production of disinfection by-products is one of the main drawbacks. Another disadvantage of chlorination is that it can change taste and odor of the water, which some people find unpleasant. This can be particularly problematic in areas where the water has a high concentration of organic matter, as the taste and odor can be more pronounced. Chlorine can also be corrosive to pipes and plumbing fixtures, which can lead to leaks and other problems over time (Stauber et al. 2004). This may cause problems in areas where the water is naturally acidic, as the combination of chlorine and acidity can accelerate the corrosion process.

Some organisms, such as cryptosporidium, can be resistant to chlorine, making it difficult to achieve complete disinfection. This is a significant concern, as cryptosporidium can cause severe gastrointestinal illness and can be particularly problematic in areas where there is a high concentration of livestock or other sources of contamination. Finally, chlorine and its by-products can have an environmental impact, such as affecting aquatic ecosystems or contributing to the formation of smog. This is a significant concern, particularly in areas where water treatment plants discharge their effluent into rivers or other bodies of water.

2.2 *Chlorination (Sodium Hypochlorite Solution)*

Sodium hypochlorite solution is a disinfectant and bleaching agent that is commonly used in water treatment, household cleaning, and industrial applications (Malik and Ahmad 2018). It is a clear, colorless, or slightly yellowish liquid that has a strong odor of chlorine. The solution is formed by dissolving sodium hydroxide and chlorine gas in water, resulting in the formation of sodium hypochlorite (NaOCl) and sodium chloride (NaCl) as by-products. The concentration of sodium hypochlorite in the solution can vary depending on its intended use, with common concentrations ranging from 5% to 15% (Vega-Méndez et al. 2017). A vast variety of microorganisms, including bacteria, viruses, and fungi, can be killed by sodium hypochlorite. It is a strong oxidizing agent that can also degrade pollutants and enhance water quality by reacting with organic substances in water (Rabêlo Braitt et al. 2013).

2.2.1 Process of Disinfection

This process typically involves several steps to ensure the microbiological quality of water (Fig. 1). The first step in disinfecting water with sodium hypochlorite solution is to determine the appropriate dosage. The amount of sodium hypochlorite needed relies upon the degree of pollutants present in the water, which can be determined through water testing and analysis (Kapoor et al. 1999).

Once the appropriate dosage has been determined, the sodium hypochlorite solution needs to be prepared. Sodium hypochlorite is typically available in liquid form and needs to be diluted to the appropriate concentration before use. The dilution process involves adding a certain amount of sodium hypochlorite to a particular volume of water to accomplish the ideal concentration.

The diluted sodium hypochlorite solution is then added to the water that needs to be disinfected. This can be done using a dosing pump or by manually pouring the solution into the water. It is important to ensure that the sodium hypochlorite solution is mixed thoroughly with the water to ensure that all parts of the water are exposed to the disinfectant. This can be achieved by using a mixing tank or by allowing the water to flow through a mixing chamber.

The water needs to be left in contact with the sodium hypochlorite solution for specific time duration to eliminate the microbial pathogens. The duration of contact time relies upon the sodium hypochlorite concentration used and the extent of water contamination. Once the contact time is complete, the sodium hypochlorite solution needs to be neutralized to prevent it from reacting with other chemicals or materials. This is typically achieved by adding sodium bisulfite or another neutralizing agent to the water. Finally, the water needs to be tested to ensure that the disinfection process is successful and that the water is fit for the drinking purpose. This is a crucial step in ensuring that the water is free from contaminants and safe to drink.

2.2.2 Disinfection by-Products

Disinfecting drinking water with sodium hypochlorite solution can create several DBPs including THMs and HAAs. THMs are formed when chlorine and NOM react in water. Long-term exposure to THMs can cause health problems, including an increased risk of cancer. HAAs are formed when chlorine reacts with humic acids and fulvic acids present in the water. Like THMs, long-term exposure to HAAs can cause health problems. As chlorine combines with inorganic molecules in the water, chlorite and chlorate are created. Chlorite and chlorate can cause health problems when consumed in high concentrations (WHO 2011). Chlorite can cause anemia, while chlorate can cause thyroid problems. Bromate is formed when bromide ions in the water react with chlorine. Bromate is a suspected carcinogen and can cause health problems when consumed in high concentrations. Long-haul exposure to bromate can build the risk of disease.

2.2.3 Advantages

Disinfecting drinking water using sodium hypochlorite solution has several advantages. Firstly, it is an effective disinfectant against a wide range of microorganisms (Khan et al. 2013). Secondly, it is very cost-effective in comparison to other widely used disinfecting procedures including UV treatment or ozonation (Momba et al. 2006). Thirdly, it is easy to use and can be applied to water in a variety of ways. Fourthly, it is a reliable disinfectant that can maintain its effectiveness over a wide range of water quality parameters. Fifthly, it can leave a residual disinfectant in the water, which can help to prevent microbial growth in the water supply. Lastly, the use of sodium hypochlorite for water disinfection has been well-established and is widely used around the world, which means that the technology is well understood and there is a wealth of knowledge and experience available to guide its use.

2.2.4 Disadvantages

One of the major concerns with the use of sodium hypochlorite solution is the unnecessary production DBPs. When sodium hypochlorite reacts with organic matter available in the water supply, it can form DBPs that can be harmful to human health if consumed in high concentrations. Therefore, it is of paramount importance to check the levels of DBPs in water on a regular basis to ensure that they are within safe limits. Another disadvantage of using sodium hypochlorite solution is that it is a corrosive chemical that can damage pipes, pumps, and other equipment used in the water treatment process. This can increase maintenance costs and reduce the lifespan of equipment, which can affect the overall efficiency of a treatment plant (UNEP, 2014).

Sodium hypochlorite is also a hazardous chemical that requires careful storage and handling to ensure the safety of human beings or the environment. It is crucial to follow strict protocols when handling sodium hypochlorite to prevent spills, leaks, or other accidents that could harm people or the environment. The environmental impact of sodium hypochlorite is also a concern. If it is released into the environment in high concentrations, it can be toxic to aquatic life and other organisms. Therefore, it is crucial to ensure that the concentration of sodium hypochlorite in water is within the safe limits before it is released into the environment (UNEP, 2014).

2.3 Chlorination (Solid Calcium Hypochlorite)

Chlorination with solid calcium hypochlorite is a method of disinfecting water by adding calcium hypochlorite, a white or pale gray powder that is used as a disinfectant and bleaching agent (Khanna et al. 2016). To perform chlorination with solid calcium hypochlorite, the powder is typically dissolved in water to create a concentrated solution, which can then be added to the water being treated (Borchardt et al. 2019). The amount of calcium hypochlorite needed relies upon the volume of water being disinfected and the level of required disinfection. Calcium hypochlorite works by releasing hypochlorous acid, which is a powerful disinfectant that can kill bacteria, viruses, and other microorganisms in the water (Snowden-Swan et al. 1998).

2.3.1 Process of Disinfection

The process of disinfection using Solid Calcium Hypochlorite involves several steps (USEPA 2003). The first step is to dissolve the solid calcium hypochlorite in water to create a chlorine solution. The chlorine solution is prepared in a mixing tank that is equipped with a mechanical mixer to ensure that the calcium hypochlorite dissolves completely. The amount of calcium hypochlorite added to the blending tank relies upon the expected concentration of chlorine in the solution. The subsequent step is to infuse the chlorine solution into the water to be cleaned. The chlorine solution is infused into the water utilizing a chemical dosing siphon that is aligned to deliver the right measure of chlorine solution to the water. The dosing siphon is constrained by a regulator that screens the chlorine concentration in the water and changes the siphon's speed likewise to keep a consistent chlorine level.

The third step is to permit the chlorine to respond to the microorganisms present in the water. The chlorine in the water reacts with the organic matter and microorganisms, damaging their cell walls and killing them. The response time relies upon the water temperature, pH, and chlorine concentration in the water. The fourth step is to guarantee that the residual chlorine concentration in the water is at a protected and viable level. The residual chlorine concentration is the amount of chlorine remaining in the water after the disinfection procedure. It is essential to keep a residual chlorine concentration in the water to guarantee that it stays safe for utilization.

2.3.2 Disinfection by-Products

The most common DBPs that originate during the disinfection procedure are THMs and HAAs. These DBPs are formed when the chlorine in the calcium hypochlorite responds to the organic matter present in the water. The formation of DBPs is affected by several factors, including the amount of chlorine used, the time duration of the procedure, and the temperature and pH of the water. THMs and HAAs are classified as carcinogenic compounds, which mean they have the potential to cause cancer (Borchardt et al. 2019). Exposure to these compounds can also cause other health problems, such as reproductive and developmental effects. To reduce the formation of THMs and HAAs, water treatment plants using Solid Calcium Hypochlorite for disinfection may use strategies such as optimizing the disinfection process to minimize the amount of chlorine used, controlling the pH of the water, and removing organic matter before disinfection (Khanna et al. 2016).

2.3.3 Advantages

Drinking water disinfection is a crucial process in ensuring the safety of freshwater consumption. One of the most commonly used disinfectants is Solid Calcium Hypochlorite, which has several advantages over other disinfectants (Snowden-Swan et al. 1998). The first advantage of using Solid Calcium Hypochlorite is that it kills a wide variety of pathogenic microbes, including parasites, viruses, and bacteria, with great efficiency. The second advantage of using Solid Calcium Hypochlorite is that it is relatively easy to use and transport. Solid Calcium Hypochlorite is stable, non-explosive, and does not require special storage conditions. It is also easy to transport, as it can be packed in small containers and transported to remote areas where access to safe drinking water is limited.

The third advantage of using Solid Calcium Hypochlorite is that it is cost-effective. The cost of Solid Calcium Hypochlorite is relatively low compared to other disinfectants, such as ozone and ultraviolet radiation. Additionally, it is simple to manufacture, making it a viable option for water treatment facilities with limited funds. The fact that Solid Calcium Hypochlorite persists in the water, providing ongoing protection against microorganisms, is the fourth advantage of using it.

2.3.4 Disadvantages

Although Solid Calcium Hypochlorite is a potent disinfectant for drinking water, its use has some drawbacks (Snowden-Swan et al. 1998). The first disadvantage of using Solid Calcium Hypochlorite is the possibility of producing DBPs. DBPs, some of which have been linked to cancer, are formed when Solid Calcium Hypochlorite is used to disinfect water. These DBPs are produced when the chemical reacts with the organic matter present in the water. These DBPs may be harmful to human health, particularly when consumed in large quantities over an extended period of time.

The second disadvantage of using Solid Calcium Hypochlorite is that it can change the odor and taste of water. This is because Solid Calcium Hypochlorite contains chlorine, which can produce a strong taste and odor in the water. This can be unpleasant for some people and may affect their willingness to drink the water. The third disadvantage of using Solid Calcium Hypochlorite is that it can corrode metal pipes and equipment. Chlorine is a highly reactive chemical, and when it comes into contact with metals, it can cause corrosion. This can result in damage to the water distribution system and may lead to leaks and water quality issues.

The fourth disadvantage of using Solid Calcium Hypochlorite is that it requires careful handling and storage. If it comes into contact with the skin or eyes, solid calcium hypochlorite is a dangerous substance that can irritate the skin and damage the eyes. It should be handled with care and kept somewhere cool, dry, and well-ventilated.

2.4 Chloramines

Chloramines are a group of disinfectants that are commonly used to treat fresh drinking water. They are formed when chlorine is added to water that already contains ammonia or other organic nitrogen compounds. The resulting chloramines are less reactive than free chlorine and have a longer-lasting residual effect, making them an effective option for maintaining water quality throughout the distribution system (Arasmith 1993).

When chlorine is added to water, three kinds of chloramines can be produced:

Monochloride (NH_2Cl): This is the chloramine that is used the most, and it is made when chlorine is added to ammonia-containing water. Disinfection by-products

(DBPs) are reduced and microbial growth is effectively controlled by monochloramine.

Dichloramine ($NHCl_2$): This less common chloramine is made when too much chlorine is added to ammonia-containing water. Dichloramine is less effective at controlling microbial growth than monochloramine, but it is more effective at controlling nitrification in distribution systems.

Trichloramine (NCl_3): This is the least commonly used chloramine and is formed when a large excess of chlorine is added to water containing ammonia. Trichloramine is not an effective disinfectant and is primarily a by-product of chloramination.

2.4.1 Process of Disinfection

According to Tech (2005), one crucial step in ensuring that the water is safe for consumption is the chloramine-based disinfection process. In most cases, the process involves a number of carefully planned steps to get rid of any pathogens or organic matter that might be in the raw water.

Pre-chlorination is the process's first step. Free chlorine is used to treat the water before adding chloramines to get rid of any pathogens or organic matter that might be in the raw water. In order to guarantee that the water does not contain any contaminants that could be harmful to human health, this step is critical. Ammonia is added to the water after the pre-chlorination process has been completed. The desired chloramine concentration and water quality parameters determine the amount of ammonia added. Chlorine and ammonia can react and form chloramines thanks to this crucial step.

Chlorine is produced when ammonia and chlorine react to form chloramines, which are allowed to grow in the water. This step is crucial to ensuring that the water is safe for consumption and effectively disinfected. After the chloramines have formed, the water is tested to make sure that the distribution system maintains the desired residual chloramine concentration. This step is very important because it makes sure that the water is safe to drink and that the disinfection process works.

After that, the water is distributed through the distribution system, and the levels of chloramine are carefully monitored to make sure they stay at the right levels to effectively disinfect the water. This step is crucial because it ensures that the water throughout the distribution system is safe to drink.

2.4.2 Disinfection by-Products

While chloramines are generally considered to produce fewer disinfection by-products (DBPs) than other disinfectants like free chlorine, they can still produce some DBPs that can have potential health effects. Nitrosamines, haloacetic acids, trihalomethanes, and N-nitrosodimethylamine (NDMA) are the most common DBPs produced when chloramines are used to disinfect drinking water.

When chloramines react with the water's organic nitrogen compounds, they can form nitrosamines, which are carcinogenic compounds. Concern should be raised because nitrosamine exposure can increase cancer risk. When chloramines react with organic matter in the water, they can form HAAs, which are also carcinogenic. Chloramines can still contribute to the formation of THMs, even though they typically produce fewer THMs than free chlorine. When chloramines react with water's organic nitrogen compounds, NDMA, a potent carcinogen, can form.

2.4.3 Advantages

Disinfecting drinking water using chloramines has several advantages, as noted by Snowden-Swan et al. (1998). First, chloramines have a longer-lasting residual effect than free chlorine, making them more stable and effective at providing disinfection throughout the distribution system. This means that water treated with chloramines can remain safe to drink for longer periods. Another advantage of chloramines is that they are generally thought to produce fewer DBPs than free chlorine. The decreased production of DBPs may contribute to a reduction in the health hazards imposed by drinking the contaminated water.

Another advantage of chloramines is their lower taste and odor threshold compared to free chlorine. This means that they can provide effective disinfection without causing significant taste and odor issues in the water. Chloramines are also less corrosive than free chlorine, which can help to reduce corrosion of pipes and other infrastructure in the distribution system. Lastly, chloramines are effective at controlling microbial growth in the distribution system, which can help prevent the growth of harmful bacteria such as *Legionella*.

2.4.4 Disadvantages

While chloramines generally produce fewer DBPs than free chlorine, they can still produce some DBPs, including nitrosamines, haloacetic acids, and trihalomethanes. These DBPs can have potential health effects and therefore must be carefully monitored and controlled. Another potential disadvantage of using chloramines is an increased risk of nitrification. In the distribution system, chloramines can encourage the growth of nitrifying bacteria, which can result in the production of nitrites and nitrates. Corrosion and other issues in the distribution system may be exacerbated by this.

Chloramines can likewise be less compelling than free chlorine at controlling particular kinds of microscopic organisms, like a few types of *E. coli*. The likelihood of microbial regrowth in the distribution system, which could be harmful to health, could rise as a result. Due to the need for additional treatment steps and monitoring, chloramine-based water treatment can be more expensive than free chlorine.

2.5 *Ozonation*

Ozonation is a water treatment procedure that utilizes ozone, a powerful oxidizing agent, to disinfect water and remove organic and inorganic contaminants. During ozonation, ozone gas is bubbled through the water, where it reacts with and destroys bacteria, viruses, and other microorganisms by oxidizing their cell walls and disrupting their metabolic processes. Ozonation can also break down organic compounds and eliminate tastes and odors in the water (Snowden-Swan et al. 1998; Forero et al. 2001).

2.5.1 Process of Disinfection

The process of ozonation is a method of disinfecting drinking water that involves several steps (Sassi et al. 2005). The first step is ozone generation, which involves the conversion of oxygen molecules in the air to ozone gas (Fig. 1). This is typically achieved by passing air through ultraviolet radiation. Once the ozone gas is generated, it is then injected into the water to be treated using a diffuser. After that, the water is allowed to come into contact with the ozone gas for a predetermined amount of time, usually a few minutes to several hours. The required contact time is determined by the desired level of disinfection and the quality of the water. After the contact time, any remaining ozone gas in the water is destroyed before it is released into the environment. This is typically achieved using activated carbon or other types of ozone destruct systems.

2.5.2 Disinfection by-Products

The water source, the amount of organic matter in the water, and the dose of ozone used all play a role in determining the specific DBPs that are produced during ozonation. Aldehydes are one type of DBP that can be made during ozonation. Aldehydes, such as formaldehyde and acetaldehyde, can be carcinogenic and irritate the eyes, nose, and throat when ozonation reacts with organic matter in the water.

Another DBP that can be formed during ozonation is bromate. Ozone can produce bromate, which is thought to be a human carcinogen and may possibly have detrimental effects on thyroid function, when it combines with bromide ions in water. In addition to bromate, ozonation can also lead to the formation of haloacetic acids (HAAs), a type of DBP that can be formed when ozone reacts with organic matter containing chlorine, such as residual chlorine in the water.

Ozonation can likewise prompt the development of trihalomethanes (THMs), which are a kind of DBP that can be framed when ozone responds with natural matter containing bromine, like normally happening bromine in the water. THMs, like

HAAs, can be harmful to health and cause skin irritation as well as cancer. Last but not least, DBPs like nitrosamines, which are potent carcinogens, can be produced when ozonation reacts with water-based compounds that contain nitrogen.

2.5.3 Advantages

Ozonation is a water treatment process that has several advantages over other disinfection methods (Lahlou 1999). One of the main advantages is its strong disinfection capabilities. Ozone is a highly effective water disinfectant due to its ability to kill a wide range of pathogens, including parasites, viruses, and bacteria, as well as its potent oxidizing power.

Another advantage of ozonation is its ability to remove organic and inorganic contaminants from water. Ozone can break down organic compounds and remove inorganic contaminants such as iron, manganese, and arsenic. Because of this, it is a treatment method that can be applied to a wide range of water treatment applications and is extremely adaptable. Unlike chlorine and other chemical disinfectants, ozonation does not leave behind any harmful residual chemicals in the water. This means that it is a safer and more environmentally friendly water treatment option.

2.5.4 Disadvantages

Ozonation is a water treatment process that has several advantages (Lahlou 1999). This process also has some limitations and potential disadvantages that should be considered. The equipment and energy required for ozonation can be expensive, making it a costly treatment option. Not at all like chlorine, ozone does not give a leftover impact to safeguard against microbial development in the distribution system. This means that water treated with ozone may require additional treatment steps to prevent regrowth of bacteria in the distribution system. Moreover, ozonation requires careful monitoring and control of ozone levels and contact time, and can be difficult to implement and maintain in some water treatment systems.

2.6 *Ultraviolet Light*

Drinking water is frequently disinfected with ultraviolet (UV) light. It is a physical process that utilizes UV radiation to inactivate microorganisms in water, including bacteria, viruses, and protozoan cysts. By exposing the water to UV light, which breaks down the DNA of bacteria and stops them from reproducing, UV disinfection kills pathogens. The amount of UV light required for disinfection is determined by the type and size of the microorganisms in the water. In most cases, a UV dose of 40 mJ/cm^2 is enough to kill 99.99% of bacteria and viruses (Sassi et al. 2005).

2.6.1 Process of Disinfection

The UV disinfection process involves several steps (Fig. 1). First, water is passed through a UV chamber containing a UV lamp. This lamp emits UV-C radiation at a wavelength of 254 nm, which is highly effective in inactivating microorganisms. As the water passes through the UV chamber, microorganisms in the water are exposed to the UV light. This exposure damages their DNA, preventing them from replicating or reproducing. Once the DNA of the microorganisms is damaged, they are considered to be inactivated and no longer pose a threat. The disinfected water is then collected for distribution or further treatment, if necessary.

2.6.2 Disinfection by-Products

Disinfectants can respond to the degradable materials in water to produce chemicals known as DBPs. In contrast to chemical disinfectants like chlorine, which result in the formation of DBPs, UV light is used to disinfect surfaces. This is due to the fact that UV disinfection does not require the addition of any chemicals to the water and does not result in the formation of any residual disinfectant.

The absence of DBPs is one of the advantages of using UV disinfection over chemical disinfection methods. DBPs can be disastrous for human beings and their health. Their formation is regulated by water quality standards. By using UV disinfection, water treatment plants can reduce the formation of DBPs and provide safe and clean drinking water to consumers.

2.6.3 Advantages

Ultraviolet (UV) light is an effective method for disinfecting water due to its ability to break down the DNA of microorganisms and prevent them from reproducing. There are several advantages to using UV disinfection, as outlined by Sassi et al. (2005). One of the primary advantages of UV disinfection is its effectiveness against a wide range of pathogenic microbes. Additionally, UV disinfection is a chemical-free method, making it environmentally friendly and safe for human consumption. Another advantage of UV disinfection is the absence of harmful DBPs. Unlike chemical disinfection methods, UV disinfection does not produce any harmful by-products that can pose a risk to human health. Furthermore, UV disinfection systems are easy to install and maintain, making them a cost-effective disinfection method.

2.6.4 Disadvantages

Disinfecting water using ultraviolet (UV) light has several advantages, making it a popular method of water treatment (Mohamad and Khayyat 1999). However, there are also some potential disadvantages to consider when using UV disinfection.

First, UV disinfection does not remove all types of contaminants, such as chemicals or particles. As a result, it is frequently used in combination with filtration and other water treatment methods. Also, while UV disinfection works on a lot of different kinds of microorganisms, it might not work on some kinds of microorganisms, like *Cryptosporidium*, which can make a protective cyst that does not like UV light.

Furthermore, UV disinfection systems require electricity to operate, which can be a disadvantage in areas with unreliable or limited electricity supply. The UV lamps used in the disinfection process have a limited lifespan and need to be replaced periodically, which can be a maintenance and cost consideration issue. The effectiveness of UV disinfection can also be affected by water quality parameters such as turbidity, color, and organic matter. High levels of these parameters can reduce the amount of UV radiation that reaches microorganisms and decrease the disinfection effectiveness. Lastly, UV radiation can be hazardous to human health and safety if not properly contained. Proper safety measures must be taken to prevent exposure to UV radiation during installation, maintenance, and operation of the UV disinfection system.

2.7 *Photocatalytic Disinfection*

Photocatalytic disinfection is a process that uses a photocatalyst, usually titanium dioxide (TiO_2), to disinfect surfaces and water by harnessing the power of light. When the photocatalyst is exposed to ultraviolet (UV) radiation, it undergoes a chemical reaction that produces reactive oxygen species (ROS), such as hydroxyl radicals ($^{\bullet}OH$) and superoxide ions ($O_2^{\bullet-}$). These ROS have strong oxidative power and can quickly destroy bacteria, viruses, and other harmful microorganisms.

2.7.1 Process of Disinfection

Photocatalytic disinfection of drinking water is a promising method that can effectively remove harmful microorganisms and make water safe for consumption (Lazar et al. 2012). The process typically involves the activation of a photocatalyst, i.e., titanium dioxide, through exposure to UV light or heat. The activated photocatalyst is then added to the water to be treated (Fig. 1).

After the photocatalyst and water have been combined, the mixture is usually exposed to ultraviolet light, usually from a UV lamp or the sun. The photocatalyst is made active by ultraviolet light, which causes the formation of reactive oxygen species (ROS). These ROS have a lot of oxidative power and can quickly kill harmful microorganisms in the water like viruses and bacteria.

The sterilization of microorganisms through the photocatalytic response successfully makes the water ok for utilization. The water can be filtered to get rid of any particles or impurities that are still present after the photocatalytic disinfection process. The treated water can then be stored and distributed for consumption.

2.7.2 Disinfection by-Products

The photocatalytic reaction's reactive oxygen species (ROS) react with the water's organic and inorganic compounds to form DBPs. THMs, HAAs, nitrosamines, and aldehydes are some common DBPs that can be produced during photocatalytic disinfection of drinking water. The photocatalytic reaction's ROS can react with the water's organic matter to form aldehydes.

2.7.3 Advantages

Due to its numerous advantages over other disinfection methods, the process of photocatalytic disinfection of drinking water is a promising water treatment method (Saravanan et al. 2017). Firstly, it is a chemical-free method of disinfection, which eliminates the need for the use of chemicals such as chlorine. This can reduce the production of harmful by-products and have a positive impact on public health. Additionally, it is effective against a wide variety of water-based microorganisms. The photocatalytic reaction's reactive oxygen species (ROS) have a lot of oxidative power, so they can quickly kill these microorganisms and make the water safe to drink. Both public health and the environment may benefit from this.

Another advantage of photocatalytic disinfection is that it can provide continuous disinfection as long as the UV light source is maintained. Unlike other disinfection methods that require periodic dosing or treatment, photocatalytic disinfection can provide ongoing protection against waterborne pathogens.

2.7.4 Disadvantages

Photocatalytic disinfection of drinking water is a promising technology that offers several advantages over other disinfection methods. First and foremost, it does not require the use of harmful chemicals like chlorine, which can result in the production of disinfection by-products and have negative effects on health. Second, it works against a wide range of microorganisms because the photocatalytically produced reactive oxygen species (ROS) have a lot of oxidative power and can quickly kill harmful bacteria, viruses, and other microorganisms in the water (Dong et al. 2015).

Thirdly, photocatalytic disinfection is cost-effective as it does not require the purchase and handling of chemicals, making it relatively inexpensive compared to other disinfection methods. Fourthly, it is a sustainable and environmentally friendly method of disinfection as it does not produce harmful by-products or generate waste. Fifthly, it provides continuous disinfection as long as the UV light source is maintained, which means that there is no need for periodic dosing or treatment. Lastly, the photocatalytic disinfection process is easy to maintain and requires minimal operator intervention, as the equipment can be automated and controlled remotely.

However, there are also some potential disadvantages of photocatalytic disinfection that must be considered. Firstly, the process may not be effective against certain types of microorganisms, such as certain viruses or parasites. Secondly, the photocatalytic process is dependent on UV light to activate the photocatalyst and produce ROS. Therefore, the process may not be effective if the UV light source is not functioning properly or is not maintained.

Thirdly, the UV lamps used in photocatalytic disinfection require a significant amount of energy to operate, which can increase the operating costs of the system. Fourthly, the process can produce DBPs when the ROS reacts with organic and inorganic compounds in the water. Fifthly, after treatment, the photocatalyst must be separated from the water, which can add additional steps and costs to the treatment process. Lastly, the photocatalytic process is sensitive to pH levels, and a lower pH can inhibit the photocatalytic reaction.

In conclusion, photocatalytic disinfection of drinking water has several advantages over other disinfection methods. Still, it also has some potential disadvantages that must be considered when choosing a water disinfection method. The benefits of photocatalytic disinfection make it a promising technology, but its limitations must be addressed to ensure safe and effective water treatment.

3 Factors Influencing the Formation of DBPs During Various Water Disinfection Processes

Disinfection is an essential step in the water treatment process to prevent the spread of diseases caused by water and eliminate harmful bacteria. However, disinfection can also result in the production of disinfection by-products (DBPs), which can harm human health (Srivastav et al. 2020). There are a number of factors that influence the formation of DBPs and must be taken into consideration when selecting a disinfection method. DBP formation can also be affected by the disinfectant used. Chlorine disinfection, on the other hand, has the potential to produce nitrogenous DBPs like monochloramine and dichloramine, while chlorine disinfection has the potential to produce trihalomethanes (THMs) and haloacetic acids (HAAs).

DBP formation is also influenced by the amount of disinfectant used and time spent in contact with water. DBPs can rise to higher levels when disinfectant dosages are increased. DBP levels can rise as a result of longer contact times. To reduce DBP formation, disinfectant dosage and contact time must be optimized. Another factor that influences the formation of DBPs is the pH of the water. THMs can be more abundant at higher pH values, while HAAs can be more abundant at lower pH values. To reduce the formation of DBP, the pH of the water must remain within a predetermined range.

Temperature is also an essential factor that affects the DBP's formations. Higher temperatures can increase the rate of chemical reactions, leading to higher levels of DBPs (Tripathy et al. 2015a). Water treatment facilities must monitor the temperature of the water and adjust the disinfection process accordingly. The formation of DBPs can also be influenced by the presence of DBP precursors, such as organic matter. The more precursors present, the more DBPs can be formed (Richardson et al. 2007).

4 Management Practices for Controlling DBPs

Disinfecting by-products (DBPs) are produced when chlorine and other disinfectants react with the organic matter in water (Benson et al. 2017). Numerous health problems, including cancer, kidney problems, and liver damage, have been linked to DBPs (Campanale et al. 2020; Therefore, it is essential to control the formation of DBPs in drinking water (Wang and Qian 2021). One way to control DBPs is to switch to alternative disinfectants like chloramines or ozone, which have been shown to reduce the formation of DBPs.

Getting rid of the water's organic matter is another option. Coagulation, sedimentation, and filtration are three methods that can be used to accomplish this (Borchardt et al. 2019). By diminishing how much natural matter is in the water, less DBPs will be framed. DBPs can be less likely to form if source water quality is improved. Another method that can be used to reduce the formation of DBPs is activated carbon filtration. According to Tripathy et al. (2015b), activated carbon can reduce the formation of DBPs by removing organic matter from the water. Standard checking of DBPs in the conveyance framework can assist with recognizing the wellspring of DBPs and consider remedial measures to be taken. The formation of DBPs can be reduced and the quality of drinking water improved by combining these management practices.

5 Conclusions

This chapter clarifies the significance of disinfection in the process of treating drinking water while also discussing the possible dangers posed by the development of DBPs. This chapter has covered the significance of disinfection, which removes harmful bacteria and microorganisms from drinking water to safeguard public health. The use of water disinfection as a public health intervention that removes pathogens from water bodies slows the spread of infections. A wide range of disinfection methods are available to meet the need for pathogen inactivation in water. Surface water treatment processes typically include a specific chemical disinfection step to prevent the spread of waterborne diseases, despite the fact that coagulation, sedimentation, filtration, lime-soda softening, and adsorption are a few physiochemical water treatment processes that can disinfect to some extent. The disinfectant chemical that is most frequently utilized is chlorine. The most popular kinds are calcium or sodium hypochlorite and chlorine gas. Alternative disinfectants include gamma radiation, ozone, chlorine dioxide, chloramines, ultraviolet light, and others. Water flow rates, application and demand rates, water pH, equipment cost, chemical availability, and maintenance issues should all be carefully taken into account before choosing the chemical or technique.

It is essential to adopt rigorous monitoring and optimization of disinfection operations in order to guarantee the safety of drinking water while minimizing the development of DBPs. This entails performing routine inspections of the disinfection system, assessing the effectiveness of different disinfectants, and maximizing operational factors including contact duration and dose. By doing this, water treatment facilities can protect the public's health by finding a balance between removing pathogens and reducing the development of DBP. This chapter also outlines ongoing research and development initiatives targeted at investigating alternate disinfection techniques and technologies to reduce DBP production in drinking water. These initiatives mark a significant advancement in tackling the problems brought on by DBPs as they look for novel strategies to efficiently disinfect water while reducing the generation of toxic by-products. These initiatives seek to either develop new disinfectants that are less likely to produce DBP or adapt current disinfection procedures to lessen DBP generation. Furthermore, cutting-edge treatment methods are being researched to successfully remove DBPs from the treated water and further improve water quality. For communities all across the world to have safer, higher-quality drinking water, ongoing work in this area are essential. We can ensure that drinking water is free from harmful microorganisms and the health risks associated with them while minimizing the presence of potentially harmful disinfection by-products by continuously improving our understanding of DBP formation mechanisms, optimizing disinfection processes, and developing cutting-edge technologies. By giving people access to safe and clean drinking water, these efforts will inevitably improve the wellbeing and quality of life for people all over the world.

The information offered in this chapter will be helpful for policymakers, water treatment experts, and researchers as the demand for safe drinking water increases globally. In conclusion, this chapter offers a thorough resource that promotes knowledge of disinfection procedures and the development of DBPs in drinking water. In order to secure the supply of clean, safe, and sustainable drinking water to fulfil the expanding requirements of our global population, it emphasizes the necessity for continual study, collaboration, and optimization.

References

Anderson SE, Meade BJ (2014) Potential health effects associated with dermal exposure to occupational chemicals. Environ Health Insights 8(Suppl 1):51–62

Arasmith S (1993) 'Introduction to small water systems', University of Alaska SE (UAS) Sitka campus & ATTAC (Alaska Training/Technical Assistance Center). ACR Publications, United States

Benson NU, Akintokun OA, Adedapo AE (2017) Disinfection byproducts in drinking water and evaluation of potential health risks of long-term exposure in Nigeria. J Environ Public Health 2017:7535797

Block MS, Rowan BG (2020) Hypochlorous acid: a review. J Oral Maxillofac Surg 78(9):1461–1466

Bond T, Goslan EH, Parsons SA, Jefferson B (2012) A critical review of Trihalomethane and Haloacetic acid formation from natural organic matter surrogates. Environ Technol Rev 1(1):1–43

Borchardt MA, Spencer SK, Kieke BA, Lambertini E, Loge FJ (2019) Viruses in alluvial aquifers of the upper Midwest United States: presence, genetic diversity, and environmental drivers. Sci Total Environ 690:1276–1285

Campanale C, Massarelli C, Savino I, Locaputo V, Uricchio VF (2020) A detailed review study on potential effects of microplastics and additives of concern on human health. Int J Environ Res Public Health 17(4):1212

Centers for Disease Control and Prevention (CDC) (2022) Water Treatment. https://www.cdc.gov/healthywater/drinking/public/water_treatment.html

Chin A, Berube PR (2005) Removal of disinfection by-product precursors with ozone-UV advanced oxidation process. Water Res 39(10):2136–2144

DeMarini DM (2020) A review on the 40th anniversary of the first regulation of drinking water disinfection by-products. Environ Mol Mutagen 61(6):588–601

Ding S, Deng Y, Bond T, Fang C, Cao Z, Chu W (2019) Disinfection byproduct formation during drinking water treatment and distribution: a review of unintended effects of engineering agents and materials. Water Res 160:313–329

Dong H, Zeng G, Tang L, Fan C, Zhang C, He X et al (2015) An overview on limitations of TiO2-based particles for photocatalytic degradation of organic pollutants and the corresponding countermeasures. Water Res 79:128–146

Erdem CU, Ateia M, Liu C, Karanfil T (2020) Activated carbon and organic matter characteristics impact the adsorption of DBP precursors when chlorine is added prior to GAC contactors. Water Res 184:116146

Eryılmaz M, Palabıyık IM (2013) Hypochlorous acid - analytical methods and antimicrobial activity. Trop J Pharm Res 12(1):123–126

Forero JE, Duque JJ, Rios F, Diaz J (2001) Ozone for phenol treatment in industrial wastewater. CT&F-Ciencia, Tecnología y Futuro 2(2):17–26

Hua G, Reckhow DA (2007) Comparison of disinfection byproduct formation from chlorine and alternative disinfectants. Water Res 41(8):1667–1678

Hua G, Reckhow DA, Kim JH (2020) Formation and control of disinfection byproducts in drinking water treatment: a review. J Environ Sci 87:147–162. https://doi.org/10.1016/j.jes.2019.05.034

Jefri UHNM, Khan A, Lim YC, Lee KS, Liew KB, Kassab YW, Choo CY, Al-Worafi YM, Ming LC, Kalusalingam A (2022) A systematic review on chlorine dioxide as a disinfectant. J Med Life 15(3):313–318

Kapoor A, Viraraghavan T, Cullimore DR (1999) Removal of contaminants in drinking water by chlorination, Chloramination, and ozonation processes. Crit Rev Environ Sci Technol 29(1):1–46

Khan MU, Malik A, Muhammad S (2013) Disinfection of drinking water: a review on the utilization of conventional methods. J Water Health 11(3):349–360

Khanna P, Lohani M, Jain R (2016) Water treatment by using various technologies- a review. Indian J Sci Technol 9(41):1–7

Khayyat AMA (2000) Study of point of use treatment methods for the disinfection of drinking water in Nepal [Doctoral dissertation]. Massachusetts Institute of Technology

Lahlou M (1999) A national drinking water clearinghouse fact sheet. Tech Brief Twelve

Lazar MA, Varghese S, Nair SS (2012) Photocatalytic water treatment by titanium dioxide: recent updates. Catalysts 2(4):572–601

Leonard LT, Vanzin GF, Garayburu-Caruso VA, Lau SS, Beutler CA, Newman AW, Mitch WA, Stegen JC, Williams KH, Sharp JO (2022) Disinfection byproducts formed during drinking

water treatment reveal an export control point for dissolved organic matter in a subalpine headwater stream. Water Res X 15:100144
Levchuk I, Márquez JJR, Sillanpää M (2018) Removal of natural organic matter (NOM) from water by ion exchange – a review. Chemosphere 192:90–104
Li X, Mitch WA (2018) Drinking water disinfection byproducts (DBPs) and human health effects: multidisciplinary challenges and opportunities. Environ Sci Technol 52(4):1681–1689
Lindsay L (2011) Chlorination. Published by The National Environmental Services Center at West Virginia University, Morgantown 4(2):1–4
Mackenzie D (2020) Ultraviolet light fights new virus. Engineering (Beijing) 6(8):851–853
Malik DJ, Ahmad S (2018) Sodium Hypochlorite. StatPearls Publishing, Treasure Island, FL. PMID: 29494080
Mao Y, Wang X, Yang H, Wang H, Xie YF (2014) Effects of ozonation on disinfection byproduct formation and speciation during subsequent chlorination. Chemosphere 117:515–520
Mishra BK, Gupta SK, Sinha A (2014) Human health risk analysis from disinfection by-products (DBPs) in drinking and bathing water of some Indian cities. J Environ Health Sci Eng 12:73
Momba MNB, Sibewu M, Onyango MS, Momba MN (2006) The effectiveness of locally manufactured sodium hypochlorite for household water treatment in rural communities in South Africa. J Environ Eng Sci 5(2):155–164
Mohamad A, Khayyat A. Study of point of use treatment methods for the disinfection of drinking water in Nepal [Thesis]. Brown University, B.Sc. Mechanical Engineering; 1999
Ngwenya N, Ncube EJ, Parsons J (2013) Recent advances in drinking water disinfection: successes and challenges. Rev Environ Contam Toxicol 222:111–170
Parvez S, Ashby JL, Kimura SY, Richardson SD (2019) Exposure characterization of Haloacetic acids in humans for exposure and risk assessment applications: an exploratory study. Int J Environ Res Public Health 16(3):471
Postigo C, Richardson SD (2014) Transformation of drugs during oxidation/disinfection processes in water treatment: a review. Environ Sci Technol 48(17):9715–9732
Pulicharla R, Proulx F, Behmel S, Sérodes J-B, Rodriguez MJ (2020) Trends in ozonation disinfection by-products—occurrence, analysis and toxicity of carboxylic acids. Water 12:756
Rabêlo Braitt G, de Almeida Rodrigues E, da Silveira Bueno CE, Braitt AH (2013) Evaluation of active chlorine releasing of sodium hypochlorite during seven days, stored at different temperatures. RSBO Revista Sul-Brasileira de Odontologia 10(2):143–148
Richardson SD, Plewa MJ, Wagner ED, Schoeny R, DeMarini DM (2007) Occurrence, genotoxicity, and carcinogenicity of regulated and emerging disinfection by-products in drinking water: a review and roadmap for research. Mutat Res/Rev Mutat Res 636(1–3):178–242
Saqib Ishaq M, Afsheen Z, Khan A et al (2019) Disinfection methods. Photocatalysts - applications and attributes. IntechOpen. https://doi.org/10.5772/intechopen.80999
Saravanan R, Gracia F, Stephen A (2017) Basic principles, mechanism, and challenges of photocatalysis. In: Nanocomposites for visible light-induced photocatalysis. Springer, Cham, pp 19–40
Sassi J, Viitasalo S, Rytkonen J, Leppakoski E (2005) 'Experiments with ultraviolet light, ultrasound and ozone technologies for onboard ballast water treatment', VTT TIEDOTTEITA. Åbo Akademi University, Turku, Finland
Snowden-Swan L, Piatt J, Lesperance A (1998) Disinfection technologies for potable water and wastewater treatment: alternatives to chlorine gas', Pacific Northwest National Laboratory (PNNL). Prepared for: US Army Forces Command Air Quality Division, United State
Srivastav AL, Patel N, Chaudhary VK (2020) Disinfection by-products in drinking water: occurrence, toxicity and abatement. Environ Pollut 267:115474
Stauber JL, Walters A, Howard G (2004) Chlorination disinfection by-products in drinking water: a review and roadmap for research. J Water Health 2(4):253–267
Tak S, Vellanki BP (2018) Natural organic matter as precursor to disinfection byproducts and its removal using conventional and advanced processes: state of the art review. J Water Health 16(5):681–703

Tech E (2005) Chlorine and alternative disinfectants guidance manual. In: Snowden-Swan L, Piatt J, Lesperance A (eds) Water stewardship-office of drinking water. Pacific Northwest National Laboratory, U.S.

Tripathy S, Mishra RK, Manjare AK (2015a) A review on formation of disinfection by-products and their control in drinking water. Chemosphere 139:734–753

Tripathy S, Saha S, Mishra S (2015b) Occurrence and health risk assessment of disinfection byproducts in drinking water: a critical review. Hum Ecol Risk Assess Int J 21(7):1861–1882

United Nations Environment Programme (2014) Handbook for the safe use of sodium hypochlorite solution for drinking water. https://www.unep.org/resources/report/handbook-safe-use-sodium-hypochlorite-solution-drinking-water

United States Environmental Protection Agency (EPA) (2021) Drinking water treatability database: Chlorination. https://iaspub.epa.gov/tdb/pages/treatment/treatmentOverview.do?treatmentProcessId=2057817286

United States Environmental Protection Agency (EPA) (2022) Surface water treatment rules. https://www.epa.gov/dwreginfo/surface-water-treatment-rules

United States Environmental Protection Agency (USEPA) (2003) Wastewater technology fact sheet, EPA 832-F-03-024

USEPA (2006) National primary drinking water regulations: stage 2 disinfectants/disinfection byproducts rule fed. Fed Regist 71(2):387–493

USEPA (2013) Controlling disinfection by-products and microbial contaminants in drinking water, United States Environmental Protection Agency, (EPA: 600-R-01-110)

Vega-Méndez ME, Sánchez-Polo M, Rivera-Utrilla J (2017) Chlorine-based disinfectants: a review on their dynamics, effectiveness and toxicity towards microorganisms and human health. J Environ Sci Health C Environ Carcinog Ecotoxicol Rev 35(1):1–30

Villanueva CM, Cordier S, Font-Ribera L, Salas LA, Levallois P (2015) Overview of disinfection by-products and associated health effects. Curr Environ Health Rep 2:107–115

Wagenet L, Darling S, Lemley A (1988) 'Chlorination of drinking water', fact sheet 5, (updated October 2005)

Wang Y, Qian H (2021) Phthalates and their impacts on human health. Healthcare (Basel) 9(5):603

Wang J, Yue W, Wang Z, Song J (2023) Removal effect of trihalomethanes (THMs) and halogenated acetic acids (HAAs) precursors in reclaimed water by polyaluminum chloride (PACl) coagulation. Water Sci Technol 87(3):672–684

Weisman RJ, Heinrich A, Letkiewicz F, Messner M, Studer K, Wang L, Regli S (2022) Estimating national exposures and potential bladder cancer cases associated with chlorination DBPs in U.S. drinking water. Environ Health Perspect 130(8):87002

World Health Organization (WHO) (2011) Guidelines for drinking-water quality. World Health Organization, Geneva, Switzerland. https://www.who.int/water_sanitation_health/publications/2011/dwq_guidelines/en/

World Health Organization (WHO) (2017) Chlorine in drinking-water. https://www.who.int/water_sanitation_health/publications/chlorine-drinking-water/en/

Yang H, Ye S, Wang J, Wang H, Wang Z, Chen Q, Wang W, Xiang L, Zeng G, Tan X (2021) The approaches and prospects for natural organic matter-derived disinfection byproducts control by carbon-based materials in water disinfection progresses. J Clean Prod 311:127799

Zazouli MA, Kalankesh LR (2017) Removal of precursors and disinfection by-products (DBPs) by membrane filtration from water; a review. J Environ Health Sci Eng 15:25

Various Disinfection Processes and Formation of Disinfection by-Products in Drinking Water

Rafael Audino Zambelli and Ítalo Waldimiro Lima de França

Abstract The increase in the urban population means that water consumption is increasing. In this way, several methods to produce drinking water have been used, such as chlorination, UV radiation, and ozonation. However, chemical reactions between disinfection methods and the different organic matter that are found in surface and groundwater lead to the formation of disinfection by-products. These by-products can, many times, present toxic and/or carcinogenic potential and have aroused the interest of researchers regarding their characterization, composition, and quantification. Some factors may favor the appearance of disinfection by-products, mainly due to the application of chlorine to eliminate bacteria, viruses, molds, and yeasts. In addition, the type of organic matter, pH of the water, level of residual chlorine, contact time, temperature, and the presence of bromine and iodine ions in the water can affect the formation of chemicals. The main by-products found in drinking water are Trihalomethanes, Halobenzoquinones, Haloketones, Haloacetic acids, and Nitrosamines. Human contamination by disinfection by-products can occur through ingestion of drinking water and through the skin while bathing or diving in chlorinated pools. This chapter addresses the different disinfection processes, their strengths, and weaknesses, as well as the characterization of the main disinfection by-products and their potential risk to human health.

Keywords Chlorination · Disinfection · Organic matter · Ozonation · Trihalomethanes

R. A. Zambelli (✉) · Í. W. L. de França
Food Engineering Department, Federal University of Ceará, Fortaleza, Brazil
e-mail: zambelli@ufc.br

S. Madhav et al. (eds.), *Drinking Water Disinfection By-products*,
https://doi.org/10.1007/978-3-031-49047-7_5

1 Introduction

The civilizations' development results in an increase in population, greater occupation of urban areas than rural areas, changes in climate, increase in industry and agriculture. These factors can compromise countries' water resources (Du et al. 2017). In this way, humanity is facing a growing water crisis at a global level, where the demand for drinking water is increasing. It is expected that climate change will increase the water stress of populations, which leads countries to be pressured to guarantee sustainable access to drinking water (Villanueva et al. 2023).

Water can be contaminated from different sources and sources, for example, nitrates, arsenic, fluorine, and chemicals. These components come from agriculture, untreated industrial effluents, and domestic sewage (Damania et al. 2019), therefore, a disinfection process is necessary.

Untreated water obtained from the surface, reservoirs, or lakes must be taken to the treatment stations so that the procedures are carried out to guarantee its quality and potability, to later be sent to consumers. In this way, some treatments can be used in the water cleaning process, such as sorting, coagulation, flocculation, sedimentation, filtration and, finally, disinfection, which eliminate microorganisms, color, turbidity, impurities, garbage, and algae (Koop and Van Leeuwen 2015; Loucks and Van Beek 2017).

The use of disinfectant agents in the processing of drinking water is nothing new, being practiced for more than 100 years and it is one of the most efficient methods in the maintenance of water quality and that collaborates with public health. Initially, chlorine was the most chosen disinfectant, with the advancement of technology, chemicals such as ozone, chlorine dioxide, and chloramines were used in water purification (Bhatt and Paneria 2017).

Disinfection is considered as a vital step to ensure the supply of clean and safe drinking water (Kali et al. 2021). Drinking water treatment has represented a great advance for the quality of life of consumers, since it has been successful in reducing health risks through disinfection processes using oxidizing agents, such as ozone, chlorine, chloramine, dioxide chlorine, and other substances (Mohan and Reckhow 2021).

The main objective of water treatment is to improve the quality of water obtained from natural sources and must be designed to target several specific contaminants. The main concern, according to WHO (2022) is biological hazards, mainly microorganisms, both in developing and developed countries. The greatest risks are linked to the ingestion of water contaminated with human or animal waste that can be a source of pathogenic bacteria, protozoa, viruses, and other parasites.

Several methods can be used to improve water quality and make it drinkable, from the more conventional ones such as chlorination and filtration, to the most innovative ones such as back-up osmosis and UV radiation.

Currently, more than 700 by-products of drinking water disinfection are cataloged in the scientific literature (Dong et al. 2019). Many of them still lack regulation for the specification and measurement of risk and toxicity.

In China, for example, more than 90% of urban water supply systems use chlorine as a disinfectant agent, this characteristic is a common link with other countries. In this way, a significant number of people are exposed to the by-products of disinfection throughout their lives, increasing the risk of developing chronic diseases such as cancer (Meng et al. 2012; Pan et al. 2014).

Thus, this chapter aims to present the main water disinfection mechanisms and map the main disinfection by-products and their harmful potential.

2 Chlorination

The chlorination process has provided safe water from a microbiological point of view in the public water supply for many years. Because of its low cost, it was introduced as a disinfection method in several rural and marginalized communities (Nielsen et al. 2022).

The chlorine application as a disinfectant during the treatment of drinking water promotes the reduction of waterborne diseases (Pichel et al. 2019).

Chlorination is the most widely used water treatment in the world, and the oldest (Llonch et al. 2023), because of its simplified operation, high oxidizing potential, and because it presents a great cost benefit for those who use it, in addition to being the treatment that developing countries can use to purify water (Mahmood 2015; Chowdhury et al. 2012).

Chlorine-treated water is used for drinking, hand washing, bathing, swimming, cleaning, or cooking (Kujlu et al. 2020), i.e., exposure is daily and uninterrupted for a large part of the world's population, which can favor the onset of chronic diseases due to exposure to toxic substances.

Chlorine has the main advantage of removing almost all types of pathogenic biological agents and the residual chlorine that remains in the water prevents microbial recontamination within the water supply chain (Hassan et al. 2010).

The use of chlorine can occur at the beginning of the treatment (known as pre-chlorination) and/or at the end of the process (known as post-chlorination). Pre-chlorination is effective in reducing contamination by microorganisms and algae growth in water treatment facilities. It was observed that pre-chlorination also promotes improvements in coagulation and flocculation processes. Post-chlorination, on the other hand, provides final disinfection and promotes the presence of free residual chlorine throughout the water distribution line (Hendricks 2016).

The by-products of chlorine disinfection consist of a mixture of substances that are formed when chlorine acts to oxidize naturally occurring anthropogenic substances (Richardson 2011). The main by-products of chlorine disinfection that can be formed are trihalomethanes, haloacetic acids, halobenzoquinones, haloketones, halonitromethanes, and halogenated acetonitriles (Mazhar et al. 2020; Tsitsifli and Kanakoudis 2018).

3 Ozonation

Ozonation has been applied in drinking water treatment for disinfection and oxidation for decades (Lim et al. 2022). Therefore, ozone can be considered a green and environmentally friendly technology, which has been a constant target of investors and companies that seek to follow the Environmental, Social, and Governance (ESG) standards established in several guidelines (Zambelli 2022).

Ozone is a molecule that can react with various contaminating particles. However, it is important to verify that its ability to oxidize other substances is selective in relation to pollutants, which can hinder the oxidation process, remove some contaminants incompletely or promote the generation of intermediate by-products that promote toxicity for consumers (Wang and Chen 2019).

The positive results of ozone's efficiency as an oxidizing agent and disinfectant have attracted interest in its use in water treatment for over a century. In recent years, this technique has shown interest in other areas, such as advanced wastewater treatment to reduce the amount of micropollutants and in the processing of drinking water (Manasfi 2021).

Ozone has a disinfectant action against pathogens that have not been inactivated using chlorine. In the ozonation process, the main by-product formed is bromate and its concentration is correlated with the presence and concentration of bromide and ammonia ions in the water (Padhi et al. 2019).

Some disadvantages are observed in the ozonation process. The first refers to the inability of ozone in the water without any type of residue remaining, which impairs secondary disinfection, that is, in the water distribution lines, which is not observed with chlorine. Furthermore, the oxidation of bromide by ozone can promote the formation of bromate, which generates potentially carcinogenic by-products (Snyder et al. 2005).

Due to the organic matter present in water, ozonation can lead to the formation of low molecular weight organic compounds such as carboxylic acids, aldehydes, and ketones which have a certain degree of toxicity. Gulde et al. (2021a) identified the formation of 227 transformation products due to the application of ozone in the water of a tributary of a water treatment plant. The reaction mechanisms for the formation of many transformation products consist of the combination of the kinetics of ozone formation and reduction with the production of hydroxyl radicals, from the oxidation of organic matter by ozone.

In the same sense, Laflamme et al. (2020) identified and studied the formation of 18 ozonation by-products (aldehydes and bromates) in drinking water systems in Canada and reported the occurrence of non-halogenated aldehydes and haloacetaldehydes. In general, the application of ozone and liquid chlorine contributed to the formation of compounds. The main non-halogenated aldehydes found were formaldehyde, glyoxal, metiglyoxal, and acetaldehyde. The authors attributed the appearance of these substances to factors such as the composition of the organic matter present and the water temperature, which favors the formation of aldehydes.

4 Thermal Disinfection

The heat application to inactivate microorganisms has been used for millennia in the production of beverages and food, being considered the oldest method of disinfection. The origin of the development of a thermal disinfection process dates to Louis Pasteur in the late nineteenth century, when he observed that the application of heat to wine and beer prevented the formation of acids, bitterness, and viscosity defects in these drinks, protecting them from spoilage. This process was defined as pasteurization and is still used today, mainly in the food industry (Currier and Widness 2018).

Thermal technology is effective against different types of microorganisms (viruses, bacteria, and protozoa), in addition to those that are resistant to some chemical treatments, such as *Escherichia coli, Salmonella typhimurium, Vibrio cholerae*, and *Clostridium* sp. (Al-Gheethi et al. 2013). The efficiency of thermal disinfection relates to three mechanisms: (a) denaturation of enzymes present inside the cells, which leads to the loss of their catalytic activity; (b) changes in the structure of proteins and fatty acids that make up the microbial cell wall; and (c) cell rupture due to expansion and consequent leakage of cell constituents (Metcalf & Eddy n.d.; Śmiech et al. 2020).

Thermal disinfection or pasteurization is associated with increasing the water temperature, through the addition of heat, to a pre-established temperature and exposure time, which allows its consumption in suitable conditions for human health. Throughout this process, microorganisms are the most affected, so problems with particulate matter in suspension and turbidity require other technologies, such as filtration, decantation, and flocculation (Abraham et al. 2015; Śmiech et al. 2020). It is important to point out that disinfection using heat is the only water treatment method that does not suffer any alteration due to the level of turbidity of the water.

This process might be conducted on batch or continuous set up. In batch processes, water tanks are heated using nonrenewable resources, such as the burning of fuels, or by using renewable energy, such as solar energy. In continuous processes, water is heated while passing through a device called a heat exchanger, which makes this process more expensive compared to the batch process. Even though, batch processes mainly require fuels and long operating time (Currier and Widness 2018).

Microorganisms exhibit a higher resistance to dry heat than to wet heat. Dry heat usually is applied on the disinfection of materials, and its mechanism is related to oxidation and protein denaturation. While wet heat is associated with microorganism inactivation by membrane injury, and has a wider application in the food industry, mainly due to its higher energy potential (Smelt and Brul 2014). The water disinfection process through heating below the boiling temperature has aroused the interest of researchers and industries because of its simple dimensioning, easy and quick implementation, and low cost (Nieuwoudt and Mathews 2005).

It is important to emphasize that thermal disinfection of drinking water does not mean exactly sterilization, once it does not kill all pathogens, it just reduces the number of microorganisms to an adequate level for human health. Some microorganisms are resistant to high temperatures for some time (Smelt and Brul 2014).

Usually, higher temperatures are associated with small-time intervals and lower temperatures with a long exposure period. These conditions might lead to a reduction in the level of microorganisms, and binomial time/temperature depends mainly on the characteristics of the pathogen, and most known microorganisms are inactivated at temperatures below boiling (Espinosa et al. 2020).

The pathogen inactivation is generally determined by the decimal reduction (D), which means the time necessary for a 1-log reduction in pathogens level (Abraham et al. 2015). It might be mathematically expressed as Eq. 1:

$$\log\left(\frac{N_t}{N_0}\right)=\frac{-t}{D} \tag{1}$$

which N_0 represents the initial level of pathogens and N_t is the level after a time interval (t).

The value of D consists of the decimal derating time at a given reference temperature. This parameter is used as a way of comparing data on resistance to heat application of some pathogens and is a function of temperature. Generally, the value of D decreases rapidly as the applied temperature increases.

Equation 1 can also be expressed for different temperatures as shown in Eq. 2:

$$\log\left(\frac{D}{D_r}\right)=\frac{-(T-T_r)}{z} \tag{2}$$

D_r is the reduction time of the number of microorganisms at a given temperature (T_r) and D is termed as the reduction time of microorganisms for a given temperature (T). The parameter z is the number of degrees required for a thermal death time curve to traverse 1 log cycle, so low values indicate considerable sensitivity to enhancement heat, specifying the correlative impact of different temperatures on microorganisms. Eqs. 1 and 2 are considerably useful, but limited to isothermal conditions, thus requiring another methodology.

The pathogen inactivation on thermal disinfection of drinking water can also be determined by the estimation of the instantaneous rate of microorganism death using a first-order rate model. Equation 3 expresses this rate model for the variation of the level of pathogens (N) during the time (t):

$$\frac{dN}{dt}=-kN \tag{3}$$

Under conditions of constant temperature, Eq. 3 can be associated with time (N_0 for the initial level of pathogens and N_t for the level after a time interval), and expressed as Eq. 4:

$$\log\left(\frac{N_t}{N_0}\right) = -kt \quad (4)$$

which K is the pseudo-first-order inactivation rate of the reactions.

Equations 1 and 4 allow a relationship between D and k, as shown in Eq. 5:

$$k = \frac{2.303}{D} \quad (5)$$

For transient conditions, Eq. 3 can be integrated into an infinitesimal interval in time (Δt) and must be performed numerically, as shown in Eq. 6:

$$\frac{dN}{dt} = \frac{\left(N_{t+\Delta t} - N_t\right)}{\Delta t} = -kN_t = \frac{2.303}{D_r}.10^{\frac{(T-T_r)}{z}.N_t} \quad (6)$$

Equation 6 can be solved by the Euler method, and it just requires some information as values for D and z for different pathogens, as well as the values for D at different temperatures (Abraham et al. 2015). These equations are essential to evaluate and project a thermal disinfection process during water treatment.

Overall, the benefits of thermal disinfection deal with impressive speed, it usually takes a few minutes and presents a considerable efficiency in the inactivation of different types of microorganisms, not requiring the presence of chemical disinfectants. It also can be applied with green energy, such as solar power, pointed out as a promising technology in this area, taking into consideration the ample solar/UV exposure in restricted infrastructure regions, such as for people living in rural fields (Abraham et al. 2015; Carielo et al. 2017; Nieuwoudt and Mathews 2005).

Despite these advantages, thermal disinfection must be carried out with caution, since it can cause corrosion during water storage equipment time, the risks of scalding injury to people operating the process, and the high temperature can deal some damage to plasticized or rubberized seals, leading to problems with pumps and other components using such seals.

5 UV Radiation

Ultraviolet light consists of the electromagnetic spectrum that corresponds to the wavelengths of visible light and X-rays (10–400 nm) and exhibits the capability to destroy RNA or DNA present in cells of pathogen microorganisms, leading them to death, and achieving disinfection of food and water. UV radiation is divided into three categories: UV-A or long-wave (black light: 320–400 nm), UV-B or mid-wave (280–320 nm), UV-C or short-wave (200–280 nm), and vacuum UV (10–200 nm), and each waveband cause a different effect on microorganisms: UV-A and UV-B are reported to present mutagenic and lethal effect, and UV-C damages the cell DNA (Hora and Heisel 2016; Wolfe et al. 1984).

UV light disinfection is seen as a promising technology for water and food treatment due to some advantages compared to chemical methods, such as it does not use chemical or toxic products, absence of the formation of by-products with toxicity, and carcinogenic or mutagenic characteristics, and it does not affect the characteristics sensory aspects of products such as taste, odor, and color (Yan et al. 2021).

UV lamps can be divided into two main categories: continuous-wave emission or pulsed emission, which allows the disinfection to be performed as a batch or continuous process. These systems use mercury lamps that have quartz tubes immersed in running water. In this way, the lamp allows the passage of UV radiation at a wavelength close to 250 nm, which leads to a minimum germicidal efficiency of 85%. However, the efficiency of these quartz tubes is limited and reduces upon continuous use, requiring cleaning methods, which can be mechanical, chemical, or ultrasonic. An alternative that has proven to be efficient is the replacement of quartz with Teflon, despite Teflon presenting a lower intensity of UV wave transmission (Mofidi et al. 2001).

The damage caused by UV light in genetic material is the main cause of pathogen inactivation at a wavelength of approximately 260 nm. DNA produces thymine and cytosine dimers during UV absorption, thus interrupting its replication and transcription, proceeding to a photochemical modification on the pathogen, consequently leading to pathogen and parasite inactivation (Hora and Heisel 2016). The pathogen inactivation is related to the UV dose ($\mu W.s.cm^{-2}$), and can be expressed as Eq. 7:

$$\frac{N_t}{N_0} = e^{-KP_d t} \quad (7)$$

Here, N_0 means the initial number of microorganisms; N_t means the final or surviving microorganism, K is the inactivation rate constant; and P_d means the UV light intensity applied to microorganisms. The required dosage ($\mu W.s.cm^{-2}$) to inactivate (90% – 1 log reduction) of microorganisms by UV on food or water varies according to their own characteristics, for example, *Salmonella typhi*, *Giardia muris*, Poliovirus, *Escherichia coli*, Hepatitis A Virus, *Pseudomonas aeruginosa*, *Rotavirus*, and *Cryptosporidium parvum* (Wolfe et al. 1984). Other factors reported to affect UV action and the required dosage are organic matter, color, suspended particles, and particle size, despite the presence of some organic and inorganic chemicals, such as phenolic and humic components, as well as iron, manganese, and lignin sulfonates. These compounds directly interfere with ultraviolet light transmission in water because of their high UV absorption coefficients. Therefore, water, which contains significant amounts of these substances, needs treatments such as sedimentation and filtration to purify it (Hora and Heisel 2016; Yan et al. 2021).

UV disinfection is widely used in vegetable and fruit surfaces and has been proposed as a potential alternative to thermal and chemical disinfection methods. However, this technology presents some limitations on these products, such as its permeability. The use of ultraviolet is particularly convenient for water disinfection, since it is considerably efficient against disease-causing agents present in groundwater sources. It is pointed out as a low-cost technology, especially when compared to chemical methods, such as chlorination, in addition, it can be applied in several areas where water treatment is inefficient or non-existent, such as small rural areas in the interior of developing countries (Disinfection 2014; Yan et al. 2021).

In general, UV light has been shown to be effective in the water disinfection process treatment, mainly in the inactivation of bacteria, viruses, and protozoa. Ultraviolet is a low-cost process that is easy and quick (usually a few seconds) to operate, does not require chemicals or affect water quality (no change in total organic carbon, pH, turbidity, and corrosivity), and there is no formation of undesirable toxic or mutagenic/carcinogenic by-products (Hora and Heisel 2016).

Despite these advantages, the use of UV is sometimes limited due to some factors: firstly, the absence of residual disinfectant action in treated water samples, formation of biofilm on the surface of the UV lamp, reduced efficiency in fluids that present high turbidity, as well as problems caused by the photoreactivation of microbial pathogens. To overcome some issues, ultraviolet technology is commonly associated with chemical methods on water treatment, such as the use of chlorine or chloramines to maintain the residual disinfection of treated water (Chen et al. 2017; Khedikar and Paidalwar 2016; Yan et al. 2021).

6 Disinfection by-Products Formation in Drinking Water

In general, surface water tends to have higher concentrations of disinfection by-products than groundwater due to the presence of organic matter in greater quantities (Abbas et al. 2014). The amount of disinfection by-products formed is regulated by the levels of precursors present in the process, such as the amount of organic and inorganic matter present in raw water. These components are transformed by disinfecting agents into other chemical species that are considered disinfection by-products (Villanueva et al. 2023).

Figure 1 shows the main precursors, disinfecting agents, and by-products.

The wastewater containing a large amount of organic matter, the chlorination process can lead to the formation of harmful by-products, especially trihalomethanes (Mallick et al. 2018).

Organic pollutants are the main natural organic matter that can be present in water. It is usually composed of a heterogeneous mixture of organic components that originate in plants and animals, in addition to chemical substances present in surface water and that are dissolved. They can be classified into three fractions: hydrophilic, hydrophobic, and transphobic (Zazouli et al. 2010). However, the hydrophobic fraction is the majority in surface waters, corresponding to approximately 75% of the total organic matter. Its composition is formed by fulvic and humic acids (Ilunga et al. 2021).

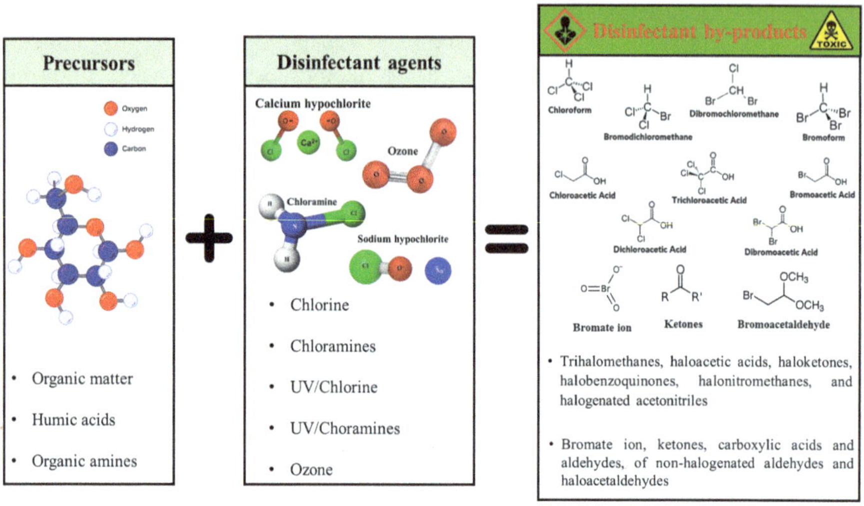

Fig. 1 Precursors, disinfecting agents, and disinfection by-products

By-products of the disinfection process are produced from the reaction of organic matter present in water with various disinfectants and chemical compounds applied in the production of water. However, this by-product formation is more associated with chlorine than with alternative disinfectants (chlorine dioxide, ozone, and chloramines (Chaukura et al. 2020).

The combination of chlorine and ozone technology in water treatment can increase the number of aldehydes in treated water than the combination of other disinfectant agents (Papageorgiou et al. 2014).

Bougeard et al. (2010) point out that the different types of disinfection by-products formed depend on the type of disinfectant applied in the process.

Significant levels of disinfection by-products are produced from favorable chemical characteristics such as the presence of bromide and iodide ions, water pH, intrinsic characteristics of organic matter, temperature, dose, and time of contact with chlorine, precursors presence, in addition to the hydrophobicity or hydrophilicity of organic matter (Chaukura et al. 2020).

The disinfection by-products production depends on some factors, one of which is the pH of the water. The formation of trihalomethanes is intensified when the pH value of the water is higher, on the other hand, at low pH, haloacetic acids are formed (Kali et al. 2021).

Zhang et al. (2017) and Mompremier et al. (2019) found a relationship between the material of the pipes through which treated water comes into contact and the formation of toxic components. For example, polyethylene, ductile iron, and stainless-steel pipes can lead to the formation of trihalomethanes. On the other hand, haloacetic acids can be formed if the piping is composed of high-density polyethylene, galvanized steel, polyvinyl chloride, or polypropylene.

In this context, Mujathel et al. (2022) observed the presence of trihalomethanes precursors in four types of chlorine-treated drinking water sources in the Aseer region of Saudi Arabia. The compounds have been found in treated surface water, treated groundwater, desalinated water, and mixed water. There was the presence of high levels of trihalomethane in the treated surface water, where only the chlorination and filtration process were applied. In the water treatment performed by reverse osmosis, the levels were minimal. In addition, it was observed that the collections carried out in the summer had a higher content of trihalomethanes, especially bromoform.

Cheshmekhezr et al. (2021) found similar results when analyzing surface water from rivers that pass through the city of Tehran, Iran and that were used as drinking water. The statistical analysis showed the presence of trihalomethanes in greater quantities in the summer, with a positive correlation between the level of trihalomethanes and the temperature.

Precursors of disinfection by-products can be removed by membranes and coagulation and flocculation processes, in addition to the application of activated charcoal (Kujlu et al. 2020).

The main factors that affect the disinfection formation by-products are:

1. Natural organic matter
2. Chlorine dose
3. pH
4. Contact time
5. Bromide presence
6. Iodide presence
7. Temperature

Figure 2 shows the main sources that influence the presence and quantity of formed disinfection by-products.

The following sections present the most common disinfection by-products.

6.1 *Halogenated Disinfection by-Products*

Large amounts of chlorine can produce halogenated disinfection by-products, which are directly related to carcinogenic, genotoxic, and cytotoxic potential, therefore, undesirable in drinking water (Wagner and Plewa 2017).

The chlorine used as a disinfectant agent produces halogenated compounds as disinfection by-products (Li et al. 2020). These substances are not regulated but have relevant toxicology and impact on public health. The formation occurs through substitutions of halogens by carbons or nitrogen. Also, a small fraction can be replaced by bromine and iodine because of the iodine and bromide ions presence in the water. They are stable structures, present in water, and difficult to detect (Craven et al. 2022).

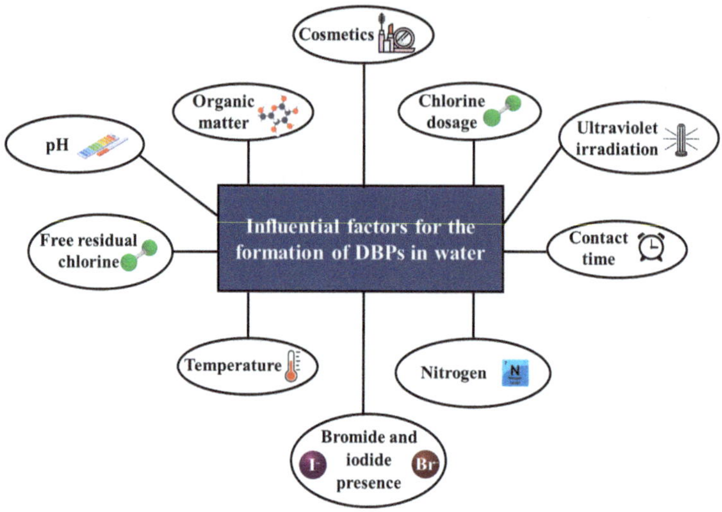

Fig. 2 The main sources that influence the presence and quantity of formed disinfection by-products

These halogenated nitrogenous disinfection by-products include substances such as halamines, haloamino acids, halonitrils, haloamides, halonitroalkanes, and halopeptides (Gao et al. 2022).

Excess residual chlorine in water distribution lines reacts with organic matter and produces disinfection by-products. The formation of these by-products can occur in two ways. The first one is based on the rapid hydrolysis of chlorine through the production of hypochlorite ions and hypochlorous acid. The substances formed react with organic matter through electron transfer and electrophilic substitution to promote the formation of by-products. In turn, halogenated disinfection products start from the hydrolysis of bromine that produces hypobromous acid and hypobromite (Duirk and Colette 2006; Langsa et al. 2017).

Dong et al. (2022) monitored halogenated disinfection by-products for 21 months in a large-scale water treatment plant in Huzhou City, Zhejiang province, China. Several halogenated disinfection by-products were found in different water sources. The largest formation took place in the summer.

In turn, Tao et al. (2020) identified 97 halogenated by-products derived from disinfection, such as 2,4-dichlorobenzoic acid, 2,6-dichloro-4-nitrophenol, and 2,6-dibromo-4-nitrophenol, in addition to brominated, iodized, and chlorinated compounds in tap water.

The use of treated water for the preparation of hot drinks, such as tea and coffee, can potentiate the production of halogenated disinfection by-products, especially if sugar is added. The study conducted by Qiu et al. (2022) obtained results that demonstrate the ability of sugar to react with residual chlorine present in tap water distribution lines. This reaction promotes the formation of halogenated by-products. The addition of sucrose can increase the total organic halogen level by up to 35%

and lead to the formation of 15 new brominated and chlorinated polar halogen disinfection by-products. The addition of disaccharides to tap water samples produced a disinfection by-products higher level formation, while addition of monosaccharides produced lesser effects due to lower molecular weight.

6.2 *Trihalomethanes*

During the chlorination process, some trihalomethane components are formed, such as chloroform, dibromochloromethane, bromodichloromethane, and bromoform, because of the reaction of chloramine and chlorine with the components of organic matter and their respective precursors, such as fulvic acid and humic acid (Thokchom et al. 2020; Benson et al. 2017), present in water.

The action of ozone can cause trihalomethanes formation and some haloacetic acids in higher amounts than conventional water treatments such as coagulation and flocculation. Álvarez-Arroyo et al. (2022) identified chloroform and dichloroacetic acid as the main components formed after treating water with ozone.

Water quality directly affects the disinfection by-products production, since the content of organic matter is high, there is a greater production of these by-products. Therefore, surface water tends to generate more trihalomethanes than groundwater.

Trihalomethanes consist of a chemical compound class that is derived from methane, where three of the four hydrogen atoms may be replaced by halogens (Mazhar et al. 2020).

This class of substance has the properties of being highly volatile, and therefore can be inhaled, in addition to being permeable to the skin. The main classes of Trihalomethanes are (Villanueva et al. 2016):

i. Bromoform ($CHBr_3$)
ii. Bromodichloromethane ($CHBrCl_2$)
iii. Chloroform ($CHCl_3$)
iv. dibromochloromethane ($CHClBr_2$)

Evlampidou et al. (2020) investigated the content of trihalomethanes in drinking water in several European countries. Cyprus, Ireland, Greece, and Spain had the highest levels. All these countries have chlorine as the main disinfectant in the treatment of drinking water, regardless of the source of the water.

Scheili et al. (2015) studied the main sources of variation that can cause an increase in trihalomethanes in water. There was a positive correlation regarding the source of water, where surface water was more prone to the formation of trihalomethanes, levels of natural organic matter, water temperature, time of contact with the color and distance from the water treatment plant.

According to Weisman et al. (2022), in the United States, a correlation was established between several cases of bladder cancer and the levels of trihalomethanes found in water. Approximately 84% of the cases come from surface water systems treated with chlorine and that presented trihalomethane contents between 40 and 80 μg/L.

Trihalomethanes were initially regulated in the United States through the United States Environmental Protection Agency (USEPA). In the regulatory aspect, trihalomethanes were considered an indicator to analyze other disinfection by-products that are in lower concentrations (Cotruvo and Amato 2019).

A study evaluated the amount of trihalomethanes in water supplied to three Canadian indigenous communities, where two of them, exceeded the limit of 100 μg/L, that is, the consumption of tap water in these communities represents a risk to human health. Furthermore, it was observed that the trihalomethanes formation occurs during the distribution of treated water to consumers' homes (Amarawansha et al. 2023).

Studies show that trihalomethanes may have carcinogenic potential in humans, especially in the bladder, kidneys, breast, colon, rectum, and liver (Mohammadi et al. 2020). Other effects are reported such as retardation of fetal growth in the uterus, premature birth, malformation, and stillbirths (Fakhri et al. 2017).

Mishaga et al. (2022) investigated the likely health risks associated with the presence of trihalomethanes in drinking water in Egypt from inhalation, ingestion, and/or dermal contact. 1667 water samples were analyzed in 23 provinces during the 3-year period. The concentrations of total trihalomethanes were always below what is allowed according to Egyptian legislation (100 μg/L). The compound in greatest presence was chloroform, while bromoform had a rare appearance in the results. According to the study, bromodichloromethane had the greatest impact on the onset of cancer, followed by chlorodibromomethane. Ingestion causes more health risks than dermal contact.

Mosaferi et al. (2021) point out that dermal exposure occurs through showering, swimming pools, and activities such as washing dishes, cars, and other equipment. Similar results were found by Helte et al. (2022), where they evaluated the association of trihalomethanes in drinking water with the risk of bladder cancer. In a sample of 58,672 individuals who were evaluated for 16 years, 831 cases of bladder cancer were observed, however, it was not possible to conclude that the presence of trihalomethanes was the main cause of the appearance of the disease.

6.3 *Halobenzoquinones*

Another class of substance that has been frequently detected as a by-product of the drinking water disinfection process is halobenzoquinones, especially 2,6-dichloro-1,4-benzoquinone, which has toxic potential for humans (Qiao et al. 2022). These components are associated with the risk of bladder cancer, damage to DNA and proteins because of their ability to induce the formation of reactive oxygen species (Wang et al. 2018). Other halobenzoquinones were identified by Hu et al. (2020): 2-chloro-6-iodo-1,4-benzoquinone; 2-bromo-6-iodo-1,4-benzoquinone; 2,6-dichloro-1,4-benzoquinone; 2,6-dichloro-3-methyl-1,4-benzoquinone;

2,6-dibromo-1,4-benzoquinone; 2,6-diiodo-1,4-benzoquinone; 2,3,6-trichloro-1,4-benzoquinone.

Halobenzoquinones can be produced from the contact of water purifying agents with aromatic amino acids, lignin hydrolysis products, humic and tannic acids, hydroquinones, as well as catechol, resorcinol, and gallic acid, which can naturally be present in water (Huang et al. 2019).

Several halobenzoquinones are formed from the chlorination process and the reaction with free amino acids present in the water, with 2,6-dichloro-1,4-benzoquinone being the dominant species. The free amino acids tryptophan and tyrosine showed a higher yield in the production of halobenzoquinones than phenylalanine, in addition, the study conducted by Zhao et al. (2022) identified ten intermediates of formed halobenzoquinones and proposed the mechanisms of electrophilic addition and substitution, oxidation, deacidification, and dehydration as ways of formation of halobenzoquinones from chlorination, and the yield of the formation of these compounds was dependent on the free chlorine content, pH, and temperature of the water.

Tu et al. (2022) performed a quantitative analysis of the relationship between structure and toxic potential of halobenzoquinone isomers in DNA. It has been shown that these substances can interact with DNA through hydrogen bonding in areas rich in guanine and that they can cause relevant chromosomal damage.

The neurotoxicity of halobenzoquinones was evaluated in in vivo models by Yang et al. (2022) and proved to be effective in reducing the locomotor activity of zebrafish larvae linked to the reduction of the neurotransmitter's dopamine and γ-aminobutyric acid. The transcriptomic analysis performed showed that exposure to halobenzoquinones affected neurodevelopmental signaling pathways, providing molecular evidence for one of the mechanisms of neurotoxicity of halobenzoquinones.

Aromatic antioxidants that are dumped into water sources are attributed to the increased risk of 2,6-dichloro-1,4-benzoquinone formation, being considered the most likely precursors of this substance from chlorination. It was observed that the application of methods before ozonation such as the use of activated carbon and quartz sand can promote the removal of precursors, which reduces the formation of halobenzoquinones (Qiao et al. 2022).

6.4 Haloketones

Haloketones consist of a disinfection by-products class that have mutagenic, carcinogenic, and toxic potential. Monitoring these substances in drinking water is essential for consumer safety (Deng et al. 2021).

These compounds are extremely volatile and do not have specific regulations for maximum limits. The formation of haloketones occurs after the use of ozone, chlorine dioxide, chlorine, and chloramine. Its detection is difficult and those found in greater amounts in water are 1,1-dichloroacetone, 1,3-dichloroketone, and

1,1,1-trichloroketone (Bougeard et al. 2010; Golfinopoulos and Nikolaou 2005). Like the trihalomethanes, the haloketones showed volatile characteristics (Xu et al. 2002).

Haloketones are considered intermediate by-products and have a long reaction time. Studies show that high temperatures and chlorine dose do not significantly influence the amount of haloketones present in the water, especially if residual chlorine is found in large quantities. However, it is important to note that other water quality parameters such as ammonia and nitrite contents influence the appearance of haloketones, as they are chlorine consumers (Yang et al. 2007; Zhou et al. 2019).

In research developed by Serrano et al. (2015), the method of microextraction in solid phase and gas chromatography coupled with mass spectrometry was used to identify 14 types of haloketones in water treated from the use of chlorine and chloramines. The results demonstrated that there was formation of substances before and after the application of the oxidation method and, in addition, the subsequent steps of water treatment were ineffective in eliminating these substances. The authors compared the amount of haloketones in pool and tap water, however, the highest concentration was observed in pool water due to the higher concentration of organic matter and residual chlorine.

6.5 Haloacetic Acids

Haloacetic acids have been identified in surface, marine, river, reservoir, groundwater, and desalinated waters in China, Cyprus, India, Qatar, Nigeria, Spain, and England (Kumar et al. 2023). Among the main contaminants produced from disinfection by-products, due to operational difficulties, haloacetic acids are substances that require greater demand and resources to carry out monitoring, quantification, and strategies to reduce their quantity in drinking water (Paull and Barron 2004).

These substances are a common by-product of water treatment and are second in incidence after trihalomethanes. Ingestion of haloacetic acids is carried out exclusively through drinking water, rather than through dermal absorption (Rodriguez et al. 2004).

Chloroacetic acid, dichloroacetic acid, trichloroacetic acid, bromoacetic acid, dibromoacetic acid, and tribromoacetic acid are the main haloacetic acids found in drinking water. Its formation occurs from unintentional chemical interaction between disinfecting agents and organic matter (Tang et al. 2013). Dong et al. (2019) state that monochloroacetic, monobromoacetic, and bromochloroacetic acids are also part of the group of haloacetic acids.

The pKa values of haloacetic acids are between 0.6 and 2.5; therefore, these acids only exist in drinking water in their protonated form and their formation takes place in an acid medium.

Kumar et al. (2023) reinforce that because of its carcinogenic potential and damage to the environment, the removal of haloacetic acids is extremely important. Several approaches or technologies such as nanofiltration, use of ozone and activated carbon, biofiltration, reverse osmosis, and adsorption can be used to reduce the amounts of haloacetic acid in water.

Nguyen et al. (2005) attribute the formation of haloacetic acids to organic carbon derived from algae, aromatic compounds, and aromatic proteins that function as precursors. In turn, Hong et al. (2009) report that amino acids generate a greater amount of haloacetic acids than trihalomethanes after the coloring process and attribute the production of acid to other organic materials such as fulvic, humic, and dicarboxylic acids.

The haloacetic acid presence in water samples in the United Kingdom was detected by Malliarou et al. (2005). The study presented a variety of levels of haloacetic acids present in water (35–95 μg/L) and peak concentration at 244 μg/L. Furthermore, there was a correlation between the presence of chloroform and dibromochloromethane. The main haloacetic acids found were dichloroacetic acid and trichloroacetic acid.

6.6 Nitrosamines

In general, the nitrosamine formation as disinfection by-products has always been associated with chlorine (Krasner et al. 2013), however, studies have recently shown that N-nitrosodimethylamine can form from the water ozonation process (Andrzejewski et al. 2008). This substance is a by-product of dyeing, tanning, rubber, and/or pesticide industries, and is found in surface or groundwater close to industries like these (Mitch et al. 2003).

According to Wang et al. (2013), disinfection with chlorine can generate the formation of nitrogenous disinfection by-products, such as N-nitrosamine, provided that the organic matter is hydrophilic and has low molecular weight.

The main nitrosamines that can be found in drinking water are N-nitrosodi-n-butylamine, N-nitrosodipropylamine, N-nitrosodiethylamine, N-nitrosomethylethylamine, N-nitrosomorpholine, N-nitrosopiperidine, N-nitrosopyrrolidine, and N-nitroso-di-phenylamine (Yoon et al. 2012).

Nitrosamines have characteristics that make them an environmental concern. This class of substances presents miscibility in water, low sorption potential, making it mobile in the environment, and susceptible to groundwater contamination (Kommineni et al. 2003; Pisarenko et al. 2012).

Oxygen plays an important role in the formation of nitrosamines during the chlorination process. The N-nitrosamine content decreases with the reduction of available oxygen for the oxidation of the components, in addition, an acidic pH limits the formation of these compounds (Alaba et al. 2017), the reaction is not fast, it occurs slowly over days, in this way, nitrosamines accumulate in the chlorinated water distribution systems (Mitch et al. 2003).

The USEPA classifies N-nitrosodimethylamine as a possible human carcinogen and presents a high risk when its concentration in water is equal to or greater than

0.7 ng/L. Its formation is correlated with the presence of dimethylamine reacting with nitrate or nitrite in an acid medium (Mirvish 1975; Yang et al. 2009).

On the other hand, if the organic matter has a limited subset of precursors containing hydrazine functional groups, ozonation can form nitrosamines in greater than 50% yield (Shen and Andrews 2011).

The presence of coagulation polymers that contain amine in their composition, pharmaceuticals, personal care and hygiene products, ion exchange resins, industrial effluents, soluble microbial products, algae, and soluble chemical products can favor the formation of nitrosamines in drinking water (Krasner et al. 2013).

The decomposition of ozone by hydroxide ions leads to the formation of hydroxyl radicals, which play an important role in the formation of nitrosamines during ozonation, mainly at alkaline pH, as nitrite is rapidly oxidized under these conditions (Staehelin and Hoigne 1982).

Significant amounts of nitrosamines are found in wastewater that has the presence of dyes containing dimethylamine group and are purified with the use of ozone for disinfection, especially if precursors hydrazine, asymmetric dimethylhydrazine, semicarbazines or N-dimethylsulfanamide are present (Andrzejewski et al. 2008 Gunten et al. 2010).

7 Conclusion

The urbanization of territories has increased the demand for drinking water in several regions of the world, especially in underdeveloped countries. In this way, to guarantee the production of drinking water, some disinfection methods have been applied, such as chlorination, ozonation, and UV radiation, the first being the most applied due to its efficiency and cost.

The water used as raw material to produce drinking water comes from natural reservoirs on the surface and from groundwater. These sources have an enormous amount of organic matter that can react with chlorine, ozone, and UV radiation, which can cause the formation of various disinfection by-products that are commonly found in drinking water.

It is important to highlight that the main by-products of disinfection found in drinking water consist of chemical substances that are potentially harmful to humans and animals, including carcinogenic and mutagenic potential. Substances such as trihalomethanes, halobenzoquinones, haloketones, haloacetic acids, and nitrosamines are the result of this interaction.

However, the amount and type of substances formed depend on factors such as the composition of the organic matter, pH of the water, dose of residual chlorine, contact time between the contaminated water and the disinfecting agent, temperature, and the presence of bromine and iodine ions.

It is extremely important to periodically check the amount of these substances in drinking water to reduce the rate of exposure to humans, as exposure occurs through ingestion and through the skin during common daily activities such as drinking water, cooking, and to have a bath.

References

Abbas S, Hashmi I, Rehman MSU, Qazi IA, Awan MA, Nasir H (2014) Monitoring of chlorination disinfection by-products and their associated health risks in drinking water of Pakistan. J Water Health 13:270–284

Abraham JP, Plourde BD, Minkowycz WJ (2015) Continuous flow solar thermal pasteurization of drinking water: methods, devices, microbiology, and analysis. Renew Energy 81:795–803

Alaba PA, Sani YM, Olupinta SF, Daud WMW, Mohammed IY, Enweremadu CC, Olubunmi O (2017) Toward N-nitrosamines free water: formation, prevention, and removal. Crit Rev Environ Sci Technol 47:2448–2489

Al-Gheethi AAS, Lalung J, Kadir MOA, Ismail N, Talib A (2013) Reduction of fecal indicators and elimination of pathogens from sewage treated effluents by heat treatment. Casp J Appl Sci Res 2:39–55

Álvarez-Arroyo R, Pérez JI, Ruiz LM, Gómez MA (2022) Chlorination by-products formation in a drinking water distribution system treated by ultrafiltration associated with pre-ozonation or coagulation/flocculation. J Water Process Eng 47:779–786

Amarawansha G, Zvomuya F, Tomy G, Farenhorst A (2023) Trihalomethanes in drinking water from three first nation reserves in Manitoba, Canada. Environ Monit Assess 195:88–95

Andrzejewski P, Kasprzyk-Hordern B, Nawrocki J (2008) N-nitrosodimethylamine (NDMA) formation during ozonation of dimethylamine-containing waters. Water Res 42:863–870

Benson R, et al. (2017) Human health screening and public health significance of contaminants of emerging concern detected in public water supplies Science of The Total Environment 5791643–1648 https://doi.org/10.1016/j.scitotenv.2016.03.146

Bhatt BV, Paneria DB (2017) Modernization in the water distribution system. New Horiz Civ Eng 8:1–6

Bougeard CMM, Goslan EH, Jefferson B, Parsons SA (2010) Comparison of the disinfection by-product formation potential of treated waters exposed to chlorine and monochloramine. Water Res 44:729–740

Carielo G, Calazans G, Lima G, Tiba C (2017) Solar water pasteurizer: productivity and treatment efficiency in microbial decontamination. Renew Energy 105:257–269

Chaukura N, Marais SS, Moyo W, Mbali N, Thakalekoala LC, Ingwani T, Mamba BB, Jarvis P, Nkambule TTI (2020) Contemporary issues on the occurrence and removal of disinfection byproducts in drinking water - A review. J Environ Chem Eng 8:659–667

Chen J, Loeb S, Kim JH (2017) LED revolution: fundamentals and prospects for UV disinfection applications. Environ Sci Water Res Technol 3:188–202

Cheshmekhezr S, Alimoradi S, Torabian A, Pardakhti A, Babaei L (2021) Investigation of the formation and effects of upstream trihalomethanes in Tehran's raw and drinking water using the spectrophotometric method. J Water Process Eng 44:426–433

Chowdhury S, Rodriguez MJ, Sadiq R, Serodes J (2012) Modeling DBPs formation in drinking water in residential plumbing pipes and hot water tanks. Water Res 45:337–347

Cotruvo JA, Amato H (2019) National Trends of bladder cancer and Trihalomethanes in drinking water: A review and multicountry ecological study. Dose-Response: An Int J 17:1–19

Craven CB, Tang Y, Carroll K, An L, Chen B, Li XF (2022) Closing the gap of known and unknown halogenated nitrogenous disinfection byproducts in water: advanced mass spectrometry techniques. TrAC - Trends Anal Chem 153:624–632

Currier RW, Widness JA (2018) A brief history of milk hygiene and its impact on infant mortality from 1875 to 1925 and implications for today: A review. J Food Prot 81:1713–1722

Damania R, Desbureaux S, Rodella AS, Russ J, Zaveri E (2019) Quality unknown: the invisible water crisis. World Bank Group, Washington DC, USA
Deng Y, Zhou X, Shen J, Xiao G, Hong H, Lin H, Wu F, Liao BQ (2021) New methods based on back propagation (BP) and radial basis function (RBF) artificial neural networks (ANNs) for predicting the occurrence of haloketones in tap water. Sci Total Environ 772:534–539
Disinfection DW (2014) Drinking water disinfection 3.1
Dong H, Qiang Z, Richardson SD (2019) Formation of iodinated disinfection byproducts (I-DBPs) in drinking water: emerging concerns and current issues. Acc Chem Res 52:896–905
Dong F, Pang Z, Yu J, Deng J, Li X, Ma X, Dietrich AM, Deng Y (2022) Spatio-temporal variability of halogenated disinfection by-products in a large-scale two-source water distribution system with enhanced chlorination. J Hazard Mater 423:113–119
Du Y, Lv XT, Wu QY, Zhang DY, Zhou YT, Peng L, Hu HY (2017) Formation and control of disinfection byproducts and toxicity during reclaimed water chlorination: a review. J Environ Sci 58:51–63
Duirk SE, Colette TW (2006) Degradation of Chlorpyrifos in aqueous chlorine solutions: pathways, kinetics, and modeling. Environ Sci Technol 40:546–551
Espinosa MF, Sancho AN, Mendoza LM, Mota CR, Verbyla ME (2020) Systematic review and meta-analysis of time-temperature pathogen inactivation. Int J Hyg Environ Health 230:113595
Evlampidou I, Fonte-Ribera L, Rojas-Rueda D, Garcia-Lavedan E, Costet N, Pearce N et al (2020) Trihalomethanes in drinking water and bladder cancer burden in the European Union. Environ Health Perspect 128:17001–17014
Fakhri Y, Mohseni-Bandpei G, Conti OH, Keramati Y, Zandsalimi N, Amanidaz R, Pouya HB, Moradi B, Bahmani L, Amirhajeloo R, Baninameh Z (2017) Health risk assessment induced by chloroform content of the drinking water in Iran: systematic review. Toxin Rev 36:342–351
Gao Y, Qiu J, Ji Y, Wawryk NJP, An T, Li XF (2022) Formation mechanism of iodinated aromatic disinfection byproducts: acid catalysis with H2OI+. Environ Sci Technol 56(3):1791–1800
Golfinopoulos SK, Nikolaou AD (2005) Survey of disinfection by-products in drinking water in Athens, Greece. Desalination 176:13–24
Gulde R, Clerc B, Rutsch M, Helbing J, Salhi E, McArdell CS, Von Gunten U (2021a) Oxidation of 51 micropollutants during drinking water ozonation: formation of transformation products and their fate during biological post-filtration. Water Res 207:117–124
Gunten UV, Salhi E, Schmidt CK, Arnold WA (2010) Kinetics and mechanisms of N-nitrosodimethylamine formation upon ozonation of N, N-dimethyl sulfonamide-containing waters: bromide catalysis. Environ Sci Technol 44:5762–5768
Hassan A, Thacker NP, Bassin J (2010) Trihalomethane formation potential in treated water supplies in urban metro cities. Environ Monit Assess 168:489–497
Helte E, Save-Soderbergh M, Ugge H, Fall K, Larsson SC, Akesson A (2022) Chlorination by-products in drinking water and risk of bladder cancer – A population-based cohort study. Water Res 214:202–210
Hendricks D (2016) Fundamentals of water treatment unit processes: physical, chemical, and biological. CRC Press, Boca Ratón, p 385
Hong H, Wong M, Liang Y (2009) Amino acids as precursors of trihalomethane and haloacetic acid formation during chlorination. Environ Contam Toxicol 56:638–645
Hora JJ, Heisel CJ (2016) Water and wastewater. The CQ Press Guide to Urban Politics and Policy in the United States
Hu S, Gong T, Zhu H, Wang J, Li Z, Chen H, Huang Z, Zhang M, Xian Q (2020) Formation and decomposition of new iodinated halobenzoquinones during chloramination in drinking water. Environ Sci Technol 54:5237–5248
Huang G, Jmaiff LK, Blackstock P, Jiang Z, Liu X, Lu X, Li F (2019) Formation, identification, and occurrence of new bromo-and mixed halo-tyrosyl dipeptides in chloraminated water. Environ Sci Technol 53:3672–3678

Ilunga AK, Mamba BH, Nkambule TTI (2021) Catalytic hydrodehalogenation of halogenated disinfection byproducts for clean drinking water production: A review. J Water Process Eng 44:402–412

Kali S, Khan M, Ghaffar MS, Rasheed S, Easeem A, Iqbal MM, Niazi MBK, Zafar MI (2021) Occurrence, influencing factors, toxicity, regulations, and abatement approaches for disinfection by-products in chlorinated drinking water: A comprehensive review. Environ Pollut 281:116–126

Khedikar IP, Paidalwar AA (2016) Overview of water Disinfection by UV technology. Research Gate 2:8

Kommineni S, Ela WP, Arnold RG, Huling SG, Hester BJ, Betterton EA (2003) NDMA treatment by sequential GAC adsorption and Fenton-driven destruction. Environ Eng Sci 2003(20):361–373

Koop SHA, Van Leeuwen CJ (2015) Assessment of the sustainability of water resources management: a critical review of the city blueprint approach. Water Resour Manag 29:5649–5670

Krasner SW, Mitch WA, McCurry DL, Hanigan D, Westerhoff P (2013) Formation, precursors, control, and occurrence of nitrosamines in drinking water: a review. Water Res 47:4433–4450

Kujlu R, Mahdavianpour M, Ghanbari F (2020) Multi-route human health risk assessment from trihalomethanes in drinking and non-drinking water in Abadan, Iran. Environ Sci Pollut Res 27:442621–442630

Kumar M, Shekhar S, Kumar R, Kumar P, Govarthanan M, Chaminda T (2023) Drinking water treatment and associated toxic byproducts: concurrence and urgence. Environ Pollut 320:121–129

Laflamme O, Sérodes JB, Simard S, Legay C, Dorea C, Rodriguez MJ (2020) Occurrence and fate of ozonation disinfection by-products in two Canadian drinking water systems. Chemosphere 260:660–669

Langsa M, Heitz A, Joll CA, Von Gunten U, Allard S (2017) Mechanistic aspects of the formation of Adsorbable organic bromine during chlorination of bromide-containing synthetic waters. Environ Sci Technol 51:5146–5155

Li Y, Jiang J, Li W, Zhu X, Zhang X, Jiang F (2020) Volatile DBPs contributed marginally to the developmental toxicity of drinking water DBP mixtures against *Platynereis dumerilii*. Chemosphere 252:611–619

Lim S, Shi JL, Gunten UV, McCurry DL (2022) Ozonation of organic compounds in water and wastewater: A critical review. Water Research 213118053-https://doi.org/10.1016/j.watres.2022.118053

Llonch L, Verdú M, Martí S, Medinyá C, Riera J, Cucurull J, Devant. (2023) Drinking water chlorination in dairy beef fattening bulls: water quality, potential hazards, apparent total tract digestibility, and growth performance. Animal 17:685–692

Loucks DP, Van Beek E (2017) Water resource systems planning and management: an introduction to methods, models, and applications. Water Resour Cyst Plann Manag Introd Methods Mod Appl 12:85–99

Mahmood K (2015) Naegleria fowleri in Pakistan-an emerging catastrophe. J Coll Physicians Surg Pak 25:159–160

Malliarou E, Collins C, Graham N, Nieuwenhuijsen MJ (2005) Haloacetic acids in drinking water in the United Kingdom. Water Res 39:2722–2730

Mallick J, Singh RA, Khan CK, Singh NB, Kahla EI, Warrag S, Islam A, Rahman P (2018) Examining the rainfall-topography relationship using non-stationary modeling techniques in the semi-arid Aseer region, Saudi Arabia. J Geosci 11:222–229

Manasfi T (2021) Ozonation in drinking water treatment: an overview of general and practical aspects, mechanisms, kinetics, and byproduct formation. Compr Anal Chem 92:85–116

Mazhar MA, Khan NA, Ahmed S, Khan AH, Hussain A, Changani F, Yousefi M, Ahmadi S, Vambol V (2020) Chlorination disinfection by-products in municipal drinking water – A review. J Clean Prod 273:159–168

Meng LP, Dong ZM, Hu JY (2012) National survey and risk assessment of haloacetic acids in drinking water in China for reevaluation of the drinking water standards. China Environ Sci 32:721–726

Metcalf & amp; Eddy I (n.d.) Wastewater engineering: treatment and reuse. Fourth edition / revised by George Tchobanoglous, Franklin L. Burton, H. David Stensel. McGraw-Hill, Boston

Mirvish SS (1975) Formation of *N*-nitroso compounds: chemistry, kinetics, and *in vivo* occurrence Toxicol. Appl Pharmacol 31:325–351

Mishaga EI, Radwan EK, Ibrahim MBM, Hegazy TA, Ibrahim MS (2022) Multi-exposure human health risks assessment of trihalomethanes in drinking water of Egypt. Environ Res 207:643–650

Mitch WA, Sharp JO, Trussell RR, Valentine RL, Alvarez-Cohen L, Sedlak DL (2003) N-nitrosodimethylamine (NDMA) as a drinking water contaminant: a review. Environ Eng Sci 20:389–404

Mofidi AA, Baribeau H, Rochelle PA, De Leon R, Coffey BM, Green JF (2001) Disinfection of cryptosporidium parvum with polychromatic UV light. J Am Water Works Assoc 93:95–109

Mohammadi A, Faraji M, Ebrahimi AA, Nemati S, Abdolahnejad A, Miri M (2020) Comparing THMs level in old and new water distribution systems; seasonal variation and probabilistic risk assessment. Ecotoxicol Environ Saf 192:285–290

Mohan A, Reckhow DA (2021) Hydrolysis and chlorination of 2,6-Dichloro-1,4-benzoquinone under conditions typical of drinking water distribution systems. Hydrolysis and chlorination of 2,6-Dichloro-1,4-benzoquinone under conditions typical of drinking water distribution systems. Water Res 200:219–229

Mompremier R, Fuentes Mariles OA, Becerril Bravo JE, Ghebremichael K (2019) Study of the variation of haloacetic acids in a simulated water distribution network. Water Supply 19:88–96

Mosaferi M, Asadi M, Aslani H, Mohammadi A, Abedi S, Mansour SN, Maleki S (2021) Temporospatial variation and health risk assessment of trihalomethanes (THMs) in drinking water (Northwest Iran). Environ Sci Pollut Res 20:8168–8180

Mujathel AM, El-Barky W, Fayed M, Aly SA (2022) Trihalomethane evaluation in chlorinated treated drinking water sources in Saudi Arabia (Aseer region a case study). Alex Eng J 61:12699–12711

Nguyen ML, Wasterhoff P, Baker L, Hu Q, Esparza-Soto M, Sommerfeld M (2005) Characteristics and reactivity of algae-produced dissolved organic carbon. J Environ Eng 131:1574–1582

Nielsen AM, Garcia LAT, Silva KJS, Sabogal-Paz LP, Hincapié MM, Montoya LJ, Galeano L, Galdos-Balzatequi A, Reygadas F, Herrera C, Golden S, Byrne JA, Fernández-Ibañez P (2022) Chlorination for low-cost household water disinfection – A critical review and status in three Latin American countries. Int J Hyg Environ Health 244:114–119

Nieuwoudt MN, Mathews EH (2005) A mobile solar water heater for rural housing in southern Africa. Build Environ 40:1217–1234

Padhi RK, Subramanian S, Mohanty AK, Satpathy KK (2019) Comparative assessment of chlorine reactivity and trihalomethanes formation potential of three different water sources. J Water Process Eng 29:769–777

Pan S, An W, Li H, Su M, Zhang J, Yang M (2014) Cancer risk assessment on trihalomethanes and haloacetic acids in drinking water of China using disability-adjusted life years. J Hazard Mater 280:288–294

Papageorgiou A, Voutsa D, Papadakis N (2014) Occurrence and fate of ozonation by-products at a full-scale drinking water. Sci Total Environ 481:392–400

Paull B, Barron L (2004) Using ion chromatography to monitor haloacetic acids in drinking water: a review of current technologies. J Chromatogr A 1046:1–9

Pichel N, Vivar M, Fuentes M (2019) The problem of drinking water access: a review of disinfection technologies with an emphasis on solar treatment methods. Chemosphere 218:1014–1030

Pisarenko AN, Stanford BD, Yan D, Gerrity D, Snyder SA (2012) Effects of ozone and ozone/peroxide on trace organic contaminants and NDMA in drinking water and water reuse applications. Water Res 46:316–326

Qiao R, Liang S, Chen C, Xiong L, Guan Q, Wang L, Fu Z, Pan Y, Liu H, Zhu J, Yechen H, Li L, Huang G (2022) 2,6-Dichloro-1,4-benzoquinone formation from chlorination of substituted aromatic antioxidants and its control by pre-ozonation in drinking water treatment plants. Chemosphere 299:134–141

Qiu C, He W, Li Y, Jiang F, Pan Y, Zhang M, Lin D, Zhang K, Yang Y, Wang W, Hua P (2022) Formation of halogenated disinfection byproducts in chlorinated real water during making hot beverages: effect of sugar addition. Chemosphere 305:417–425

Richardson SD (2011) Disinfection by-products: formation and occurrence in drinking water. Encycl Environ Health 85:110–136

Rodriguez MJ, Sérodes JB, Levallois P (2004) Behavior of trihalomethanes and haloacetic acids in a drinking water distribution system. Water Res 38:4367–4382

Scheili A, Rodriguez MJ, Sadiq R (2015) Seasonal and spatial variations of source and drinking water quality in the small municipal system of two Canadian regions. Sci Total Environ 508:514–524

Serrano M, Silva M, Gallego M (2015) Determination of 14 haloketones in treated water using solid–phase microextraction and gas chromatography–mass spectrometry. J Chromatogr A 1407:208–215

Shen R, Andrews SA (2011) Demonstration of 20 pharmaceuticals and personal care products (PPCPs) as nitrosamine precursors during chloramine disinfection. Water Res 45:944–952

Smelt JPPM, Brul S (2014) Thermal inactivation of microorganisms. Crit Rev Food Sci Nutr 54:1371–1385

Śmiech KM, Kovács T, Wildschut RF, Criado Monleon AJ, de Vries-Onclin B, Bowen JG, Agostinho LLF (2020) Thermal disinfection of hospital wastewater in a pilot-scale continuous-flow system. Appl Water Sci 10:1–12

Snyder SA, Vanderford BJ, Rexing DJ (2005) Trace analysis of bromate, chlorate, iodate, and perchlorate in natural and bottled waters. Environ Sci Technol 39:4586–4593

Staehelin J, Hoigne J (1982) Decomposition of ozone in water: rate of initiation by hydroxide ions and hydrogen peroxide. Environ Sci Technol 16:676–681

Tang S, Wang XM, Yang HW, Xie YF (2013) Haloacetic acid removal by sequential zero-valent iron reduction and biologically active carbon degradation. Chemosphere 90:1563–1567

Tao D, Wang R, Shi S, Yun L, Tong R, Peng Y, Guo W, Liu Y, Hu S (2020) The identification of halogenated disinfection by-products in tap water using liquid chromatography–high resolution mass spectrometry. Sci Total Environ 740:888–895

Thokchom B, Radhapyari K, Dutta S (2020) Occurrence of trihalomethanes in drinking water of Indian states: a critical review. Disinfection by-products in drinking water - Detection and Treatment 8:83–107

Tsitsifli S, Kanakoudis V (2018) Disinfection impacts to drinking water safety - a review. Multidisciplinary Digital Publishing Institute Proceedings 2:603–615

Tu N, Liu H, Li W, Yao S, Liu J, Guo Z, Yu R, Du H, Li J (2022) Quantitative structure-toxicity relationships of halobenzoquinone isomers on DNA reactivity and genotoxicity. Chemosphere 309:763–769

Villanueva CM, Gracia-Lavedan E, Bosetti C, Righi E, Molina AJ, Martin V (2016) Colorectal cancer and long-term exposure to trihalomethanes in drinking water: a multicenter case-control study in Spain and Italy. Environ Health Perspect 125:56–65

Villanueva CM, Evlampidou I, Ibrahim F, Donat-Vargas C, Valentin A, Tugulea AM, Echigo S, Jovanovic D, Lebedev AT, Lemus-Pérez M, Rodriguez-Susa M, Luzati A, Nery TCS, Pastén PA, Quiñones M, Regli S, Weisman R, Dong S, Ha M (2023) Phattarapattamawong, S., Kogevinas, M. Global assessment of chemical quality of drinking water: the case of trihalomethanes. Water Res 230:568–579

Wagner ED, Plewa MJ (2017) CHO cell cytotoxicity and genotoxicity analyses of disinfection by-products: an updated review. J Environ Sci 58:64–76

Wang J, Chen H (2019) Catalytic ozonation for water and wastewater treatment: recent advances and perspective. Sci Total Environ 704:249–258

Wang C, Shi H, Adams CD, Timmons T, Ma Y (2013) Investigation of removal of N-nitrosamines and their precursors in water treatments using activated carbon nanoparticles. Int J Environ Technol Manag 16:34–48
Wang C, Yang X, Zheng Q, Moe B, Li XF (2018) Halobenzoquinone-induced developmental toxicity, oxidative stress, and apoptosis in zebrafish embryos. Environ Sci Technol 52:10590–10598
Weisman RJ, Heinrich A, Letkiewicz F, Messner M, Studer K, Wang L, Regli S (2022) Estimating National Exposures and potential bladder cancer cases associated with chlorination DBPs in U.S. Drinking Water. Environ Health Perspect 130:87002–87010
Wolfe RL, Ward NR, Olson BH (1984) Inorganic chloramines as drinking water disinfectants: A review. J Am Water Works Assoc 76:74–88
World Health Organization (WHO) (2022) Guidelines for drinking-water quality: fourth edition incorporating the first and second addenda. WHO Graphics, Geneva, Switzerland
Xu X, Mariano TM, Laskin JD, Weisel CP (2002) Percutaneous absorption of trihalomethanes, haloacetic acids, and haloketones. Toxicol Appl Pharmacol 184:19–26
Yan H, Cui ZK, Manoli T, Zhang H (2021) Recent advances in non-thermal disinfection technologies in the food industry. Food Sci Technol Res 27:695–710
Yang X, Shang C, Westerhoff P (2007) Factors affecting formation of haloacetonitriles, haloketones, chloropicrin and cyanogen halides during chloramination. Water Res 41:1193–1200
Yang L, Chen Z, Shen J, Xu Z, Liang H, Tian J, Ben Y, Zhai X, Shi W, Li G (2009) Reinvestigation of the nitrosamine-formation mechanism during ozonation. Environ Sci Technol 43:5481–5487
Yang X, Wang C, Yang L, Zheng Q, Liu Q, Wawryk NJP, Li XF (2022) Neurotoxicity and transcriptome changes in embryonic zebrafish induced by halobenzoquinone exposure. J Environ Sci 117:129–140
Yoon S, Nakada N, Tanaka H (2012) A new method for quantifying N-nitrosamines in wastewater samples by gas chromatography - triple quadrupole mass spectroscopy. Talanta 97:256–261
Zambelli RA (2022) Principles of ozonation and its equipment. In: Jafari SM, Therdhtai N (eds) Non-thermal food processing operations. Woodhead Publishing, Cambridge, pp 17–54
Zazouli MA, Ulbricht M, Naseri S, Susanto. (2010) Effect of hydrophilic and hydrophobic organic matter on amoxicillin and cephalexin residuals rejection from water by nanofiltration. Iran J Environ Heal Sci Eng 7:15–24
Zhang Y, Chu W, Yao D, Yin D (2017) Control of aliphatic halogenated DBP precursors with multiple drinking water treatment processes: formation potential and integrated toxicity. J Environ Sci 58:322–330
Zhao J, Hu S, Zhu L, Wang W (2022) Formation of chlorinated halobenzoquinones during chlorination of free aromatic amino acids. Sci Total Environ 825:904–912
Zhou X, Zheng L, Chen S, Du H, Gakoko BM, Song Q, Wu F, Chen J, Lin H, Hong H (2019) Factors influencing DBPs occurrence in tap water of Jinhua region in Zhejiang Province, China. Ecotox Environ Saf 171:813–822

Disinfection by-Products (DBPs) and their Toxicological Risk on Human Wellbeing: A Public Health Concern

Dharmeswar Barhoi, Sweety Nath Barbhuiya, Arpana Rabha, and Sarbani Giri

Abstract Disinfection is an essential step in any water or wastewater treatment facility that results in the production of Disinfection by-products (DBPs) and when these disinfectants interact with dissolved organic matter and a few other inorganic compounds in water, leads to the production of a unique type of substance called DBPs in trace amount. Based on the potential health hazards and impairments associated with exposure to DBPs, USEPA has established threshold levels for certain DBPs like THMs, HAAs, HKs, and HANs. DBPs possible toxicological effects on human health have led to their recognition as a public health concern and they are generally exposed through ingestion, inhalation, and skin contact while participating in daily activities including drinking water, taking a shower, and swimming. Exposure to DBPs in humans and other animals has a variety of negative health consequences which include higher chances of Bladder cancer, cardiovascular diseases, neurological disorders, reproductive problems, and skin lesions. With a focus on their effects on public health, this chapter is aimed to discuss in detail about DBPs, their formation, toxicological potential, and effect on the development and aggravation of human disorders.

Keywords Disinfection by-products · Health hazards · Wastewater treatment

---

D. Barhoi · A. Rabha
Department of Zoology, The Assam Royal Global University, Guwahati, India

S. N. Barbhuiya (✉)
Department of Zoology, Patharkandi College, Karimganj, India

S. Giri
Molecular and Cell Biology Laboratory, Assam University, Silchar, India

S. Madhav et al. (eds.), *Drinking Water Disinfection By-products*,
https://doi.org/10.1007/978-3-031-49047-7_6

1 Introduction

Consuming contaminant-free food and beverages is crucial for maintaining good health. The most crucial element needed for sustainable living is water. The quality of water from natural sources decreased with fast industrialization and other anthropogenic activity, making it unfit for human consumption. To obtain water that is safe to drink, chemicals must therefore continue to be used to decontaminate the water. Pathogenic bacteria that are present in water are inactivated by a procedure called disinfection to stop the spread of various diseases to consumers. Disinfection/decontamination is an indispensable step in any water reservoir facility as well as in wastewater treatment, and during this decontamination procedure, humans may be exposed. Disinfection is essential for preserving the water's microbiological safety as it helps to inactivate microorganisms that can induce a variety of ailments upon exposure (Gomez-Alvarez et al. 2016). The most widely used disinfection, such as chlorine, chloramines, and ozone as shown in Fig. 1, produce a variety of reactive chemical by-products and intermediates that may be hazardous to humans and other animals, raising concerns across the globe. DBPs are created as a consequence of the procedure used to detoxify drinking water using various chemicals, which pose a concern to a sizable population in developed countries. People are mostly exposed to disinfection by-products (DBPs) through drinking water. The modes of human exposure to DBPs include cutaneous absorption during bathing and swimming, drinking DBP's contaminated water, and inhalation of DBP's contaminated air. Chlorination, which is widely employed in both developed and developing nations, is one of the most popular methods for decontaminating water. Additional means of disinfecting water include exposing it to UV rays and ozone treatment, among others. They are also much more common and practical than the earlier methods. UV disinfection is the most effective in terms of effectiveness for eradicating germs, yet running and maintaining such a system only requires a small number of resources (Tak et al. 2020). DBPs may be influenced by several precursors, including organic matter available in the environment (water body), algal organic matter (AOM), man-made pollutants created within the treatment facility, such as pesticides, drugs, and detergents, compounds containing bromine and chlorine, upstream sewage discharge, and drinking water treatment (DWT) (Alexandrou et al. 2018). DBPs are classified as "hazardous for human health," so the existence of these materials in resources of untreated water necessitates additional monitoring and work on the part of water companies (WHO 2017a). The current chapter's objective is to describe the health issues caused by the recently identified DBPs that are produced during water treatment steps in the drinking water reservoirs.

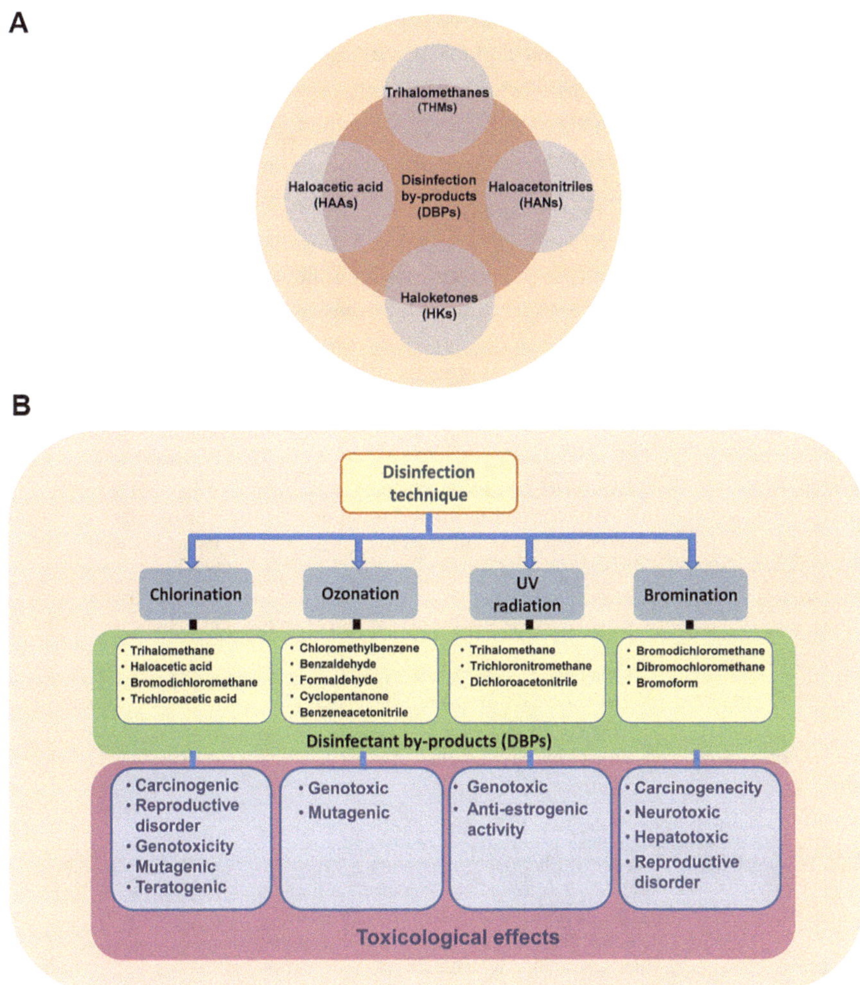

Fig. 1 Disinfection by-products (DBPs) and their toxicological implication on human health upon exposure. **(a)** Common types of DBPs. **(b)** Different techniques are used to decontaminate or purify the water reservoirs for drinking the water

2 What Are DBPs?

Disinfection by-products (DBPs) are a major group of emerging contaminants (ECs). They are distinct from other categories because, although being the outcome of anthropogenic activity, they are created naturally as opposed to being manufactured by people. During the sanitization step of water purification, when disinfectants interact with soluble organic material and a little other inorganic material in the water, a special kind of material known as DBPs is produced in trace amounts (Richardson and Postigo 2012). According to Rivera-Nez et al. (2018), DBPs are often created when various species of free chlorine interrelate with naturally stirring organic stuff in water bodies. DBPs have a long half-life in the human body and, when exposed to them, can induce an extensive variety of aberrant physiological and morphological symptoms in people and other animals. DBPs are likewise recognized as long-lasting dangerous chlorinated chemicals because of these characteristics (Alexandrou et al. 2018).

Based on the possible health hazards and impairments coupled with DBPs contact, the United States Environmental Protection Agency (USEPA) has indicated the threshold levels for certain DBPs. The permissible limit of different DBPs is shown in Table 1. Those DBPs for which the USEPA has not set or enforced threshold

Table 1 Permissible limits of microbial pollutants and disinfection by-products in drinking water

Sl. no.	Microbes/DBPs	Permissible limit	References
1	Bacteria	As per WHO, USEPA, and Indian Standards, bacterial contamination should not be at detectable range 1dL of water	WHO (2017b), USEPA (2003), Indian Standard 10500 (2012)
2	Viruses	The permissible limit for viruses is such that no virus should be detectable in samples between 100 and 1000 liters of directly reclaimed drinking water. As per USEPA (2002) and India (IS 110500:2012) the permissible limit for viruses is zero	USEPA (2003), Indian Standard 10500 (2012).
3	Vibrio cholera	No relaxation has been set.	WHO (2011, 2017a), USEPA (2003), Indian Standard 10500 (2012).
4	Trihalomethanes (THMs)	As per USEPA, the concentration of THMs should not exceed 80 micrograms/liter for total THMs. THMs include many compounds of chlorine and bromine. Whereas, in case of WHO, the highest impurity level for individual THMs are100 micrograms/L for bromoform, 300 micrograms/L chloroform, 100 micrograms/L for dibromochloromethane and 60 micrograms/L for bromodichloromethane. However, as per European Union (EU), the maximum level is 100 micrograms/L for total THMs.	WHO (2022), EU (2020) USEPA (1998).
5	Haloacetic acid (HAAs)	According to Canadian drinking water quality rules, the highest permissible quantity of HAAs in water that is suitable for consumption is 80 micrograms/L (GCDWQ). According to the USEPA, the maximum HAA contamination level is 60 micrograms per liter.	EPA (2016a, 2016b).
6	Haloketones (HKs)	Not set by USEPA	EPA (2002)
7	Haloacetonitriles (HANs)		
8	Bromate	EPA established 0.010 mg/L as the highest limit of bromated contamination in water	EPA (2002)
9	Chlorite	As per suggested by EPA, the greatest amount of chlorite in drinking water is 1.0 mg/L	EPA(2002)

constraints are referred to as unregulated DBPs, whereas those for which such limits have been established are regarded as regulated DBPs (Richardson 2011). Yet, more DBPs than the allowable limit is present in the water bodies in many different nations. As a result, drinking water is the main way that humans and other animals get exposed to these DBPs. This exposure to humans and other animals has a variety of negative health consequences on the cardiovascular system, and the reproductive system, and also increases the risk of cancer. In this chapter, the negative consequences of these DBPs on the development and aggravation of human disorders were discussed.

3 How Are DBPs Generated?

DBPs are synthesized in the water body by going through several chemical reactions with the components found in the water bodies. When a water body is infected with microbes and other chemical compounds, some compounds were being synthesized known as a disinfectant which is commonly used to clean the infected/contaminated water. These disinfectants like chlorine, chloramine, and peracetic acid react with the other organic components of the water and synthesize by-products in the reaction (Fig. 1). These compounds newly synthesized in the treated water body are regarded as DBPs. The most commonly used disinfectants are described in the subsequent heading.

4 Disinfectants Used to Treat Water

Water must be disinfected by removing pathogenic organisms such as bacteria, viruses, protozoa, worms, and larvae to utilize for other household tasks. Chemicals called disinfectants are used to purify water (Srivastav et al. 2020). At the municipal board level of government, chlorine (Cl_2), chloramines (NH_2Cl, $NHCl_2$), and chlorine dioxide are the most often utilized disinfectants (ClO_2). In addition, ozonization and ultraviolet (UV) radiation are the other two methods most frequently employed to disinfect potable water (Chhipi-Shrestha et al. 2018).

4.1 Ozone

Ozone is a powerful oxidizing agent with a tremendous redox capability of around 2.06 V. Because of its thermodynamic instability, it easily breaks down into oxygen. As a result, ozone is recognized as a powerful disinfectant with positive environmental effects (Nghi et al. 2018). Due to its high ability to oxidize a range of pollutants, ozone has the ability to be used as a strong oxidant for wastewater treatment as well as a broad-spectrum disinfectant (Wei et al. 2017). To disinfect microorganisms and organic materials in water, the complex ozonation process uses two main mechanisms: (i) the generation of free radicals as the consequence of the interaction with $OH^-$and (ii) the oxidation of bacteria as carbon-based chemicals (Rosero-Moreano 2018). It is especially superior to chlorine at eliminating protozoan cysts and oocysts. Ozone attacks pathogens directly at first, weakening the cell wall and causing surface ruptures that let the contents of the cell escape. The interaction between free radicals generated by ozone depletion and the nucleic acids causes enough damage to DNA or RNA to result in inactivation. Ozone's power as an oxidant makes it beneficial for reducing colors and odors, improving the removal of suspended particles, conditioning sludge, and oxidizing pesticides and herbicides in addition to its obvious advantage over chlorine as a disinfectant (Masten and Davies 1994). Ozone has certain disadvantages as well, the three main ones being the high initial cost, the requirement for on-site production, and the requirement for highly qualified employees to keep the ozone plant running smoothly (Masten and Davies 1994). Moreover, Ozone makes the water more biodegradable, which can be problematic for renewal in distribution networks but advantageous for biological treatment when combined with activated carbon granulation (Servais et al. 1994).

4.2 *Ultraviolet Radiation*

Ultraviolet radiation (UV) light is electromagnetic radiation with wavelengths greater than X-rays but less than illumination (Turtoi 2013). As UV disinfection is ideally suited for wastewater disinfection and it is well known that UV radiation is particularly effective against harmful bacteria, this is of great importance in the water treatment industry (Hijnen et al. 2006). An organism's DNA and RNA are exposed to electromagnetic radiation from a mercury arc lamp via a UV disinfection system, which affects an organism's ability to reproduce or replicate when it reaches the cell wall. The benefit of employing ultraviolet light as a disinfectant is that it is a physical procedure as opposed to a chemical one, and it has no enduring consequence that could be hazardous to people or aquatic life. UV light is used to clean the drinking water contaminated with microorganisms, hence the duration of time exposed to these microorganisms is related to the cleanliness of the drinking water. Besides, the intensity and the features of the wastewater also depend on the efficacy of the UV light interrogated disinfection of the drinking water (Turtoi 2013).

4.3 *Chlorine*

Chlorine is a very efficient disinfectant that is uncomplicated to handle, has minimal installation costs, is economical, is straightforward to dose, monitor, and regulate, and has a great persistent impact. No other disinfectants have yet been discovered that can match chlorine's general adaptability, even though there are likely other disinfectants that are as effective or even superior to chlorine. Gaseous chlorine, chloramines, and sodium hypochlorite are the three main types of chlorine (Li et al. 2017). The organic matter concentration, pH, and water temperature are some of the determining factors. Chlorine gas speedily catalyzes hydrolysis into hydrochloric and hypochlorous acid when dissolved in water (Lin 2014).

4.4 Peracetic Acid

In comparison to chlorine or chlorine dioxide, peracetic acid (PAA) has a larger oxidation potential and is a powerful disinfectant and oxidant (Gehr and Cochrane 2002). The effectiveness of peracetic acid as a purifier has been demonstrated, and in the majority of situations, it provided disinfection on par with that achieved with chlorine at equivalent mass concentrations. Veschetti et al. (1998) discovered that sodium hypochlorite and peracetic acid had a similar mechanism of action to disinfect fecal and total coliforms, albeit peracetic acid was less efficient against fecal *streptococci*. Peracetic acid, on the other hand, was found to be at least as effective as chlorine in killing these microorganisms, and in many situations, it was significantly more effective (Nozaic 2004).

4.5 Chloramine

Free chlorine has long been used as a disinfectant in drinking water due to its quick inactivation of the majority of pathogenic bacteria. Nevertheless, when chlorine interacts with the organic materials present in water, trihalomethanes (THMs), which are thought to be carcinogenic, are created. Due to chloramines' reduced propensity to produce halogenated organic by-products, many water utilities have shifted from free chlorine (in the form of $HOCl/OCl^-$) to them to reach this low level of trihalomethanes. Monochloramine is less effective for initial disinfection because its disinfection action is slower than that of free chlorine. Yet, because it is more stable than free chlorine, a distribution system can maintain residual disinfection over a greater distance, potentially saving money. Monochloramine can penetrate the biofilm and kill sessile biofilm bacteria, such as certain *Pseudomonas* spp., more efficiently than free chlorine (Kool et al. 1999).

5 Types of DBPs

DBPs have garnered a lot of attention in the field of water treatment since Rook found trihalomethanes (THMs) in chlorinated water in the 1970s. Different types of DBPs are generated depending on the disinfection process used; for instance, trihalomethanes, haloacetic acids, haloacetonitriles, haloketones (HKs), chloral hydrate, chloropicrin, chlorinated furanone, and cyanogen halides are frequently generated during chemical disinfection using chlorine, chloramines, or chlorine dioxide (Serrano et al. 2015). The four main families of the more than 700 identified species of DBPs are the haloacetic acids, halomethanes, haloketones, and haloacetonitriles (Xue et al. 2020; Rosero-Moreano 2018). Trihalomethanes and haloacetic acids (HAAs) were the two DBP types that were found in chlorinated water at the greatest levels and most often (Chen et al. 2008).

5.1 Trihalomethanes (THMs)

Trihalomethanes (THMs), which include Tribromochloromethane, Bromodichloromethane (BDCM), Dibromochloromethane (DBCM), and Chloroform, are the most common chlorinated DBPs in the controlled DBP category (TBM). According to Sadiq and Rodriguez (2011), bromide and dissolved organic substances such as humic acid in drinking water encourage the growth of THM and, when exposed to people and other animals, cause a number of lesions in the skin and other bodily systems.

5.2 Haloacetic Acid (HAAs)

The second-largest class of DBPs after THMs is haloacetic acid (HAA). When drinking water that is contaminated with organic material is chlorinated, these very stable, non-volatile chemicals are created. One of the nine unique components that HAAs often contain is dichloroacetic acid (DCAA), while the other two are trichloroacetic acid (TCAA). Other HAAs including monochloroacetic acid, dibromoacetic acid, bromochloroacetic acid, and monobromoacetic acid are also present in lower quantities. Drinkable water frequently has greater concentrations of DCAAs and TCAAs (Aslani et al. 2017).

5.3 Haloketones (HKs)

Volatile DBPs known as haloketones (HKs) are uncontrolled and can cause damage even in low quantities. Water that has been treated has been discovered to contain up to 14 HKs from uncontrolled DBPs, with 1,1-dichloroacetate (1,1-DCA), 1,3-dichloroacetate (1,3-DCA), and 1,1,1-trichloroacetone being the most common (1,1,1-TCA). Recent studies have measured chloroacetone, 1,1-DCA, 1,3-DCA, 1,1,1-TCA, 1,1,3-trichloroacetic, 1,1,3,3-tetrachloroacetone, and 1,1,1,3-tetrachloroacetone at values ranging from 0.13 to 14 g/L in swimming pool water. Low amounts of the most prevalent HKs are usual in drinking water (Serrano et al. 2015).

5.4 *Haloacetonitriles (HANs)*

When disinfectants like chlorine, chloramine, chlorine dioxide, and ozone react with precursors that include dissolved nitrogen, including amino acids found in drinking water, haloacetonitriles (HANs), which are DBPs, are produced (Muellner et al. 2007). Many brominated and chlorinated chemicals are created when chlorine compounds react with amino acids. Chloroacetonitrile (CAN), dichloroacetonitrile (DCAN), and trichloroacetonitriles are the three chlorinated nitrogenous chemicals (TCAN). Similar to this, bromoacetonitrile (BAN), dibromo acetonitrile (DBAN), and tribromoacetonitriles(TBAN) are chemicals that contain brominated nitrogen. The most common HAN species found in drinking water is DCAN (Komaki et al. 2014).

6 Formation of Disinfection by-Product

The disinfection process has consistently eradicated and rendered inactive pathogens discovered in water during treatment (Krasner 2009). There are numerous ways to disinfect surfaces, including using chloramines, chlorine dioxide, ozone, UV light irradiation, potassium permanganate, and nanofiltration (Dubey et al. 2020). The most popular method is chlorination, which, by leaving residual chlorine in the distribution networks, kills pathogens at the end of treatment and prevents bacterial regrowth. However, frequent chlorination has resulted in unintended chemical reactions between organic precursors (NOM, algal organic material, organic material in industrial wastewaters), inorganic components (bromate, chlorate, chlorite, iodate, and nitrate), chemical disinfectants (chlorine, chlorine dioxide, chloramines, and ozone), and other organic disinfectants to create regulated DBPs (Sadiq and Rodriguez 2011; Mazhar et al. 2020). A number of factors, including pH, temperature, different source fluids, disinfection type, and residence time, totally regulate the formation of these DBPs. Furthermore, the temperature significantly affects the concentrations of DBPs over time and geography, as do the physicochemical properties of water, the kind of disinfection procedure, the types of distribution networks, and so on (Amy et al. 2000; Nieuwenhuijsen et al. 2000).

The use of alternate disinfections with the same or better properties than chlorine or chloride derivatives has been raised in light of the health risks associated with DBPs that are present in water supplies as a consequence of the interaction between NOM and chlorine in water treatment plants (WTP), as well as the immediate need is to control the action of microbial pathogens in the water sample because of unregulated wastewater discharges into the catchment water. Ozone, hydrogen peroxide, UV radiation, potassium permanganate, peroxyacetic acid, and others are examples of potential substitutes (Rosero-Moreano 2018). Nevertheless, the emergence of a new class of DBPs referred to as emergent or unregulated (UR-DBPs) as a consequence of this nonconventional disinfection is raising fresh worries in different parts of the world (Legay et al. 2015). When employed as secondary disinfection, chloramine is preferred to chlorine (Liu et al. 2016), yet it has been discovered that chlorination produces a major class of nitrogenous DBPs known as N-nitrosamines (Le Roux et al. 2010). Ozone is the second strong main disinfectant employed in the cleaning of drinkable water systems (Washington State 2004). It increases the quantity of organic matter that can be digested, which reduces the production of THMs, promotes microbial renewal in distribution networks, and creates a variety of organic and inorganic DBPs (Sadiq and Rodriguez 2011).

6.1 *Trihalomethanes (THMs)*

Trihalomethanes (THMs) are a significant class of DBPs that are known to be harmful to human health and are produced when drinking water is chlorinated (Liang and Singer 2003). It is created when free chlorine reacts with naturally occurring organic material (NOM) for a long time, as it does during ordinary chlorination while treating water (Di Cristo et al. 2013). Other physical parameters that have been linked to THM generation include the temperature of the water system, pH, and duration of contact within the water body (Rasheed et al. 2018). It is now recognized that bromide makes a substantial contribution to the creation of THMs. When bromide is exposed to chlorine, it is oxidized to intermediates (presumably Br_2, HOBr, and OBr-, with a little amount of BrCl), which seem to have a far larger impact on the halogenation phase of the THM reaction sequence than chlorine does. The effects of bromide ions on the production of THMs have been thoroughly researched since brominated analogs are riskier than their chlorinated equivalents (Richardson et al. 2007). Due to the fact that brominated analogs are riskier than their chlorinated counterparts, the effects of bromide ions on the generation of THMs have been extensively studied (Richardson et al. 2007). Active chlorine quickly converts bromide to active bromine (hypobromous acid, HOBr, and hypobromite ion, OBr) during the disinfection process, which starts the substitution reaction that produces brominated analogs (Liu et al. 2018). Active bromine and chlorine species then react with organic molecules to create halogenated intermediates, which serve as a precursor to halogenated THMs, free chloride ions, or bromide ions that are later released as the intermediates decompose (Westerhoff et al. 2004).

6.2 *Haloacetic Acid (HAAs)*

Both the chlorination procedure used to disinfect drinking water and the presence of inorganic species like bromide in surface and ground waters have been connected to the generation of haloacetic acid (HAA). Unlike the atmosphere, where it naturally develops through the photochemical breakdown of chlorinated solvents, where HAAs are naturally created, HAAs are produced when disinfectants with halogen species oxidize the NOMs present in water (Frank et al. 1994). The number of

HAAs likewise increases when the chlorine level rises. Hypobromous acid is produced during the disinfection process by the reaction of bromide and chloride, which interacts with NOM roughly 25 times more quickly than hypochlorous acid to create HAAs (Chang et al. 2001). Nevertheless, a number of factors, such as pH, temperature, contact duration, season, bromide and chloride concentrations, and chlorination affect how they grow (Kim et al. 2003).

6.3 Haloacetonitriles (HANs)

Haloacetonitrile disinfection by-products (HAN-DBPs) are produced when disinfectants including chlorine, chloramine, chlorine dioxide, and ozone react with dissolved nitrogen-containing precursors such as amino acids in drinking water (Muellner et al. 2007). Chloramines significantly raise HAN concentrations in drinking water. After disinfection, the following HANs are the most common ones found in drinking water: bromoacetonitrile (BAN), chloroacetonitrile (CAN), dichloroacetonitrile (DCAN), trichloroacetonitrile (TCAN), and bromochloroacetonitrile (BCAN) (Komaki et al. 2014). Algal blooms cause urban wastewater effluents to have high quantities of organic nitrogen, which leads to the production of HANs, an unregulated class of nitrogen-containing DBPs (Shah and Mitch 2012). Moreover, a number of disinfectants used to lower regulated DBPs also increase HAN production (Anglada et al. 2011).

6.4 Bromide

Several investigations demonstrated that the CuO-catalyzed disproportionation of the hypobromous acid (HOBr) route can produce considerable bromate (BrO_3^-) (Liu and Croué 2016). Anionic and acidic forms of chlorine are produced during chlorination. The presence of bromide ions causes chlorine disinfection, which produces bromide DBPs. After that, chlorine reacts with the Br ion to create HOBr, which then interacts with the natural organic matter NOM to produce Br-DBPs. HOBr can also react with water's ammonia and chloramines to create NH2Br and NHClBr, which can then interact with NOM to produce Br-DBPs (Sundaram 1990).

7 Impact of DBPs on Human Health

DBPs that are synthesized due to the treatment of water bodies with infections pose serious health implications upon exposure. Humans are exposed to these byproducts when they use the treated water for their daily household activities like drinking treated water, bathing, and showering. The DBPs cause health hazards upon its exposure to human that includes skin lesions, cardiovascular diseases, reproductive diseases, neurological disorders, and cancer. Scientists and researchers all over the world have investigated the impact of exposure to DBPs on human health. However, the research is still going on to confirm the reports of its implications on human health as controversy exists on it. Some reports indicated that the DBPs potentially cause cancer (discussed in subsequent headings) and some reports suggest a positive correlation of it with an increased risk of cancer. In this chapter, we have collated both positive and negative findings that have been investigated so far.

7.1 Cancer

Continued prolonged exposure to chlorinated water has been related to an increased risk of cancer, notably Cancer of the colorectum as well as the urinary bladder, according to several epidemiological studies (Richardson et al. 2007). In addition to DBPs, drinking water may also include radionuclides and arsenic, both of which have the potential to cause cancer. THM levels, however, have historically been linked to a higher risk of bladder cancer (Cantor 1997; Villanueva et al. 2007). One investigation discovered a link between the water's mutagenicity and a greater prevalence of bladder and kidney cancer, which may be attributed to concentrations of the chlorinated furanone and/or other mutagenic DBPs (Koivusalo et al. 1994). There is evidence from several types of research that the presence of THM bromoform increases the chance of getting rectal cancer (Bove et al. 2007). Following an epidemiologic investigation evaluating risk by route of exposure (Cantor et al. 2006; Villanueva et al. 2007), a substantial number of the bladder cancer threat linked to chlorinated water shows up to be due to taking showers, bathing, and diving and the risk may be highest for those who carry the GSTT1-1 gene. These findings imply a potential link between cutaneous and inhalation exposure, as well as inherited susceptibility, and the chance of getting cancer.

An Ontario-based Canadian study found some evidence suggesting an increased risk of bladder cancer in people who were exposed to THMs over a lengthy period (greater than 35 years) (King and Marrett 1996). According to studies relating intestinal cancer with DBPs in water, consuming chlorinated water increases the risk of developing intestine cancer. King and Marret looked for the incidence of bladder, intestine, or anal cancer in 5000 patients in Ontario (Canada), and among them, 950 patients had the risk due to the use of drinking water contaminated with DBPs. In another study, the risk of intestinal cancer was 1.5 times higher in individuals exposed to DBPs with levels of 50 g/L or higher (King and Marrett 1996). Controversial reports also exist in the science community that report no risk upon exposure to DBPs. According to research done in Iowa (USA) between 1986 and 1989 using data from patients with anal and intestinal cancer, there is no increased risk of developing intestinal cancer following prolonged exposure to trihalomethanes or chlorinated drinking water. The majority of research does, however, point to a relationship between the length of exposure to DBPs and the risk of cancer development and progression. Also, people who consume diets low in fiber are far more likely to develop cancer when exposed to DBP. Together with not exercising, DBP exposure increases the risk of anal cancer (Hildesheim et al. 1998). The health complications due to DBP exposure are listed and compared in Table 2.

7.2 *Reproductive Disease*

In both human and animal investigations, some epidemiological findings have suggested a connection between DBP exposure and poor reproductive outcomes (Keegan et al. 2001). Epidemiological research on disinfection by-product exposure and its effects on pregnancy and birth defects are scarce. However, some studies show a connection between trihalomethane exposure and spontaneous abortion, abnormal birth, and developmental retardation (Dodds and King 2001). The results were based on the use of highly chlorinated drinking water, and it was found that the study's participants who had been exposed to a high level of natural organic matter during their pregnancies had positively correlated with spontaneous abortion, abnormal birth, and developmental anomalies. Dodds and King conducted a cohort study in 2001 to assess the risk of birth defects related to exposure to various trihalomethanes in public water sources. They examined the birth defects that included chromosomal abnormalities, cleft defects, cardiovascular disorders, and neural tube defects. Previous studies reported that exposure to chloroform was not related to an increased incidence of neural tube anomalies but exposure to bromodichloromethane at concentrations of 20 g/L or above drastically induced neural tube defects. In separate research from 2012, Levallois et al. looked at the connection between maternal DBP exposure and the chance of having a small-for-gestational-age infant. With HAA concentrations above the fourth quartile and THM or HAA concentrations exceeding current water guidelines, the risk of small-for-gestational-age babies rose (Levallois et al. 2012). Additionally, a Bradford, England research that included 7438 newborns discovered a link between TTHM exposure during pregnancy and lower birth weight (Smith et al. 2016). Also, a study of 2460 stillbirth cases in Massachusetts between the years of 1997 and 2004 revealed that exposure to dichloroacetic acid and chloroform was linked to stillbirths (Rivera-Nunez et al. 2018). The individuals exposed to TTHM encountered lower sperm counts and serum testosterone levels in men (Zeng et al. 2013). It was found that the GSST1 gene is associated with sperm parameters in individuals. Investigations also revealed that a GSTT1 polymorphism changed upon exposure to Bromo-THM which resulted in decreased sperm motility.

Table 2 Disinfection by-products (DBPs) and their toxicological implications on human health

Sl. no.	Disinfection by-product	Exposure route	Exposure source	Toxicological implications	References
1.	Trihalomethane	Drinking water	These compounds are produced due anthropogenic sources. When water is decontaminated by chlorination techniques, these products are getting generated in the treated water body and mineral water plants. Upon drinking mineral water and supplying water from such treated bodies, humans get exposed to it.	Exposure to Trihalomethane produced due to application of chlorine at conc. of 2–5 mg/L produces cancer initiation. Besides, trihalomethane could induce toxicities in reproductive organs.	Bond et al. (2012)
2.	Bromodichloromethane	Drinking water		Exposure to brominated chloromethane showed DNA instability among the individuals exposed to it. Along with its genotoxic effect, it also elevates the anomalies in the reproductive organs leading to congenital anomalies in the infant of exposed women.	Leavens et al. (2007)
3.	Trichloroacetic acid	Drinking water and Spa		Trichloroacetic acids have immense potential to elevate chromosomal damage, in the reproductive tissues and other cells of the body. TCA also causes reproductive toxicities among the individuals exposed to it.	Mashau et al. (2018)
4.	Halo acetic acid	Drinking water and Spa	Both natural and anthropogenic sources are responsible for the production of these DBP's. Chlorination techniques in water purification lead to the production and exposure of humans to these DBPs.	Exposure to halo acetic acid is associated with DNA damage, stillbirth, spontaneous abortion, neural defect, and congenital effects.	Bond et al. (2012)
5.	Iodoacetic acid	Drinking water		Exposure to Iodoacetic acid causes exclusive DNA damage among the individuals exposed to it.	Wei et al. (2013)
6.	Di-iodoacetic acid		–	Di-iodoacetic acid causes toxicity to cells upon exposure.	Sayess et al. 2017

Sl. no.	Disinfection by-product	Exposure route	Exposure source	Toxicological implications	References
7.	Iodoform		–	Iodoform exhibited both DNA damaging potential as well as cytotoxic potential when exposed to humans.	Wei et al. (2013)
8.	Bromochloromethane	Through secondary treatment (combination exposure)	Man-made sources in combination with two or more disinfection techniques	Upon exposure to Bromochloromethane, the individuals were reported to suffer from reproductive anomalies.	Yang et al. (2017)
9.	Bromoform	Sources other than drinking water	–	Bromoform DBPs are toxic to a variety of cell types. Bromoform exposure causes toxicity to nervous systems as well as hepatocytes. Anomalies in nerve impulse and liver functions in individuals exposed to Bromoform are common.	Risher (2005)
10.	Dibromochloromethane	Sources other than drinking water	–	Dibromochloromethane has extensive cancer-inducing potential.	Risher (2005)
11.	Formaldehyde, benzaldehyde, cyclopentanone, benzeneacetonitrile, chloromethylbenzene	Drinking water and spa	Anthropogenic sources. Water purification techniques lead to the formation of such DBPs.	Exposure to these DBPs exhibits genotoxic and mutagenic effects among the exposed individuals.	Papageorgiou et al. (2014)

7.3 Cardiovascular Diseases

Not only is the reproductive system, but the cardiovascular system is also affected upon contact with DBPs. Exposure to DBPs altered the physiology of the cardiovascular system leading to de-normalize the lives of the exposed individuals. The use of chlorine dioxide in the chlorination process, in particular, seems to be associated with elevated jeopardy of congenital cardiac defects as well as an increase in the overall THM content. In a study, the composition of municipal drinking water was collected from 58,669 women using the health database in Swedish. In the study, 753 infants of women drinking water contaminated with DBPs had cardiac conditions among the other infants (Cnattingius et al. 1990). Various techniques have been used to quantify the trihalomethanes and nitrate in the ground and surface water and their role to increase heart defects was also investigated. When crude rates were analyzed, groundwater was associated with an increased risk of heart abnormalities; however, following appropriate modifications, it was discovered that this excess rate was generated by chlorination procedures that integrated chlorine dioxide throughout water treatment (Cedergren et al. 2002). Chlorine dioxide has an adjusted hazard ratio of 1.61, indicating that it may be a risk factor for cardiac issues in and of itself. Trihalomethane concentrations increased the incidence of cardiac anomalies (Cedergren et al. 2002). Nitrate concentrations were linked to higher-than-average cardiac anomalies; however, this increased risk is not statistically significant as the reports suggested (Blaisdell et al. 2019). Despite the minimal individual risks for congenital heart problems brought on by trihalomethanes and chlorine dioxide, the attributable risk for cardiac malformations may be high due to the vast population exposed to public drinking water (Cedergren et al. 2002).

Monobromoacetic acid (MBAA) exposure in water increased the risk of cardiovascular illnesses, according to research by Randall et al. (CVDs). According to research done in Massachusetts between 2000 and 2004, atrial septal abnormalities (ASDs) and ventricular septal defects (VSDs) accounted for 45 of the total birth malformations evaluated (Wright et al. 2017). According to 1995 research by Christ et al. on rats, exposure to bromochloroacetonitrile led to CVDs, levocardia, a right-sided aortic arch, and ductus arteriosus (Christ et al. 1995). Chlorine dioxide DBPs, including chlorate and chlorite, did not, however, reveal an increased incidence of VSDs in more recent research (Righi et al. 2012). A cohort research done in New Jersey found that women exposed to high levels of THMs had a higher chance of developing significant cardiac problems (Bove et al. 1995). Chlorination DBPs, halogenated acids dichloroacetic acid, and trichloroacetic acid were shown to cause cardiac malformation (TCAA). An elevated risk of cardiovascular malformation was seen in pregnant Long-Evans rats after dichloroacetic acid (DCAA) induction.

7.4 Skin Lesion

When a combinatory treatment strategy is approached, it causes the production of considerably fewer chlorinated by-products (CBP) than chlorination alone. Chlorine alone will create a substantially larger quantity of DBPs by interacting with the organic content in the water, making chlorine-only swimming pools generally more harmful than those treated with ozone or UV. The most typical symptoms of CBPs in swimmers and pool staff are irritated eyes, noses, throats, and respiratory issues (Liviac et al. 2010). Among the chlorinated by-products, NCl_3 produces the most irritative symptoms (Parrat et al. 2012). There was evidence that irritating ocular, nasal, and throat symptoms had a significant concentration-response connection with exposure levels of NCl_3 (Massin et al. 1998). Water and CBPs have detrimental effects on skin cells since they dilute or drain natural moisturizing away. As NCl_3 concentrations rose, the likelihood of skin irritation increased (Parrat et al. 2012). Earlier reports suggest that chlorination water causes temporary but considerable alterations in the skin surface qualities among recreational swimming compared to women with healthy skin (Gardinier et al. 2009). The pool attendants are also suffering from skin dryness, irritation, and erythema which are expected to occur due to water treatment by chlorination techniques (Dalmau et al. 2012; Fernandez-Luna et al. 2016). There have also been reports of allergic contact urticaria and allergic contact dermatitis brought on by chlorinated pool water. Dermal symptoms (rash, widespread itching, skin infection) have been linked to swimming in public pools or spas upon use for several weeks (Najnin et al. 2014). In addition, compared to other pool employees, lifeguards, and trainers have been found to have greater verrucas, mycosis, eczema, and rash that might have occurred due to their exposure to chlorination treated water (Najnin et al. 2014).

7.5 Neurological Disorders

Activities intended to make the water more potable impair the water's quality. Microbial growth is facilitated by biofilm formation in drinking water distribution networks (Liu et al. 2016). The exercise of chlorine and its derivatives is one way to disinfect. The interaction of these components with the minuscule quantities of organic molecules in water results in the creation of chlorinated organic, which is persistent and potentially hazardous. It has been demonstrated that mammalian cells are vulnerable to chlorinated wastewater (Du et al. 2017). When chlorine and ammonia are used to disinfect water, copper pipe erodes, perhaps raising the metal content of the drinking water. The pH level affects how much chlorine will corrode copper pipes (Lytle and Liggett 2016). Another potentially hazardous element added to water to precipitate organic material and so clear it is alumina, which is used to coagulate particle pollutants. As a result, the quantity of aluminum in water may rise (Gopal et al. 2004), which might have serious negative effects on human health. Aluminum levels in drinking water are linked to the emergence of Alzheimer's disease (AD), according to an epidemiological research (Rondeau et al. 2009). Laboratory studies on animals exposed to aluminum amounts in drinking water equivalent to those seen in some home sources corroborate this claim (Campbell et al. 2004; Bondy 2016). In a recent research, the consequences of pregnant women's exposure to the disinfection by-products contained in tap water on their developing brains were examined. Although there was a link between girls' mental scores and exposure to disinfection by-products at age one, the difference disappeared between ages 4 and 5 years (Villanueva et al. 2018).

Flint, Michigan had one of the most recent cases of lead contamination of drinking water. To save money from the Detroit Water and Sewerage Department, the city of Flint switched to Flint River Water Treatment at the Flint Water Service Center (Zahran et al. 2017). Following testing, levels of lead over the permissible threshold were found in several water samples (Edwards 2015). Lead levels in children's blood were noticeably higher (Hanna-Attisha et al. 2016) than the permissible standards. Lead can affect neurodevelopment as well as other brain functions, causing abnormalities. Amyloid beta (Aβ) protein senile plaques, which are formed from an amyloid precursor protein, are a defining feature of Alzheimer's disease. Fetal exposure to lead could result in long-lasting repercussions since it can interfere with cellular processes connected to aluminum and Alzheimer's disease (Mazumdar et al. 2012). A rising amount of information showing that some drinking water contains dangerously high levels of aluminum is being produced through epidemiological and laboratory investigations. Studies on animals have shown that exposure to these levels over an extended period of time may result in neuroinflammation and oxidative stress (Campbell et al. 2004; Becaria et al. 2006). Alzheimer's and other neurological diseases may develop more quickly as a result of aluminum exposure. Reports suggested higher concentrations of aluminum in the post-mortem brain tissues of individuals suffering from Alzheimer's (Perl and Brody 1980; Mirza et al. 2017). Many DBPs, such as dibromacetic acid present in water sources, have been investigated for their potential neurotoxicity (Moser et al. 2004). Parkinson's disease has been linked to trichloroethylene (TCE) exposure. Animals exposed to TCE may have striatal degeneration and the onset of Parkinson's symptoms. The production of chloral from TCE and subsequent synthesis of 1-trichloromethyl-1,2,3,4-tetrahydro-beta-carboline (TaClo), a particular dopaminergic neurotoxic chemical, have both been linked to mitochondrial malfunction (Zaheer and Slevin 2011).

8 Conclusion

The DBPs have immense potential to cause toxicological implications among the exposed individuals. Drinking water is the principal source of human exposure to these disinfection by-products, while there are a few small sources outside of drinking water. The exposed individuals are suffering from various sorts of health complications when these DBPs encountered them. The common effects include adverse functioning of the reproductive, cardiovascular, and nervous systems. The different types of skin lesions are the major impact of disinfection by-product exposure. These DBPs are synthesized in the water reservoir that is treated to disinfect the contamination. The compounds used for disinfection of the water body release the by-products. Therefore, the concerned statuary body must look for alternate and/or appropriate strategies to clean the water reservoirs in such a way that humans should not get exposed to the products generated due to such treatment procedures. One might employ the following strategies by which, the production of DBPs can be minimized and thus its exposure to humans and other animals.

- **By reducing the organic load:** The organic components that are present in the water body react with chlorine and other chemicals that are used as disinfectants. The generation of disinfection by-products in the water body will be stopped by a reduction in organic compounds.
- **Reducing the contact time of chlorine during the chlorination process:** The amount of time that chlorine is in contact with a substance during chlorination directly relates to the number of chlorinated disinfection by-products produced. Less chlorinated DBPs will be produced as a result of reduced exposure to chlorine.
- **Elimination of areas with stagnant water:** The stagnant water areas are high with algal blooms and organic load. Therefore, by reducing the growth and deposition of algal blooms and organic load by eliminating the areas with stagnant water, the production of DBPs could be reduced.
- **Use of combinatory and different treatments:** A combinatory treatment strategy reduces the production of DBPs. Moreover, frequent changing of disinfectants will also decrease the synthesis and deposition of DBPs.

Abbreviations

AD	Alzheimer's Disease
AOM	Algal organic matter
ASD	Atrial septal disease
BAN	Bromoacetonitrile
BCAN	Bromochloroacetonitrile
CAN	Chloroacetonitrile
CBP	Chlorinated by-product
CVDs	Cardiovascular diseases
DBAN	Dibromoacetonitrile
DBCM	Dibromochloromethane
DBPs	Disinfection by-products.
DCAA	Dichloroacetic acid
DCAN	Dichloroacetonitrile
DWT	Drinking water treatment
ECs	Emerging contaminants.
EPA	Environmental Protection Agency
HAAs	Haloacetic acids
HAN-DBPs	Haloacetonitrile disinfection by-product
HANs	Haloacetonitriles
HKs	Haloketones
MBAA	Monobromoacetic acid
NOM	Natural organic matter
PAA	Peracetic acid

TBAN	Tribromoacetonitrile
TCAA	Trichloroacetic acid
TCAN	Trichloroacetonitrile
TCE	Trichloroethylene
THMs	Trihalomethanes
TTHM	Total Trihalomethane
Ur-DBPs	Unregulated disinfection by-products
USPEA	United states environmental protection agency
VSDs	Ventricular septal disease
WHO	World Health Organization
WTP	Water treatment plant

References

Alexandrou L, Meehan BJ, Jones OA (2018) Regulated and emerging DBPs in recycled waters. Sci Total Environ 637:1607–1616

Amy G, Bull R, Craun GF, Pegram RA, Siddiqui M, World Health Organization (2000) Disinfectants and disinfectant by-products. World Health Organization

Anglada Á, Urtiaga A, Ortiz I, Mantzavinos D, Diamadopoulos E (2011) Boron-doped diamond anodic treatment of landfill leachate: evaluation of operating variables and formation of oxidation by-products. Water Res 45(2):828–838

Aslani H, Nasseri S, Nabizadeh R, Mesdaghinia A, Alimohammadi M, Nazmara S (2017) Haloacetic acids degradation by an efficient ferrate/UV process: byproduct analysis, kinetic study, and application of response surface methodology for modeling and optimization. J Environ Manag 203:218–228

Becaria A, Lahiri DK, Bondy SC, Chen D, Hamadeh A, Li H, Taylor R, Campbell A (2006) Aluminum and copper in drinking water enhance inflammatory or oxidative events specifically in the brain. J Neuroimmunol 176(1–2):16–23

Blaisdell J, Turyk ME, Almberg KS, Jones RM, Stayner LT (2019) Prenatal exposure to nitrate in drinking water and the risk of congenital anomalies. Environ Res 176:108553

Bond T, Templeton MR, Graham N (2012) Precursors of nitrogenous disinfection by-products in drinking water—a critical review and analysis. J Hazardous Materials 235:1–6

Bondy SC (2016) Low levels of aluminum can lead to behavioral and morphological changes associated with Alzheimer's disease and age-related neurodegeneration. Neurotoxicology 52:222–229

Bove FJ, Fulcomer MC, Klotz JB, Esmart J, Dufficy EM, Savrin JE (1995) Public drinking water contamination and birth outcomes. Am J Epidemiol 141(9):850–862

Bove GE, Rogerson PA, Vena JE (2007) Case control study of the geographic variability of exposure to disinfectant byproducts and risk for rectal cancer. Int J Health Geogr 6:1–12

Campbell A, Becaria A, Lahiri DK, Sharman K, Bondy SC (2004) Chronic exposure to aluminum in drinking water increases inflammatory parameters selectively in the brain. J Neurosci Res 75(4):565–572

Cantor KP (1997) Drinking water and cancer. Cancer Causes Control 8(3):292–308. Find this article online

Cantor KP, Villanueva C, Garcia-Closas M, Silverman D, Real FX, Dosemeci M, Malats N, Yeager M, Welch R, Chanock S, Tardon A (2006) Bladder cancer, disinfection byproducts, and markers of genetic susceptibility in a case-control study from Spain. Epidemiology 17(6):S150

Cedergren MI, Selbing AJ, Löfman O, Källen BA (2002) Chlorination byproducts and nitrate in drinking water and risk for congenital cardiac defects. Environ Res 89(2):124–130
Chang EE, Lin YP, Chiang PC (2001) Effects of bromide on the formation of THMs and HAAs. Chemosphere 43(8):1029–1034
Chen C, Zhang XJ, Zhu LX, Liu J, He WJ (2008) DBPs and their precursors in a water treatment plant in North China: seasonal changes and fraction analysis. Sci Total Environ 397(1–3):140–147
Chhipi-Shrestha G, Rodriguez M, Sadiq R (2018) Unregulated DBPs in drinking water in Quebec: A meta analysis. J Environ Manag 223:984–1000
Christ SA, Read EJ, Stober JA, Kate Smith M (1995) The developmental toxicity of bromochloroacetonitrile in pregnant long-Evans rats. Int J Environ Health Res 5(2):175–188
Cnattingius S, Ericson A, Gunnarskog J, Källén B (1990) A quality study of a medical birth registry. Scand J Soc Med 18(2):143–148
Dalmau G, Estela Martínez-Escala M, Gázquez V, Pujol-Montcusí JA, Canadell L, EsponaQuer M, Pujol RM, Vilaplana J, Gaig P, Giménez-Arnau A (2012) Swimming pool contact dermatitis caused by 1-bromo-3-chloro-5, 5-dimethyl hydantoin. Contact Dermatitis 66(6):335–339
Di Cristo C, Esposito G, Leopardi A (2013) Modelling trihalomethanes formation in water supply systems. Environ Technol 34(1):61–70
Dodds L, King WD (2001) Relation between trihalomethane compounds and birth defects. Occup Environ Med 58(7):443–446
Du Y, Wu QY, Lu Y, Hu HY, Yang Y, Liu R, Liu F (2017) Increase of cytotoxicity during wastewater chlorination: impact factors and surrogates. J Hazard Mater 324:681–690
Dubey S, Gusain D, Sharma YC, Bux F (2020) The occurrence of various types of disinfectant by-products (trihalomethanes, haloacetic acids, haloacetonitrile) in drinking water. In: DBPs in drinking water. Butterworth-Heinemann, pp 371–391
Edwards M (2015) Our sampling of 252 homes demonstrates a high lead in water risk: Flint should be failing to meet the EPA Lead and copper rule. Flint water study, September, 8
EPA (2002) National primary drinking water regulations: long term 1 enhanced surface water treatment rule. Final rule. Fed Reg 67(9):1811–1844
EPA (2016a) Conventional treatment. U.S. Environmental Protection Agency. https://tdb.epa.gov/tdb/treatmentprocess?treatmentProcessId=1934681921. Accessed 17 Oct 2016
EPA (2016b) Six-year review 3 Compliance Monitoring Data (2006–2011). U.S. Environmental Protection Agency. Last updated: 12/16. https://www.epa.gov/dwsixyearreview/six-year-review-3-compliance-monitoring-data-2006-2011
European Parliament and the Council of the European Union (2020) Directive (EU) 2020/2184 of the European Parliament and of the council of 16 December 2020 on the quality of water intended for human consumption (recast). Off JL 435:1–62
Fernández-Luna Á, Burillo P, Felipe JL, Del Corral J, García-Unanue J, Gallardo L (2016) Perceived health problems in swimmers according to the chemical treatment of water in swimming pools. Eur J Sport Sci 16(2):256–265
Frank H, Scholl H, Renschen D, Rether B, Laouedj A, Norokorpi Y (1994) Haloacetic acids, phytotoxic secondary air pollutants. Environ Sci Pollut Res 1:4–14
Gardinier S, Guéhenneux S, Latreille J, Guinot C, Tschachler E (2009) Variations of skin biophysical properties after recreational swimming. Skin Res Technol 15(4):427–432
Gehr R, Cochrane D (2002, January) Peracetic acid (PAA) as a disinfectant for municipal wastewaters: encouraging performance results from physicochemical as well as biological effluents. In Disinfection and Reuse Symposium 2002 (pp 182–198). Water Environment Federation
Gomez-Alvarez V, Pfaller S, Pressman JG, Wahman DG, Revetta RP (2016) Resilience of microbial communities in a simulated drinking water distribution system subjected to disturbances: role of conditionally rare taxa and potential implications for antibiotic-resistant bacteria. Environ Sci Water Res Technol 2(4):645–657
Gopal K, Srivastava SB, Shukla S, Bersillon JL (2004) Contaminants in drinking water and its mitigation using suitable adsorbents: an overview. J Environ Biol 25(4):469–475

Hanna-Attisha M, LaChance J, Sadler RC, Champney Schnepp A (2016) Elevated blood lead levels in children associated with the Flint drinking water crisis: a spatial analysis of risk and public health response. Am J Public Health 106(2):283–290
Hijnen WAM, Beerendonk EF, Medema GJ (2006) Inactivation credit of UV radiation for viruses, bacteria and protozoan (oo) cysts in water: a review. Water Res 40(1):3–22
Hildesheim ME, Cantor KP, Lynch CF, Dosemeci M, Lubin J, Alavanja M, Craun G (1998) Drinking water source and chlorination byproducts II. Risk of colon and rectal cancers. Epidemiology 9(1):29–35
Indian Standard 10500 (2012) Drinking water [FAD 25: Drinking Water] https://law.resource.org/pub/in/bis/S06/is.10500.2012.pdf
Keegan T, Whitaker H, Nieuwenhuijsen MJ, Toledano MB, Elliott P, Fawell J, Wilkinson M, Best N (2001) Use of routinely collected data on trihalomethane in drinking water for epidemiological purposes. Occup Environ Med 58(7):447–452
Kim J, Chung Y, Shin D, Kim M, Lee Y, Lim Y, Lee D (2003) Chlorination by-products in surface water treatment process. Desalination 151(1):1–9
King WD, Marrett LD (1996) Case-control study of bladder cancer and chlorination by-products in treated water (Ontario, Canada). Cancer Causes Control 7:596–604
Koivusalo M, Jaakkola JJ, Vartiainen T, Hakulinen T, Karjalainen S, Pukkala E, Tuomisto J (1994) Drinking water mutagenicity and gastrointestinal and urinary tract cancers: an ecological study in Finland. Am J Public Health 84(8):1223–1228
Komaki Y, Mariñas BJ, Plewa MJ (2014) Toxicity of drinking water disinfection byproducts: cell cycle alterations induced by the monohaloacetonitriles. Environ Sci Technol 48(19):11662–11669
Kool JL, Carpenter JC, Fields BS (1999) Effect of monochloramine disinfection of municipal drinking water on risk of nosocomial Legionnaires' disease. Lancet 353(9149):272–277
Krasner SW (2009) The formation and control of emerging DBPs of health concern. Philos Trans R Soc A Math Phys Eng Sci 367(1904):4077–4095
Leavens TL, Blount BC, DeMarini DM, Madden MC, Valentine JL, Case MW, Silva LK, Warren SH, Hanley NM, Pegram RA (2007) Disposition of bromodichloromethane in humans following oral and dermal exposure. Toxicol Sci 99(2):432–445
Legay C, Levallois P, Aranda-Rodriguez R, Dabeka L, Hnatiw J, Rodriguez MJ (2015) Variability of non-regulated DBPs in distribution systems: impact of the storage tank. In: Recent advances in DBPs. American Chemical Society, pp 341–362
Levallois P, Gingras S, Marcoux S, Legay C, Catto C, Rodriguez M, Tardif R (2012) Maternal exposure to drinking-water chlorination by-products and small-for-gestational-age neonates. Epidemiology:267–276
Li HY, Osman H, Kang CW, Ba T (2017) Numerical and experimental investigation of UV disinfection for water treatment. Appl Therm Eng 111:280–291
Liang L, Singer PC (2003) Factors influencing the formation and relative distribution of haloacetic acids and trihalomethanes in drinking water. Environ Sci Technol 37(13):2920–2928
Lin SD (2014) Water and wastewater calculations manual. McGraw-Hill Education
Liu C, Croué JP (2016) Formation of bromate and halogenated disinfection byproducts during chlorination of bromide-containing waters in the presence of dissolved organic matter and CuO. Environ Sci Technol 50(1):135–144
Liu S, Gunawan C, Barraud N, Rice SA, Harry EJ, Amal R (2016) Understanding, monitoring, and controlling biofilm growth in drinking water distribution systems. Environ Sci Technol 50(17):8954–8976
Liu R, Tian C, Hu C, Qi Z, Liu H, Qu J (2018) Effects of bromide on the formation and transformation of DBPs during chlorination and chloramination. Sci Total Environ 625:252–261
Liviac D, Wagner ED, Mitch WA, Altonji MJ, Plewa MJ (2010) Genotoxicity of water concentrates from recreational pools after various disinfection methods. Environ Sci Technol 44(9):3527–3532

Lytle DA, Liggett J (2016) Impact of water quality on chlorine demand of corroding copper. Water Res 92:11–21

Mashau F, Ncube EJ, Voyi K (2018) Drinking water disinfection by-products exposure and health effects on pregnancy outcomes: a systematic review. J Water Health 16(2):181–196

Massin N, Bohadana AB, Wild P, Hery M, Toamain JP, Hubert G (1998) Respiratory symptoms and bronchial responsiveness in lifeguards exposed to nitrogen trichloride in indoor swimming pools. Occup Environ Med 55(4):258–263

Masten SJ, Davies SH (1994) The use of ozonation to degrade organic contaminants in wastewaters. Environ Sci Technol 28(4):180A–185A

Mazhar MA, Khan NA, Ahmed S, Khan AH, Hussain A, Changani F, Yousefi M, Ahmadi S, Vambol V (2020) Chlorination disinfection by-products in municipal drinking water – a review. J Clean Prod 273:123159

Mazumdar M, Xia W, Hofmann O, Gregas M, Sui SH, Hide W, Yang T, Needleman HL, Bellinger DC (2012) Prenatal lead levels, plasma amyloid β levels, and gene expression in young adulthood. Environ Health Perspect 120(5):702–707

Mirza A, King A, Troakes C, Exley C (2017) Aluminium in brain tissue in familial Alzheimer's disease. J Trace Elem Med Biol 40:30–36

Moser VC, Phillips PM, Levine AB, McDaniel KL, Sills RC, Jortner BS, Butt MT (2004) Neurotoxicity produced by dibromoacetic acid in drinking water of rats. Toxicol Sci 79(1):112–122

Muellner MG, Wagner ED, McCalla K, Richardson SD, Woo YT, Plewa MJ (2007) Haloacetonitriles vs. regulated haloacetic acids: are nitrogen-containing DBPs more toxic? Environ Sci Technol 41(2):645–651

Najnin N, Forbes A, Sinclair M, Leder K (2014) Risk factors for community-based reports of gastrointestinal, respiratory, and dermal symptoms: findings from a cohort study in Australia. J Epidemiol 24(1):39–46

Nghi NH, Cuong LC, Dieu TV, Ngu T, Oanh DTY (2018) Ozonation process and water disinfection. Vietnam J Chem 56(6):717–720

Nieuwenhuijsen MJ, Toledano MB, Eaton NE, Fawell J, Elliott P (2000) Chlorination disinfection byproducts in water and their association with adverse reproductive outcomes: a review. Occup Environ Med 57(2):73–85

Nozaic DJ (2004) Chlorine: is it really so bad and what are the alternatives? Water SA 30(5):18–24

Papageorgiou A, Voutsa D, Papadakis N (2014) Occurrence and fate of ozonation by-products at a full-scale drinking water treatment plant. Sci Total Environ 481:392–400

Parrat J, Donzé G, Iseli C, Perret D, Tomicic C, Schenk O (2012) Assessment of occupational and public exposure to trichloramine in Swiss indoor swimming pools: a proposal for an occupational exposure limit. Ann Occup Hyg 56(3):264–277

Perl DP, Brody AR (1980) Alzheimer's disease: X-ray spectrometric evidence of aluminum accumulation in neurofibrillary tangle-bearing neurons. Science 208(4441):297–299

Rasheed S, Hashmi I, Kim JK, Zhou Q, Campos LC (2018) Species-specific interaction of trihalomethane (THM) precursors in a scaled-up distribution network using response surface methodology (RSM). Environ Technol 39(3):346–355

Richardson SD (2011) Disinfection by-products: formation and occurrence in drinking water. Encyclopaedia Environ Health 2:110–136

Richardson SD, Postigo C (2012) Drinking water DBPs. Emerging organic contaminants and human health, pp 93–137

Richardson SD, Plewa MJ, Wagner ED, Schoeny R, DeMarini DM (2007) Occurrence, genotoxicity, and carcinogenicity of regulated and emerging DBPs in drinking water: a review and roadmap for research. Mutat Res Rev Mutat Res 636(1–3):178–242

Righi E, Bechtold P, Tortorici D, Lauriola P, Calzolari E, Astolfi G, Nieuwenhuijsen MJ, Fantuzzi G, Aggazzotti G (2012) Trihalomethanes, chlorite, chlorate in drinking water and risk of congenital anomalies: a population-based case-control study in northern Italy. Environ Res 116:66–73

Risher J (2005) Toxicological profile for bromoform and dibromochloromethane. Agency for toxic substances and disease registry
Rivera-Núñez Z, Wright JM, Meyer A (2018) Exposure to disinfectant by-products and the risk of stillbirth in Massachusetts. Occup Environ Med 75(10):742–751
Rondeau V, Jacqmin-Gadda H, Commenges D, Helmer C, Dartigues JF (2009) Aluminum and silica in drinking water and the risk of Alzheimer's disease or cognitive decline: findings from 15-year follow-up of the PAQUID cohort. Am J Epidemiol 169(4):489–496
Rosero-Moreano M (2018, November) New trends in chemical analysis of DBPs. In: Disinfection. IntechOpen, London, UK, pp 57–77
Roux JL, Gallard H, Croué JP (2010) Formation of NDMA by chloramination of nitrogenous pharmaceuticals. Water Pract Technol 5(4):wpt2010084
Sadiq R, Rodriguez MJ (2011) Empirical models to predict DBPs in drinking water
Sayess R, Khalil A, Shah M, Reckhow DA, Godri Pollitt KJ (2017) Comparative cytotoxicity of six iodinated disinfection byproducts on nontransformed epithelial human colon cells. Environ Sci Technol Lett 4(4):143–148
Serrano M, Montesinos I, Cardador MJ, Silva M, Gallego M (2015) Seasonal evaluation of the presence of 46 DBPs throughout a drinking water treatment plant. Sci Total Environ 517:246–258
Servais P, Billen G, Bouillot P (1994) Biological colonization of granular activated carbon filters in drinking-water treatment. J Environ Eng 120(4):888–899
Shah AD, Mitch WA (2012) Halonitroalkanes, halonitriles, haloamides, and N-nitrosamines: a critical review of nitrogenous disinfection byproduct formation pathways. Environ Sci Technol 46(1):119–131
Smith RB, Edwards SC, Best N, Wright J, Nieuwenhuijsen MJ, Toledano MB (2016) Birth weight, ethnicity, and exposure to trihalomethanes and haloacetic acids in drinking water during pregnancy in the born in Bradford cohort. Environ Health Perspect 124(5):681–689
Srivastav AL, Patel N, Chaudhary VK (2020) DBPs in drinking water: occurrence, toxicity and abatement. Environ Pollut 267:115474
Sundaram A (1990) Journal of environmental science and health. Part B, pesticides, food contaminants, and agricultural wastes. Journal of Environmental Science and Health. Part B, Pesticides, Food Contaminants, and Agricultural Wastes 25(3):309
Tak S, Vellanki BP, Ahuja S (2020) A review on disinfection and disinfection byproducts. In: Contaminants in our water: identification and remediation methods, pp 105–117
Turtoi M (2013) Ultraviolet light treatment of fresh fruits and vegetables surface: a review. J Agroaliment Process Technol 19(3):325–337
USEPA (1998) National primary drinking water regulations: disinfectants and disinfectionbyproducts. Fed Regist 63(241):69406–69407
USEPA (2003) National Primary Drinking Water Regulations; announcement of completion of EPA's review of existing drinking water standards. Fed Reg 68(138):42907
Veschetti E, Cutilli D, Bonadonna L, Libera SD, Ottaviani M, Martini C, Alaimo C, Pizzari P Anastasi P (1998) Preliminary results on the possibility of using peracetic acid as disinfectant of wastewater. AWT98-advanced wastewater treatment recycling and reuse, Milano, pp 14–16
Villanueva CM, Cantor KP, Grimalt JO, Malats N, Silverman D, Tardon A, Garcia-Closas R, Serra C, Carrato A, Castano-Vinyals G, Marcos R (2007) Bladder cancer and exposure to water DBPs through ingestion, bathing, showering, and swimming in pools. Am J Epidemiol 165(2):148–156
Villanueva CM, Gracia-Lavedan E, Julvez J, Santa-Marina L, Lertxundi N, Ibarluzea J, Llop S, Ballester F, Fernández-Somoano A, Tardón A, Vrijheid M (2018) Drinking water DBPs during pregnancy and child neuropsychological development in the INMA Spanish cohort study. Environ Int 110:113–122
Washington State D (2004) Alternate disinfectants. Washington State Department of Health. https://www.doh.wa.gov/portals/1/Documents/Pubs/331-252.pdf

Wei X, Wang S, Zheng W, Wang X, Liu X, Jiang S, Pi J, Zheng Y, He G, Qu W (2013) Drinking water disinfection byproduct iodoacetic acid induces tumorigenic transformation of NIH3T3 cells. Environ Sci Technol 47(11):5913–5920

Wei C, Zhang F, Hu Y, Feng C, Wu H (2017) Ozonation in water treatment: the generation, basic properties of ozone and its practical application. Rev Chem Eng 33(1):49–89

Westerhoff P, Chao P, Mash H (2004) Reactivity of natural organic matter with aqueous chlorine and bromine. Water Res 38(6):1502–1513

World Health Organization (2022) Guidelines for drinking-water quality: incorporating the first and second addenda. World Health Organization

WHO (2011) Guidelines for drinking water quality.4th edn. World Health Organization, Geneva, Switzerland. http://apps.who.int/iris/bitstream/10665/44584/1/9789241548151_eng.pdf

World Health Organization (WHO) (2017a) Guidelines for drinking-water quality, 4th edition. ISBN 978–92–4-54815-1. pp 541

World Health Organization (WHO) (2017b) Guidelines for drinking-water quality, 4th edn. Incorporating the first addendum. WHO, Geneva, Switzerland

Wright JM, Evans A, Kaufman JA, Rivera-Núñez Z, Narotsky MG (2017) Disinfection by-product exposures and the risk of specific cardiac birth defects. Environ Health Perspect 125(2):269–277

Xue B, Dai K, Zhang X, Wang S, Li C, Zhao C, Yang X, Xi Z, Qiu Z, Shen Z, Wang J (2020) Low-concentration of dichloroacetonitrile (DCAN) in drinking water perturbs the health-associated gut microbiome and metabolic profile in rats. Chemosphere 258:127067

Yang P, Zhou B, Cao WC, Wang YX, Huang Z, Li J, Lu WQ, Zeng Q (2017) Prenatal exposure to drinking water DBPs and DNA methylation in cord blood. Sci Total Environ 586:313–318

Zaheer F, Slevin JT (2011) Trichloroethylene and Parkinson disease. Neurol Clin 29(3):657–665

Zahran S, McElmurry SP, Sadler RC (2017) Four phases of the Flint water crisis: evidence from blood lead levels in children. Environ Res 157:160–172

Zeng Q, Li M, Xie SH, Gu LJ, Yue J, Cao WC, Zheng D, Liu AL, Li YF, Lu WQ (2013) Baseline blood trihalomethanes, semen parameters and serum total testosterone: a cross-sectional study in China. Environ Int 54:134–140

Route of Exposure, Mechanism of Action, and Toxicological and Human Health Impacts of Disinfection By-Products

Astha Pathak and Avinash Pratap Gupta

Abstract Disinfection of water is considered as the greatest public health achievement of the twentieth century since it greatly decreased waterborne sickness and raised life expectancy. However, an unforeseen result was the production of DBPs, which are created when disinfectants (such as chlorine, chloramine, ozone, chlorine dioxide, and UV) react with organic matter that is both naturally occurring and man-made, bromide and iodide. Treatment of water for purification process also invites a lot of harmful diseases due to the by-products that are formed during the process. Trihalomethanes are considered to be dominant by-product which is associated with non-neoplastic toxic effects. High dose of the four most common trihalomethanes, i.e., chloroform, bromodichloromethane, dibromochloromethane, and bromoform, has also shown lethal to rodents in chronic investigations of high doses. These by-products have also shown kidney toxicities affecting the vital functions. As a salt, dichloroacetic acid present in potable water is a strong by-product of chlorination which has shown to harness neural, hepatic, reproductive, and developmental growth of human beings. Other by-products like tricholoroacetic acid and brominated haloacetic acid have also shown various toxicological effects. The spread of illnesses like infertility, asthma, stillbirth, and other types of cancer has also been attributed to the discharge of toxins. Reverse osmosis, ultrafiltration, and nanofiltration are three innovative strategies that have been discovered to be effective in offsetting the production of harmful by-products. The toxicological consequences of DBPs and various strategies to ameliorate them are the major topics of this chapter. The health risks brought on by various DBPs are examined, as well as the numerous removal methods.

A. Pathak
Department of Pharmacy, Sant Gahira Guru University, Sarguja, Chhattisgarh, India

A. P. Gupta (✉)
Department of Environmental Science, Sant Gahira Guru University, Sarguja, Chhattisgarh, India

S. Madhav et al. (eds.), *Drinking Water Disinfection By-products*,
https://doi.org/10.1007/978-3-031-49047-7_7

Keywords Disinfection · Chlorination · Toxicological consequence · Carcinogenic · Dichloroacetic acid · Trihalomethane · Tricholoroacetic acid · Brominated haloacetic acid

1 Introduction

Disinfecting water involves removing, inactivating, or eliminating bacteria that could potentially cause disease. Microorganisms are either eliminated or rendered inactive, which brings an end to their capacity for growth and reproduction. When germs are not eliminated from drinking water, the consumption of drinking water can make people sick. Both chemical and physical disinfectants can be utilized to accomplish the task of sterilization. In addition, the agents eliminate organic pollutants from the water, which might provide microbes with either nourishment or a place to hide. It is not enough for disinfectants to simply eliminate microbes. It is necessary for disinfectants to have a residual impact, which indicates that they keep working even after the water has been treated with them. After the water has been disinfected, the plumbing should be protected from the growth of pathogenic bacteria by a disinfectant, which would otherwise cause the water to become re-contaminated.

The end products of the disinfection can be chemical, organic, or inorganic that are generated when a disinfectant combines with naturally existing organic materials in the water. These substances can be either chemical or organic. During the disinfection process, DBPs are mostly generated when disinfectants react with dissolved organic matter (DOM) and a small number of inorganic compounds already present in the water (Mazhar et al. 2020). In the underdeveloped world, diarrheal infections caused by a lack of water sanitation continue to be a primary cause of death, especially among children. Shortly after the year 1900, chemical disinfection of municipal water supplies began. By the 1940s, outbreaks of cholera, typhoid fever, and other illnesses associated with water were effectively eradicated throughout the developed world. The benefits of chlorine disinfection were significant because they occurred before the advent of vaccinations and antibiotics.

The widespread use of chlorine to disinfect water for drinking is a significant victory for public health. In the process of treating water to make it suitable for human consumption, the disinfection phase is without a doubt the most essential one. Concerns regarding the potential adverse long-term effects of disinfectants and DBPs on human health should not lead to a reduction in the standards for the microbiological purity of drinking water. The danger of becoming sick or dying as a consequence of being exposed to pathogens in drinking water is much, much higher than the risk of getting sick or dying as a result of being exposed to disinfectants or DBPs. In situations where it is necessary to make a choice between microbiological limits and limitations for disinfectants and DBPs because of local circumstances, the quality of the microbiological standards must always take precedence. Never, under any circumstances, should the level of disinfection be compromised.

2 Synthesis of Disinfectant Derivatives

It is possible for disinfection by-products to be produced as a result of a response among disinfectants, like chlorine and natural chemical compounds that are found in water. The synthesis of said products primarily occurs through reactions involving organic compounds, namely humic and fulvic acids. The decomposition of organic plant matter results in the discharge of these chemicals into aquatic ecosystems. In 1971, an American scientist named Bellar made the revelation that chloroform had not been detected in the water from the Ohio River that was used for the manufacturing of drinking water. Nonetheless, it was observed that chloroform was detected in potable water sourced from water treatment facilities. This provides evidence for the creation of disinfection by-products as a result of the chlorination process. The chemical structures of humic acids and fulvic acids are not widely documented. The process by which disinfection by-products are formed remains unclear. Conducting research in the field of organic matter can be challenging due to the vast array of substances involved.

The following factors all contribute to the process of defining the many kinds of by-products from disinfection that are produced:

- The disinfectant used
- The amount used
- The resulting residue

There will be a greater number of disinfection by-products produced if both the dose and the residue of the disinfectant are increased. Today, different disinfectants are used in place of halogen-based ones so that the by-products of halogen-based disinfection do not form. However, there is still the possibility of the formation of disinfection by-products.

2.1 Circumstances that Cause Disinfection

1. *Reaction time*

   Shortening the period of a reaction can lead to more trihalomethanes (THM) & halogenic acetic acids (HAA). Tribromine acetic acid and bromoform, both of which are transient disinfection by-products, may be formed if the reaction time for disinfection is sufficiently lengthy. The decomposition of haloacetonitriles (HAN) and haloketones (HK) is observed.
2. *Temperature*

   When temperatures rise, the rate at which reactions take place increases, this means that a higher quantity of chlorine is necessary for effective disinfection. This results in an increased production of halogenic disinfection by-products. The breakdown of tribromine acetic acids, HAN and HK, is facilitated by an increase in temperature, which also has the effect of speeding up the process.

3. *pH*

   An increase in pH levels results in the generation of a greater number of hypochlorite ions, leading to a reduction in the efficacy of chlorine disinfection. The formation of THM increases with higher pH values, while the formation of HAA is enhanced at lower pH values. The hydrolysis of HAN and HK occurs at elevated pH levels due to an augmentation in hydrolysis reactions.

3 Mechanism of Action of Disinfection By-products

In vitro tests have shown that several disinfection by-products are genotoxic, and animal studies have shown that they cause cancer (Richardson et al. 2009). The "International Agency for Research on Cancer" (IARC) of the World Health Organization found that chloroform and other common disinfection by-products were carcinogenic in animals (Lyon: IARC 1999). There is experimental data demonstrating fetotoxic effects, delayed foetal growth, spermato-toxicity, delayed sexual maturity, alterations in reproductive organs and the placenta, and skeletal implications from studies conducted on animals (Tardiff et al. 2006). While neurotoxicity has been studied less thoroughly, other findings linked to DBP exposures in animal experiments have been observed. When DBP was given to the animals, same effects occurred (Guariglia et al. 2011; Moser et al. 1999; Moser et al. 2004). Certain biochemical theories for DBP harmful effects have already been put forth based on research using transgenic microorganism's human and animal cultured cell lines and rodents (Nieuwenhuijsen et al. 2009). Both chloroform and trichloroacetic acid have been linked to cytotoxicity, which is a disturbance in the integrity of the cell that can occur with or without DNA being damaged (TCAA) (Testai et al. 1996). Phosgene and dichloromethyl are considered as intermediary compounds of chloroform subsequent to in vivo metabolic processes facilitated by cytochrome P450 2E1 (CYP2E1) (Gemma et al. 2003). Phosgene exhibits high reactivity and has the ability to bind with proteins, phospholipids, and reduced glutathione cellular components (Cantor et al. 2010). A higher likelihood of bladder cancer has been associated with interaction with dibenzopyrones (DBPs) when particular polymorphism of the enzyme CYP2E1 is present in human populations (Larson and Bull 2002). The mode of action of TCAA is distinct, as it can either be eliminated without alteration owing to its polar association with plasma proteins or undergo a reductive dechlorination process that generates free radicals, resulting in the formation of dichloroacetic acid, a more potent lipoperoxidative compound (Ni et al. 1996). Furthermore, it is plausible that alternative compounds could potentially serve as precursors for the production of TCAA, such as chloral hydrate or trichloroethanol (Pegram et al. 1997). In in vitro models, it has been shown that brominated trihalomethanes are metabolized into intermediate carbonyl-reactive chemicals that target DNA via glutathione S-transferases such as glutathione S-transferase theta-1 (GSTT1) (Ross and Pegram 2004; Ross and Pegram 2003; Landi et al. 1999). Likewise, the expression of GSTT1 is a prerequisite for the detection of genotoxicity in cultures of human cells (Landi et al. 2003; Coffin et al. 2000). In the realm of epidemiology, research carried out on human populations has shown that variations in GSTT1 can have an effect on the risk that is associated with DBP exposure. Individuals who are GSTT1-positive have been found to exhibit greater susceptibility (Larson and Bull 2002).

Modifications in epigenetic patterns are just one example of the many molecular pathways that have been uncovered and reported. In rodent models, the phenomenon of lower methylation amounts in DNA on a worldwide basis and in numerous proto-oncogenes as a result of being exposed to THM and HAA is being documented (Tao et al. 2004; Ge et al. 2001; Pereira et al. 2001; Salas et al. 2014). Evidence pertaining to humans that is currently accessible addressing changes in DNA methylation in retrotransposons (RT) or other areas of DNA is limited (Geter et al. 2004). Finally, it has been observed that folate metabolism may have a significant impact on the development of colorectal cancer, as folate deficiency has been linked to heightened preneoplastic lesion formation in rodents exposed to DBPs (Geter et al. 2005).

4 Water DBPs: What Affects Their Makeup and Levels

The formation of DBP is determined not only by the overall quality of the water but also by the conditions that are present throughout the water's treatment and distribution, as will be demonstrated in the following section.

4.1 The Quality of the Source Water

A significant contributor to the development of DBP is the kind and quality of the raw water that is brought into the treatment plant. The amount of DBP's precursors that are present in the untreated drinking water supply, specifically the NOM and bromide ion, is an important component that plays a role in the synthesis of DBP. However, the NOM will have a propensity to have an impact on the whole amount of DBPs. The proportion of bromide to the overall organic carbon would have a major impact on the differentiation of certain DBPs, while the NOM will have a significant impact on the speciation of other DBPs (Cantor et al. 2010).

The composition of the precursors might be affected by the characteristics of the water supply. Groundwater supplies the remaining 30% of the United Kingdom's drinkable water supply, while surface water provides around 70% of the country's potable water supply. Surface water sources in the United Kingdom include upland lakes, rivers, and reservoirs (Cantor et al. 2010). The amount of natural organic matter present in sources of surface water shows variability, whereas in groundwater, it typically manifests as low. Groundwater is safeguarded from the discharge of non-organic matter due to certain reasons (Ross and Pegram 2004). The discharge of water from agricultural activities, specifically from areas with a significant amount of peat soil, constitutes a significant origin of precursors for total organic carbon (TOC) and disinfection by-products (DBP) (Ross and Pegram 2003). Wetland areas have the ability to boost total organic carbon, or TOC, due to the abundance of vegetative biomass as well as the leaching of peat soils that lie beneath the surface of the wetland (Ross and Pegram 2003). Typically, surface water sources are associated with higher concentrations of DBP compared to groundwater sources.

In the presence of high amounts of bromide in the source water, the creation of disinfection by-products (DBPs) tends to favor the generation of brominated forms of trihalomethanes (THMs), haloacetic acids (HAAs), and haloacetonitriles (HANs) (Testai et al. 1996). Elevated concentrations of bromide in groundwater can be attributed to the intrusion of saltwater, while in low-lying surface waters, the presence of industrial and agricultural chemicals may contribute to such levels. Conversely, bromide levels in high-altitude surface waters are typically low. Studies have demonstrated that groundwater yields a greater proportion of brominated trihalomethanes (THMs) in comparison with mixed and surface waters. DCBM and DBCM, two forms of THMs which include both bromine and chlorine, are likely to be present at higher quantities in sources of water that are mixtures of several types of water in comparison with surface or underground water sources.

5 The Treatment Process

The method of treatment, the pH level, the temperature, the kind and amount of disinfectant used, the amount of time required for the reaction, and the stage of the treatment process at which the disinfecting agent is added are all elements that have the potential to affect the formation of DBP. Organic material from the raw water can be eliminated using treatment techniques such as flocculation, coagulation, sedimentation, and filtering before the water is disinfected. It has been demonstrated that disinfection is an efficient approach for reducing the possibility of DBP development; hence, these processes ought to be utilized (Ross and Pegram 2004). It has been proven that the utilization of pre-ozonation, before to the usage of chlorine treatment, leads to reduction in the rise of THMs, HAAs, dihaloacetonitriles, and TOX, but a rise in the production of haloketones and chloropicrin (CP). This is the case because pre-ozonation causes an increase in the synthesis of haloketones. It has been established that pre-ozonation produces an alteration in the distribution of species towards a greater proportion of brominated species. This movement can be attributed to the presence of more brominated species (Landi et al. 1999).

The creation of DBP is made more likely by increasing the amount of chlorine present (Ross and Pegram 2004). It has been shown that pH has a considerable impact on the process of DBP production. Increasing the pH results in a higher concentration of THM (Landi et al. 2003). A high pH is related to lower quantities of HAAs, HANs, haloketones, and chloral hydrate. This is because a high pH favors the destruction of these chemicals, which is why a high pH is favorable. It's possible that this is a consequence of a conflicting process between formation and decay (Landi et al. 2003). Observations have shown that an increase in temperature favorably affects the rates at which DBPs are formed. It has been demonstrated that as the response time increases, the concentrations of both THM and HAA rise as well (Landi et al. 2003).

At low pH, the concentrations of chloral hydrate and DCAN rise with increasing reaction time; however, this trend is reversed when the pH level is raised (Landi et al. 2003).

There are many different options available today that can be utilized as substitutes for disinfectants in the process of treating drinking water. It has been established that chloramination resulted in a decreased formation of THMs compared to chlorination (Coffin et al. 2000). The THM regulatory agency has found that treatment facilities that make use of chloramination are better equipped to achieve the standards that they have set (Tao et al. 2004). In order to maintain compliance with the requirements set for regulated DBPs, the water sector is currently observing a trend toward converting to chloramination as a means of disinfection. On the other hand, chloramination may hasten the production of certain of the newly discovered by-products like DBPs that are iodinated and nitrogen containing DBPs like nitrosodimethylamine (NDMA) considering a claim that chloramination is not the primary contributor to the production of these by-products. (Gemma et al. 2003; Ge et al. 2001).

6 During Distribution

A noteworthy phenomenon is the variation in the concentration of disinfection by-products (DBPs) that can be found throughout a water distribution network. The generation of disinfection by-products (DBPs) in the system of distribution is affected by a variety of elements, such as the amount of unbound residual chlorine, the amount of residence time, the pH of the water, and the temperature. It is a standard procedure to keep a residual level of chlorine across the distribution system. This is done with the goal of preventing the regeneration of microbes that can cause disease. The existence of residue can make the creation of disinfection by-products (DBPs) during the flow of the water throughout the distribution chain easier, assuming that there is additionally sufficient DBP precursors present in the water itself for the intent of reaction. This is the case only if there are adequate DBP precursors available in the water. Generally speaking, the length of residence is correlated with an increase in THM concentrations, which show an upward trend (Pereira et al. 2001; Salas et al. 2014; Geter et al. 2004).

The prolonged duration of water residence is positively correlated with an increase in residual chlorine consumption, which subsequently leads to a greater production of THMs. This phenomenon is the primary explanation for the elevated concentrations of THMs that are detected at the endpoints of water distribution networks (Geter et al. 2004). Nonetheless, it has been demonstrated that higher residence time is associated with a reduction in the levels of HANs, HKs, CP, and HAAs (Pereira et al. 2001). It has been demonstrated that HAAs undergo degradation as they near the extremities of the system (Geter et al. 2004).

In the context of treatment facilities, it is generally observed that the levels of disinfection by-products (DBPs) tend to rise with a corresponding increase in temperature. The promotion of additional THM formation is typically favored by an increase in pH, although the converse outcome may be apparent for alternative DBPs, such as HAAs (Landi et al. 2003).

6.1 Distribution System Spatial Variation

The continuous interaction among chlorine residue and DBP precursors in the water supply system is responsible for the spatial dispersion of disinfection by-products (DBPs) in that system; this interaction is also dependent on the amount of time that water spends residing across the system (Salas et al. 2014).

During the process of transporting water along the distribution system, there is a normal tendency for there to be an increase in the amount of THMs present. The highest concentrations can typically be found in the regions that are situated at the most remote portions of the system. On the other hand, the concentrations of HAAs have a tendency to fall as they make their way within the distribution system. As a consequence of this, the amounts of HAAs are at the highest level when the water is being discharged from the treatment plant, and they are at their lowest at the system's extremes (Geter et al. 2004).

6.2 Temporal and Seasonal Variation in DBP Amounts

A specific phenomenon known as periodic fluctuation of DBP values has been detected on numerous occasions. The levels of THMs exhibit a seasonal tendency, with peak concentrations happening during the summer and fall months and a decline occurring during the winter season. This trend can be seen in both the facilities for treatment and the distribution system (Pereira et al. 2001). The probable cause for this phenomenon can be attributed to fluctuations in temperature and variations in the quality of raw water over different seasons. Trihalomethanes (THM) along with additional disinfection by-products (DBP) are more likely to be produced when temperatures are higher throughout the summer season. This phenomenon results in a more rapid decline of residual chlorine levels. To ensure that an appropriate residual chlorine concentration is maintained within the distribution system, it is necessary to administer higher doses of chlorine during the summer months (Ross and Pegram 2004).

The augmentation of biological activity during the summer season and the shedding of leaves in autumn are expected to result in elevated levels of natural organic matter (NOM) in untreated water during these seasons. This, in turn, is likely to promote the formation of trihalomethanes (THMs) (Cantor et al. 2010).

The study has demonstrated that chloroform exhibits significant temporal variation over a shorter duration, such as from month-to-month, whereas brominated THMs tend to be more stable (Cantor et al. 2010; Salas et al. 2014).

7 Natural Events

The term "natural organic matter" (NOM) refers to the pollution that is caused by non-point sources, and run-off is a major contributor to this type of pollution. An increase in the amount of precipitation leads to an increase in the flow of surface water, which in turn leads to an increase in the quantity of organic matter from nature (NOM) in reservoirs water that is untreated as well as an increase in the potential for the creation of disinfection-related by-products (DBPs). It has been discovered that the level of THM is higher during instances of precipitation as compared to conditions where the weather is dry. In addition to this, research has shown that the influence of rainy weather on the possibility for the production of THM remains even after an event of precipitation has stopped (Geter et al. 2005). The occurrence of floods has the potential to result in an increase in the total amount of organic matter that is natural (NOM) in bodies of water, which may then lead to higher degrees of disinfection by-product production potential (Geter et al. 2005). On the other hand, during times of drought, there have been a likelihood of decreased run-off, which can lead to a fall in the amounts of organic matter that occurs naturally (NOM) in untreated water sources. This can be attributed to the fact that less water is being carried away by rivers and streams. This, in turn, can lead to a reduction in the possible creation of disinfection by-products (DBPs). In the event of such a scenario, the proportionate involvement of brominated entities may escalate as a result of heightened concentrations of bromide in the originating water body, for instance, owing to the intrusion of saltwater (Ross and Pegram 2003).

8 Routes of Exposure to DBPs

Intake, breathing, and dermal absorption are the basic ways that DBPs can get into a human body. However, other routes, such as injection, are also possible. The unique chemical features of every single disinfectant by-product (DBP) are what dictate the pathways that can be taken to be exposed to that particular DBP. When assessing the route of exposure, one of the most important factors to consider is how the 27 different DBPs are separated as volatile and non-volatile molecules.

8.1 Ingestion

When drinking tap water or consuming foods and beverages which have been prepared using tap water, it is possible to swallow disinfection by-products (DBPs), which are naturally occurring contaminants that are found in tap water. It has been reported that the amount of water that pregnant women drink from the tap on a daily basis ranges anywhere from 0.6 to 3.4 liters (Egorov et al. 2003; Kaur et al. 2004; Shimokura et al. 1998; Ershow et al. 1991, Barbone et al. 2002; Forssen et al. 2007; Zender et al. 2001). It has been estimated that pregnant women should drink between 2 and 2.7 liters of water per day, regardless of the source of the water (Kaur et al. 2004; Shimokura et al. 1998; Ershow et al. 1991). This range accounts for the overall water intake from all sources. The state of being pregnant is linked with notable elevations in the consumption of water (Ershow et al. 1991; Zender et al. 2001). The authors Forssen et al. (Forssen et al. 2007) have demonstrated that there are variations in tap water consumption among pregnant women based on their ethnicity. In addition, a recent study that was carried out at the request of the Inspectorate of Water for Drinking (DWI) of the United Kingdom found that there is a positive correlation between age and the use of water from the tap (DWI. National Tap Water Consumption Study 2008).

Before being consumed, the amounts of disinfectant by-products (DBPs) that are present in water, drinks, or food products may undergo changes as a result of operations such as filtering, boiling, or storage. These processes are examples of pre-consumption steps. These modifications have the potential to impact the quantity of DBPs ingested by an individual. Water filtration systems, such as tap-mounted filters or filter jugs, have been found to effectively remove trihalomethanes (THMs) by 31% to 100%, haloacetic acids (HAAs) by 40% to 95%, and haloacetonitriles (MX) by 60% to 99%. The amount of contamination that can be removed is proportional to the filter's capacity as well as its age (Egorov et al. 2003; Weinberg et al. 2006; Levesque et al. 2006; Eslinger and Weinberg 2003; Gibbons and Laha 1999; Ma 2008). Studies have demonstrated that filters exhibit greater efficacy in eliminating disinfection by-products (DBPs) that contain bromine, owing to their relatively lower solubility and polarity as compared to DBPs that do not contain bromine (Weinberg et al. 2006).

The concentration of THMs can be reduced by boiling tap water, with the extent of reduction ranging from 34% to 98% depending on the duration of boiling (Weinberg et al. 2006; Ma 2008; Krasner and Wright 2005; Wu et al. 2001; Batterman et al. 2000). Previous studies have reported that boiling for 2 min can result in an average reduction of 92% in TTHM concentration (Ma 2008; Krasner and Wright 2005). The decrease in quantity can be attributed to the process of THM volatilization. Consequently, it is recommended that beverages and food items made using boiled water should contain only minimal amounts of THMs. The formation of THMs can occur through the chemical reactions between free chlorine and organic compounds found in consumable items such as tea and other food products (Huang and Batterman 2009). The process of boiling results in a reduction of TCAA concentration and an increase in DCAA concentration (Levesque et al. 2006; Krasner and Wright 2005). Boiling has been demonstrated to decrease the concentrations of total haloacetic acids (HANs), haloalkanes (HKs), and haloaldehydes (Krasner and Wright 2005). People who live in areas where the quality of the microbiological water is below standard may have a restricted intake of water that has not been boiled, which results in a minimum exposure to THMs through the process of ingestion (Egorov et al. 2003). Research has demonstrated that the presence of trihalomethanes (THMs) can be mitigated through the storage of water, particularly when the container is not covered. The results indicate that storage for a duration of 48 h leads to a mean reduction of 61% in THM levels. However, no significant impact on HAA concentrations was observed following storage, possibly due to their non-volatile nature in contrast to THMs (Levesque et al. 2006).

Undoubtedly, a considerable number of individuals opt for bottled water as a potential replacement for their tap water intake, either partially or entirely. The utilization of bottled water for drinking purposes was reported by 46% of households surveyed in the UK during the year 2008. However, it remains ambiguous whether this practice was adopted as a substitute for tap water or as a supplement to it ((DWI) National Tap Water Consumption Study 2008). More than 80% of women who were pregnant from the UK reported drinking water from bottled containers at some point during their pregnancies (Kaur et al. 2004). According to estimates, the average daily consumption of bottled water by pregnant women falls somewhere between 0.6 and 0.9 liters (Kaur et al. 2004; Forssen et al. 2007). A recent study that was carried out in the USA (Saleh et al. 2008) involved the sampling of thirty-five different brands and varieties of bottled water and in the results, it was seen that no THMs were detected. It is commonly believed that bottled water does not contain any DBPs.

Dibutyl phthalates (DBPs) have the potential to be orally consumed through dietary intake. Research has demonstrated that the utilization of drinking water containing THMs for washing or cooking meat and vegetables can lead to the absorption of THMs into these food items (Huang and Batterman 2010). It is possible that the level of disinfectant by-products (DBPs) that have been detected in the supply of water for drinking can have an effect on the level of DBPs which are absorbed by foodstuffs during the process of preparing meals in a household environment. As a consequence of the widespread use of chlorine in commercial food processing methods, a chemical compound known as trihalomethanes (THMs) may be found in the finished product. For example, the use of chlorinated wash in the processing of the meat can potentially lead to increased quantities of chloroform in chicken (Robinson et al. 1981).

8.2 Absorption Via the Skin and the Lungs

Common water-using activities including bathing, taking showers, and diving, as well as a variety of household duties, (DBPs), have the possibility to be inhaled and infiltrate the skin.

(i) *Showering and bathing*

There exist multiple factors that could potentially impact the inhalation and dermal absorption of disinfection by-products (DBPs) while engaging in showering and bathing activities. The concentration of disinfection by-products (DBPs) in tap water is significantly influenced by the temperature of the water (Mazhar et al. 2022). The process of heating water, such as in a household boiler, has been observed to expedite the creation of certain disinfection by-products (DBPs) in the presence of residual chlorine. The study conducted by Weisel and Chen revealed that the concentration of THM in water increased twofold when heated to 65 °C for 8 h. On the contrary, the levels of DCAN and TCP both declined when the temperature was increased, whereas the concentration of DCP first elevated and then decreased as the temperature climbed (Weisel and Chen 1994).

It is necessary for DBPs to exist either gas-like or an aerosolized condition for them to be able to be inhaled. The level of vaporization that occurs with DBPs and consequently affect inhalation exposure is contingent upon the variations in volatility among DBP species. Because chloroform is more volatile than haloketones, the proportion of chloroform that is released into the air from shower water is significantly larger than that of haloketones (Xu and Weisel 2005). Under standard shower water temperatures, it is anticipated that HAAs will remain non-volatile. As a consequence of this, inhalation of the aerosols produced by the shower, which are subsequently inhaled through the respiratory system, is the most common way that people are exposed to HAAs when they are bathing (Xu and Weisel 2003).

The transdermal absorption of disinfection by-products (DBPs) is subject to modulation by their capacity to permeate the human skin barrier. Within the THMs, bromoform was found to have the maximum penetration through the skin of humans, followed by chloroform, then DBCM, and finally DCBM, according to the findings of an investigation on dermal absorption that was carried out by Xu et al. It was determined that the penetration of haloketones was roughly ten times less compared to THMs, whereas the penetration of HAAs was determined to be two orders of magnitude less than that of THMs. This was a significant difference in terms of the magnitude of the difference (Xu et al. 2002). According to the calculations made by the authors, the dermal dosage that results from every day's bathing activities accounts for between 40% and 70% of the everyday ingesting dose of THMs, whereas the dermal dose that results from daily haloketone exposure accounts for only 10% of the each day ingestion dose. Nevertheless, this amount would be just a tiny portion of the everyday intake dose of HAAs that would be required to produce the same effect. Therefore, it is imperative to incorporate dermal exposure assessment for THMs and HKs, while excluding HAAs ((Xu et al. 2002). In accordance with Weisel et al.'s findings, cutaneous exposure to chloroform, such as that which occurs when showering, and inhalation both have the potential to have an equal impact on the overall body burden (Weisel and Jo 1996).

The rate of water flow and the ventilation conditions of a bathroom, such as the state of a bathroom window (open or closed) and the operation of an extractor fan, have been observed to have an impact on the level of exposure during showering. This is due to the demonstrated influence of these factors on the concentration of THMs in the air (Weisel and Chen 1994; Jo et al. 1990; Kerger et al. 2000).

The length of time spent showering or bathing is a significant factor in determining one's risk of being exposed to disinfection by-products. There is a positive correlation between extended periods of bathing and an increase in dermal chloroform exposure, as indicated by measurements obtained from breath samples (Weisel and Jo 1996). The concentrations of THMs in the air are comparatively higher during a standard shower lasting 12 min, as opposed to a bath that lasts 20 min. Nonetheless, shorter durations of both showers and baths have been linked to elevated average concentrations of THMs in the air, as well as greater fluctuations in THM concentrations between occurrences. The study indicates that relying solely on event duration for exposure assessment may not comprehensively capture the range of exposure variability (Kerger et al. 2000). The duration of daily showers for pregnant women has been estimated to range between 13.1 and 15.5 min (Egorov et al. 2003; Shimokura et al. 1998). Research has demonstrated variations in the duration of showering and bathing practices across different ethnicities (Williams et al. 2001), smoking statuses, and socio-economic statuses (SES), with a decline in duration observed with increasing age (Forssen et al. 2007).

The levels of disinfection by-products (DBPs) present in the air of a bathroom may persist at elevated concentrations for a period of time subsequent to the cessation of showering or bathing activities (Backer et al. 2008). The study determined that the half-life of shower aerosols varied between approximately 5 to 15 min subsequent to the cessation of showering, contingent upon the size of the aerosol (Xu et al. 2002). The duration of time spent in the bathroom following bathing and showering is a factor that can affect inhalation exposure.

The act of hand-washing has the potential to be a contributing factor to exposure to disinfection by-products (DBPs). There is a significant increase in ambient air TTHM concentration ranging from 4 to 11-fold subsequent to hand-washing activity. The THM biomarkers (breath or blood) exhibited a comparatively moderate increase as a result of hand-washing, in contrast to the significant increase observed due to showering or bathing (Nuckols et al. 2005). This observation may be attributed to the comparatively brief duration of hand-washing activities in contrast to the lengthier periods of showering and bathing. Nevertheless, in cases where hand-washing occurs frequently, such as in certain professions like food preparation or nursing, it could conceivably have a more noteworthy impact on exposure to disinfection by-products.

(ii) *Swimming*

The TTHM concentrations in London's swimming pools were discovered to vary between 57–222.5 μg/l (Chu and Nieuwenhuijsen 2002), whereas the chloroform concentrations in swimming pools located in the northern region of Italy ranged from 9–179 μg/l (Aggazzotti et al. 1998; Aggazzotti 1995). The primary contributor to the concentration of trihalomethanes (TTHMs) in both water and air at swimming pools is chloroform. There exists a positive correlation between the concentrations of chloroform present in the ambient air surrounding the swimming pool and the levels of chloroform detected in the pool water (Aggazzotti 1995). The reported range of chloroform levels in ambient air within indoor swimming pools is between 16–853 μg/m3. The concentration of THMs in swimming pool water and air is subject to the influence of the number of swimmers present. This is due to the turbulence generated by swimming, which can trigger the release of THMs from water into the air (Chu and Nieuwenhuijsen 2002; Aggazzotti et al. 1998; Aggazzotti 1995). Furthermore, there exists a noteworthy positive correlation between the quantity of individuals swimming in the pool and the levels of total organic carbon (TOC), indicating that the likelihood of forming disinfection by-products (DBPs) amplifies with an increase in the number of swimmers.

The uptake of DBP at the swimming pool can also be influenced by individual behavior. The findings of biomonitoring studies indicate a significant positive correlation between the duration of pool exposure and plasma chloroform levels. Furthermore, heightened physical activity in the pool is associated with elevated levels of chloroform in both plasma and alveolar air, indicating a corresponding increase in uptake with increasing physical activity. (Aggazzotti 1995). A research study was conducted to examine the phenomenon of water ingestion during swimming. The results indicated that, on average, adult swimmers ingested 16 ml (with a range of 0–53 ml) of water from the swimming pool during a 45-min session of active swimming (Dufour et al. 2006).

(iii) *Other household activities*

Research has demonstrated that indoor air chloroform levels are notably elevated by domestic water-related tasks, including manual dishwashing, automatic dishwashing, and laundry washing with bleach (Nuckols et al. 2005; Gordon et al. 2006).

The utilization of blood and breath biomarkers for quantifying uptake has demonstrated a correlation between these activities and heightened levels of THM in both blood and breath (Nuckols et al. 2005). Research has demonstrated that hand dishwashing following showering and bathing results in the most significant average elevation of blood and exhaled breath THM levels in comparison with other household water usage activities and tap water ingestion. However, it is important to note that there is a considerable degree of variation among participants (Nuckols et al. 2005; Ashley et al. 2005, 73).

(iv) *Comparative significance of exposure pathways*

The significance of various exposure pathways is contingent upon the category and inherent characteristics of the disinfection by-products under consideration. The impact of showering and bathing on the overall dose of THMs, with a focus on chloroform, has been demonstrated to be significant in terms of dermal exposure and inhalation, whereas ingestion has a comparatively minor effect (Nuckols et al. 2005; Gordon et al. 2006; Whitaker et al. 2003; Backer et al. 2000). In the case of haloketones, the role of inhalation exposure during showering is comparatively less significant than that of THMs (Xu and Weisel 2005). The uptake of heterocyclic aromatic amines (HAA) through inhalation and dermal routes is anticipated to have minimal contribution (Xu and Weisel 2003; Xu et al. 2002) This implies that ingestion is the primary route for HAA exposure.

9 Health and Toxicological Impact of By-Products

9.1 *Cancer-Inducing By-Products*

Numerous epidemiological studies have been conducted to evaluate the risk of cancer following the discovery of chloroform in drinkable water, as well as the cytotoxic nature of chloroform in rats. These studies have found that chloroform can cause cancer in rodents (Villanueva et al. 2015). Case–control studies were conducted which reflects positive relationships between rates of mortality and locations with higher amounts of disinfection by-products (DBPs) in drinking water. These findings were confirmed by case–control research that was conducted after the ecological research (Villanueva et al. 2015). On the other hand, in light of certain methodological shortcomings (such as the utilization of certificate data, limits in evaluation of exposure, and a lack of control measures for complicating variables), more thoroughly constructed studies have been carried out in recent years. The various types of cancer caused due to disinfection by-products are as under:

- Bladder cancer (Costet et al. 2011; Hrudey 2012; Bull 2012).
- Colon and rectal cancer (King 2001; Rahman et al. 2010).
- Cancers of other locations (including the pancreatic, kidney, brain, breast, esophagus, and lung) in addition to leukemia, melanoma, and skin cancers that are not melanoma (IARC 2012).

9.2 *Reproductive Impact of By-products*

9.2.1 Fertility

The effect of disinfectant by-products (DBPs) on the fertility of men has gotten much less focus than its influence on the result of pregnancy, despite the potential relevance that animal studies have suggested it may have (Villanueva et al. 2015). Epidemiological examinations possess, have for the most part, been carried out on smaller populations and have been based on a varied range of population groupings.

For example, partners in couples who are subfertile or, less frequently, allegedly fertile men have been the subjects of these investigations. After accounting for other potential factors, research has demonstrated that being exposed to DBP had a negative effect on the normal morphology of sperm (Fenster et al. 2003), as well as on concentrations of sperm (Luben et al. 2007), but does not have an effect on sperm motility (Xie et al. 2011). Chloroform, total trihalomethanes (THMs), and brominated THMs in public drinking water were not associated with low sperm quality, as determined by the results of a large case–control study conducted in the United Kingdom (Iszatt et al. 2013). In a cross-sectional study done in China, researchers looked at the levels of TCAA in the urine of almost 2000 males who had gone to a fertility clinic. There was an inverse relationship between TCAA levels in the urine and sperm quality parameters; however, the study did not find a dose-response relationship. This information was reported in literature (Zeng et al. 2014). Over the course of six months, urine steroid metabolites were used to collect information regarding menstrual parameters in a study group of 403 women. Trihalomethane (THM) levels and water use patterns at participants' homes were also recorded. A shorter cycle and shorter follicular phase were linked to higher levels of total THM, with the link being greater for brominated chemicals (Villanueva et al. 2015). It was seen that these parameters dropped off in a monotone fashion. Women in the USA who were exposed to high amounts of THMs through their diet had a longer period before becoming pregnant, as documented by MacLehose et al. Pregnancy was consequently delayed as a result (MacLehose et al. 2008).

9.2.2 Fetal Loss

An increased risk of spontaneous abortion was seen in women who drank a minimum of 5 glasses of cold water from the tap per day containing at least 75 g/liter of total THMs, according to a study being conducted in California. The risk is even greater when considering the presence of bromodichloromethane (Villanueva et al. 2015). A subsequent study was carried out in North Carolina, which involved a more precise exposure characterization. However, the findings did not validate the aforementioned results (Savitz et al. 2006). For the purpose of determining the likelihood of a stillbirth, a comprehensive review of five separate research was carried out. The findings indicated that cases with high THM exposure had an overall increase of 9% (2–17%) in the risk of stillbirth (Nieuwenhuijsen et al. 2010).

9.2.3 Fetal Growth

Numerous scholarly investigations on fetal growth were disseminated prior to the year 2000, and the findings have been consolidated in various literature reviews (Villanueva et al. 2015). Although several research studies have suggested a correlation with low birth weight, the available evidence has been deemed inconclusive. Subsequent to that time, several investigations employing meticulous evaluation of individual exposure have been carried out in regions characterized by notable THM exposure (Villanueva et al. 2015). In general, there was no noticeable escalation in the likelihood of neonates being small-for-gestational-age (SGA). Notwithstanding, certain regions with elevated levels of THMs (≥80 μg/L) in their tap water (Yulin et al. 2020; Levallois et al. 2012), instances with substantial absorption during showering (Costet et al. 2012), and vulnerable groups, such as neonates harboring a CYP2E1 gene genetic variation (Infante-Rivard 2004), were observed as outliers. According to a meta-analysis conducted in 2009, which included eight studies, there was a marginal elevation in the risk of small-for-gestational-age (SGA) with exposure to trihalomethanes (THMs) (Yulin et al. 2020). Recent research investigations that used urine TCAA as an indicator of exposure revealed that the biggest exposure class had a detrimental effect on fetal growth. These studies were conducted with the goal of determining the relationship between fetal growth and exposure levels (Infante-Rivard 2004; Pandian et al. 2022).

9.2.4 Congenital Malformations

In order to evaluate the possible connection between the DBP exposure and a variety of birth abnormalities, meta-analyses of fifteen separate research were conducted. According to the findings from the research there was an increased risk of ventricular septal defect as well as an increase of 17% in the total number of congenital defects. The authors did not deem the outcomes pertaining to the group "all congenital anomalies" as significant due to the diverse phenotype and likely causation when examining all anomalies as a collective.

The heightened probability of ventricular septal defects was derived from a limited pool of three studies, with only one of them demonstrating indications of a consistent exposure-response correlation. Additionally, it is important to highlight that ventricular septal abnormalities are considered one of the oddities that are one of the most likely to have discordant diagnoses and reporting within conventional anomaly registries (Levallois et al. 2012). The present meta-analysis did not reveal any significant association between hypospadias and the variables under investigation (four studies). According to the findings of a recent investigation that was carried out on the nationwide Birth Deformity Prevention Study in the United States, there are certain heightened risks (specifically gastroschisis) that are linked to the duration of showering (Costet et al. 2012). It is important to note that in countries in which chlorine dioxide is used as the major disinfectant for water, a relationship has been discovered between the levels of chlorite or chlorate in water used for drinking and it shows higher probability of urinary tract infections and birth abnormalities. This is something that should be brought to attention. This observation has been documented in various studies (Grellier et al. 2010).

9.2.5 Toxicological Evidence

Multiple analyses of the published research in animal toxicology point to the possibility that exposure to substantial amounts of chloroform and other HAAs and HANs may induce retarded fetal development, which may be quantified by a decrease in fetal length and weight (Zhou et al. 2012; Righi et al. 2012; Tardiff et al. 2006). The primary takeaways from these investigations are summarized here:

Trihalomethanes: Six different research have been conducted to investigate the effects that chloroform has on reproductive results in rats, rabbits, and mice. In all of these studies, it was observed that high doses of chloroform resulted in impaired fetal development, as evidenced by reduced fetal weight. Additionally, some studies also reported a decrease in fetal length (Smith 2011). Nevertheless, the aforementioned negative consequences on fetal growth were observed alongside maternal toxicity. There were a number of unfavorable effects observed, such as fetal resorption, fetotoxicity, skeletal abnormalities, cleft palate, increased fetal death, and decreased pregnancy incidence (Smith 2011). Among the various research investigations conducted, the minimum level of observed adverse effect, commonly known as LOAEL, was found to be 20 mg/kg/day (Smith 2011).

There were a total of six investigations into the effects of DCBM on the fetal development of either rats or rabbits. The results of all six studies indicated that there were no negative effects of DCBM on fetal growth (Smith 2011; Christian et al. 2002a, b; Christian et al. 2001; Narotsky et al. 1997). Nonetheless, DCBM has been linked to additional unfavorable reproductive consequences such as minor skeletal anomalies, postponed sexual development, and fetal resorption (Christian et al. 2002a, b). In addition, in vitro research has been conducted to investigate the effect that DCBM has on human cytotrophoblast cells. According to the findings, DCBM has the capacity to lower the production of chorionic gonadotrophin and also impair the growth of placenta syncytiotrophoblast cells in a way that is dependent on the dosage that is delivered (Christian et al. 2001; Narotsky et al. 1997). This potential is predicated on the fact that DCBM has the ability to do both of these things. The in vitro study reported a minimum adverse effect level of 0.2 nM (approximately 0.0321 μg). According to Borzelleca and Carchman (1982) research, DBCM was linked to a minor decrease in fetal weight, as well as a decrease in litter size and pup viability in mice when administered in high doses (Delaney et al. 1997). Employing reduced dosages, the study conducted by Ruddick and colleagues did not reveal any negative impacts on the reproductive system of rats due to exposure to DBCM. The available evidence does not support the notion that bromoform has any impact on fetal growth or other reproductive outcomes. In rats and mice, no harmful effects have been observed even at doses as high as 200 mg/kg/day (National Toxicology Program 1998).

Haloacetic acids: A research study has examined the impact of DCAA on fetal growth in rats. The study has found a correlation between DCAA and decreased fetal weight and length, as well as a link to fetal resorption and cardiovascular malformations. Additionally, the study has reported a dose-response relationship between DCAA and these adverse outcomes (Chen et al. 2004). The study determined that the level at which no adverse effects were observed (NOAEL) was 14 milligrams per kilogram per day.

A research study has examined the impact of TCAA on reproductive outcomes in rats. The study has revealed a correlation between TCAA and decreased fetal weight and length, as well as links to embryolethality, fetal resorption, cardiovascular, soft tissue, and skeletal malformations. Furthermore, the study has identified a dose-response relationship between TCAA and these adverse effects (Chen et al. 2003). The Lowest Observed Adverse Effect Level (LOAEL) was determined to be 330 milligrams per kilogram per day. There have been no reported negative impacts on fetal growth due to DBAA at doses of up to 50 mg/L in rats (Borzelleca and Carchman 1982; Gulati et al. 1989).

Haloacetonitriles: The administration of chloroacetonitrile (CAN) at a dosage of 55 mg/kg/day was observed to have a correlation with decreased fetal weight in both mice (Smith et al. 1992) and rats (Smith et al. 1989a, b). The Lowest Observed Adverse Effect Level (LOAEL) was determined to be 12.5 milligrams per kilogram per day.

According to two studies, the exposure to dichloroacetonitrile (DCAN) was linked to decreased fetal weight, fetal resorption, and diverse malformations in rats (Smith et al. 1992; Christian et al. 2002a, b). The greatest amount of a substance that was tested and shown to have no adverse effects was 15 milligrammes per kilogramme per day (NOAEL). There were a total of three research projects that were carried out to investigate the connection between trichloroacetonitrile (TCAN) and the development of the fetal organs in rats. All three studies reported a negative correlation between TCAN exposure and fetal weight, while one of the studies also provided evidence for a negative correlation between TCAN exposure and fetal length (Klinefelter et al. 2004). Several negative reproductive outcomes were noted to be linked with exposure to TCAN, including fetal resorption, reduced litter size, and diverse malformations (Smith et al. 1992; Klinefelter et al. 2004; Smith et al. 1987). Two studies have reported an association between BCAN and decreased fetal weight and length in rats (Smith et al. 1992; Ahmed et al. 2008). The Lowest Observed Adverse Effect Level (LOAEL) was determined to be 5 milligrams per kilogram per day. The research findings suggest that there is a correlation between DBAN and decreased fetal weight in rats (Smith et al. 1992).

Chlorophenols: Exon et al. (Smith et al. 1989a, b; Borzelleca and Carchman 1982; Christ et al. 1996) have reported no negative impact on fetal development and growth in rats and mice, respectively, for 2-CP, 2,4-DCP, 2,4,6-TCP, and 2,4-DCP. Exon and colleagues reported a reduction in litter size among rats exposed to 2-chlorophenol (2-CP), 2,4-dichlorophenol (2,4-DCP), and 2,4,6-trichlorophenol (2,4,6-TCP). In addition, they found a correlation between exposure to 2-CP and a reduction in the number of pups that survived to adulthood, an increase in the number of conceptions, and a rise in the number of stillbirths (Smith et al. 1992).

Chloral hydrate: According to the findings of Kallman et al., the treatment of chloral hydrate failed to have any adverse effects on the expansion and development of the fetuses that were being carried by mice (Smith et al. 1998). The NOAEL was found to be 205 milligrammes per kilogramme per day after research was conducted. To summarize, there is some toxicological evidence suggesting that animals used for research may experience harmful effects on the growth and development of their fetuses as a result of being exposed to DBCM, TCAA, chloroform, HANs, and DCAA. However, it is challenging to project these findings to humans.

10 Treatment of DBPs

Water treatment plants use a variety of techniques to control DBPs in drinking water. These techniques include changing the disinfectant used, adjusting the pH of the water, and removing organic matter from the water. Some newer treatment techniques, such as ozonation and membrane filtration, have been shown to be effective in reducing DBP levels in drinking water.

11 Conclusion

During the process of water treatment, disinfection by-products can be produced, which can represent a risk to the general population's health. Exposure to DBPs has been linked to various health impacts and toxicological effects, including cancer, reproductive and developmental effects, and neurological effects. However, there are measures that can be taken to mitigate exposure to DBPs, including reducing their formation, removing them from water, and using alternative sources of drinking water. Additional study is required to improve our understanding of the health implications and toxic effects of DBPs, as well as to create more effective techniques for mitigating their effects.

References

Aggazzotti G, Fantuzzi G, Righi E, Predieri G (1998) Blood and breath analyses as biological indicators of exposure to trihalomethanes in indoor swimming pools. Sci Total Environ 217(1–2):155–163

Aggazzotti G (1995) Environmental and biological monitoring of chloroform in indoor swimming pools. J Chromatogr 710(1):181

Ahmed AE, El-Mazar HM, Nagy AA, Bdel-Naim AB (2008) Chloroacetonitrile induces intrauterine growth restriction and musculoskeletal toxicity in fetal mouse. Toxicol Ind Health 24(8):511–518

Ashley DL, Blount BC, Singer PC, Depaz E, Wilkes C, Gordon S et al (2005) Changes in blood trihalomethane concentrations resulting from differences in water quality and water use activities. Arch Environ Occup Health 60(1):7–15

Backer LC, Ashley DL, Bonin MA, Cardinali FL, Kieszak SM (2000) Household exposures to drinking water disinfection by-products: whole blood trihalomethane levels. J Expo Anal Environ Epidemiol 10(4):321

Backer LC, Lan Q, Blount BC, Nuckols JR, Branch R, Lyu CW et al (2008) Exogenous and endogenous determinants of blood trihalomethane levels after showering. Environ Health Perspect 116(1):57–63

Barbone F, Valent F, Brussi V, Tomasella L, Triassi M, Di LA et al (2002) Assessing the exposure of pregnant women to drinking water disinfection byproducts. Epidemiology 13(5):540–544

Batterman S, Huang A-T, Wang S, Zhang L (2000) Reduction of ingestion exposure to trihalomethanes due to volatilization. Environ Sci Technol 34:4418–4424

Borzelleca JF, Carchman RA (1982) Effects of selected organic drinking water contaminants on male reproduction. EPA-600/1–82-009, NTIS PB82–259847. U.S. Environmental Protection Agency, Health Effects Research Laboratory, Research Triangle Park, NC. Ref Type: Report

Bull R (2012) Toxicological evaluation of experimental data that informs the magnitude of cancer risk from DBPs. In: Hrudey SE, Charrois JW (eds) Disinfection by-products and human health. IWA Publishing, London, pp 179–212

Cantor KP, Villanueva CM, Silverman DT et al (2010) Polymorphisms in GSTT1, GSTZ1, and CYP2E1, disinfection by-products, and risk of bladder cancer in Spain. Environ Health Perspect 118(11):1545–1550

Chen J, Douglas GC, Thirkill TL, Lohstroh PN, Bielmeier SR, Narotsky MG et al (2003) Effect of Bromodichloromethane on chorionic gonadotrophin secretion by human placental trophoblast cultures. Toxicol Sci 76(1):75–82

Chen J, Thirkill TL, Lohstroh PN, Bielmeier SR, Narotsky MG, Best DS et al (2004) Bromodichloromethane inhibits human placental trophoblast differentiation. Toxicol Sci 78(1):166–174

Christ SA, Read EJ, Stober JA, Smith K (1996) Developmental effects of Trichloroacetonitrile administered in corn oil to pregnant Long-Evans rats. J Toxicol Environ Health 47(3):233–247

Christian MS, York RG, Hoberman AM, Diener RM, Fisher LC (2001) Oral (drinking water) developmental toxicity studies of Bromodichloromethane (BDCM) in rats and rabbits. Int J Toxicol 20(4):225–237

Christian MS, York RG, Hoberman AM, Fisher LC, Brown WR (2002a) Oral (drinking water) two-generation reproductive toxicity study of Bromodichloromethane (BDCM) in rats. Int J Toxicol 21(2):115–146

Christian MS, York RG, Hoberman AM, Frazee J, Fisher LC, Brown WR et al (2002b) Oral (drinking water) two-generation reproductive toxicity study of Dibromoacetic acid (DBA) in rats. Int J Toxicol 21(4):237–276

Chu H, Nieuwenhuijsen MJ (2002) Distribution and determinants of trihalomethane concentrations in indoor swimming pools. Occup Environ Med 59(4):243–247

Coffin JC, Ge R, Yang S et al (2000) Effect of trihalomethanes on cell proliferation and DNA methylation in female B6C3F1 mouse liver. Toxicol Sci 58(2):243–252

Costet N, Garlantezec R, Monfort C et al (2012) Environmental and urinary markers of prenatal exposure to drinking water disinfection by-products, fetal growth, and duration of gestation in the PELAGIE birth cohort (Brittany, France, 2002–2006). Am J Epidemiol 175(4):263–275. The first study on pregnancy out- comes using a biomarker of DBP exposure (urinary trichloroacetic acid)

Costet N, Villanueva CM, Jaakkola JJ et al (2011) Water disinfection by products and bladder cancer: is there a European specificity? A pooled and meta-analysis of European case–control studies. Occup Environ Med 68(5):379–385. The most recent pooled analysis on bladder cancer and disinfection by-products

Delaney JC, Wolfe GW, Kaiser L, Lanning L, Klinefelter G, Hunter ES et al (1997) Short term reproductive and developmental effects of chlorodibromomethane in S-D rats211 when administered in the drinking water. Toxicologist 36(1, Part 2):257

Dufour AP, Evans O, Behymer TD, Cantu R (2006) Water ingestion during swimming activities in a pool: a pilot study. J Water Health 4(4):425–430

DWI (2008) National tap water consumption study DWI 70/2/217 Phase two final report. (http://www.dwi.gov.uk/research/reports/DWI70_2_217.pdf) 2008 Available from: URL: http://www.dwi.gov.uk/research/reports/DWI70_2_217.pdf

Egorov AI, Tereschenko AA, Altshul LM, Vartiainen T, Samsonov D, LaBrecque B et al (2003) Exposures to drinking water chlorination by-products in a Russian city. Int JHyg Environ Health 206(6):539–551
Ershow AG, Brown LM, Cantor KP (1991) Intake of tapwater and total water by pregnant and lactating women. Am J Public Health 81(3):328–334
Eslinger SA, Weinberg HS (2003) Estimating average daily exposure to disinfection by-products in drinking water by examining alternate ingestion pathways. Presented at the AWWA Water Quality Technology Conference of the American Water Works Association 2-6 November 2003
Fenster L, Waller K, Windham G et al (2003) Trihalomethane levels in home tap water and semen quality. Epidemiology 14(6):650–658
Forssen UM, Herring AH, Savitz DA, Nieuwenhuijsen MJ, Murphy PA, Singer PC et al (2007) Predictors of use and consumption of public drinking water among pregnant women. J Expo Sci Environ Epidemiol 17(2):159–169
Ge R, Yang S, Kramer PM et al (2001) The effect of dichloroacetic acid and trichloroacetic acid on DNA methylation and cell proliferation in B6C3F1 mice. J Biochem Mol Toxicol 15(2):1000–1006
Gemma S, Vittozzi L, Testai E (2003) Metabolism of chloroform in the human liver and identification of the competent P450s. Drug Metab Dispos 31(3):266–274
Geter DR, Chang LW, Hanley NM et al (2004) Analysis of in vivo and in vitro DNA strand breaks from trihalomethane exposure. J Carcinog 3(1):2
Geter DR, Moore TM, George MH et al (2005) Tribromomethane exposure and dietary folate deficiency in the formation of aberrant cryptfoci in the colons of F344/N rats. Food Chem Toxicol 43(9):1405–1412
Gibbons J, Laha S (1999) Water purification systems: a comparative analysis based on the occurrence of disinfection by-products. Environ Pollut 106(3):425–428
Gordon SM, Brinkman MC, Ashley DL, Blount BC, Lyu C, Masters J et al (2006) Changes in breath trihalomethane levels resulting from household water-use activities. Environ Health Perspect 114(4):514–521
Grellier J, Benett J, Patelarou E et al (2010) Exposure to disinfection by-products and adverse birth outcomes related to fetal growth and prematurity—a systematic review and meta-analysis. Epidemiology 21(3):300–313
Guariglia SR, Jenkins Jr EC, Chadman KK et al (2011) Chlorination byproducts induce gender specific autistic-like behaviors in CD-1 mice. Neurotoxicology 32(5):545–553
Gulati DK, Hope E, Barnes LH, Russell S, Poonacha KB (1989) Bromoform: reproduction and fertility assessment in Swiss CD-1 mice when administered by gavage. NTP-89-068, NTIS PB89–169254. National Toxicology Program, Research Triangle Park, NC
Hrudey SE (2012) Epidemiological inference and evidence on DBPs and human health. In: Hrudey SE, Charrois JW (eds) Disinfection by-products and human health. IWA Publishing, London, pp 213–282
Huang AT, Batterman S (2009) Formation of trihalomethanes in foods and beverages. Food Addit Contam Part A Chem Anal Control Expo Risk Assess 26(7):947–957
Huang AT, Batterman S (2010) Sorption of trihalomethanes in foods. Environ Int 36(7):754–762
IARC (1999) International Agency for Research on Cancer monographs on the evaluation of carcinogenic risks to humans. Some chemicals that cause tumors of the kidney or urinary bladder in rodents and some other substances. In: International Agency for Research on Cancer. Some chemicals that cause tumors of the kidney or urinary bladder in rodents and some other substances, vol 73, pp 131–182
IARC (2012) Trichloroethylene and other chlorinated agents. IARC monographs on the evaluation of carcinogenic risks to humans, vol 106. IARC Scientific publications, Lyon, France
Infante-Rivard C (2004) Drinking water contaminants, gene polymer-phisms, and fetal growth. Environ Health Perspect 112(11):1213–1216
Iszatt N, Nieuwenhuijsen MJ, Bennett J et al (2013) Chlorination by- products in tap water and semen quality in England and Wales. Occup Environ Med 70(11):754–760

Jo WK, Weisel CP, Lioy PJ (1990) Routes of chloroform exposure and body burden from showering with chlorinated tap water. Risk Anal 10(4):575–580
Kaur S, Nieuwenhuijsen MJ, Ferrier H, Steer P (2004) Exposure of pregnant women to tapwater related activities. Occup Environ Med 61(5):454–460
Kerger BD, Schmidt CE, Paustenbach DJ (2000) Assessment of airborne exposure to Trihalomethanes from tap water in residential showers and baths. Risk Anal 20(5):637–652
King W (2001) Epidemiological studies of disinfection by-products and cancer risk. In: Craun GF, Hauchman FS, Robinson DE (eds) Microbiological pathogens and disinfection by-products in drinking water: health effects and management of risks. ILSI Press, Washington, DC, pp 243–254
Klinefelter GR, Strader LF, Suarez JD, Roberts NL, Goldman JM, Murr AS (2004) Continuous exposure to Dibromoacetic acid delays pubertal development and compromises sperm quality in the rat. Toxicol Sci 81(2):419–429
Krasner SW, Wright JM (2005) The effect of boiling water on disinfection by-product exposure. Water Res 39(5):855–864
Landi S, Hanley NM, Warren SH et al (1999) Induction of genetic damage in human lymphocytes and mutations in Salmonella by trihalomethanes: role of red blood cells and GSTT1-1 polymorphism. Mutagenesis 14(5):479–482
Landi S, Naccarati A, Ross MK et al (2003) Induction of DNA strand breaks by trihalomethanes in primary human lung epithelial cells. Mutat Res 538(1–2):41–50
Larson JL, Bull RJ (2002) Metabolism and lipoperoxidative activity of trichloroacetate and dichloroacetate in rats and mice. Toxicol Appl Pharmacol 115(2):268–277
Levallois P, Gingras S, Marcoux S et al (2012) Maternal exposure to drinking-water chlorination by-products and small-for-gestational- age neonates. *Epidemiology* 23(2):267–276. One of the most recent studies on pregnancy outcomes and DBP exposure, based on a robust methodology and very detailed exposure assessment
Levesque S, Rodriguez MJ, Serodes J, Beaulieu C, Proulx F (2006) Effects of indoor drinking water handling on trihalomethanes and haloacetic acids. Water Res 40(15):2921–2930
Luben TJ, Olshan AF, Herring AH et al (2007) The healthy men study: an evaluation of exposure to disinfection by-products in tap water and sperm quality. Environ Health Perspect 115(8):1169–1176
Ma A (2008) Investigation into different tap water household handling treatments to determine the effects on disinfectant by-products (MSc thesis–Imperial College London); 2008
MacLehose RF, Savitz DA, Herring AH et al (2008) Drinking water disinfection by-products and time to pregnancy. Epidemiology 19(3):451–458
Mazhar MA, Khan NA, Ahmed S, Khan AH, Hussain A, Rahisuddin, Changani F, Yousefi M, Ahmadi S, Vambol V (2020) Chlorination disinfection by-products in municipal drinking water—a review. J Clean Prod 273:123159
Mazhar M, Ahmed S, Husain A, Uddin R (2022) Monitoring of trihalomethanes and its cancer risk assessment in drinking water of Delhi City, India. Pollution 8(3):830–843
Moser VC, Phillips PM, Levine AB et al (2004) Neurotoxicity produced by dibromoacetic acid in drinking water of rats. Toxicol Sci 79(1):112–122
Moser VC, Phillips PM, McDaniel KL et al (1999) Behavioral evaluation of the neurotoxicity produced by dichloroacetic acid in rats. Neurotoxicol Teratol 21(6):719–731
Narotsky MG, Pegram RA, Kavlock RJ (1997) Effect of dosing vehicle on the developmental toxicity of Bromodichloromethane and carbon tetrachloride in rats. Fundam Appl Toxicol 40(1):30–36
National Toxicology Program. Short Term Reproductive and Developmental Toxicity of Bromodichloromethane (CAS No. 75-27-4) Administered in the Drinking Water to Sprague-Dawley Rats. http://ntp.niehs.nih.gov/indexcfm?objectid=070EC97E-B871-16C6-1C7596B55DF73B20 (1998 Available from: URL:http://ntp.niehs.nih.gov/index.cfm?objectid=070EC97E-B871-16C6-1C7596B55DF73B20

Ni YC, Wong TY, Lloyd RV et al (1996) Mouse liver microsomal metabolism of chloral hydrate, trichloroacetic acid, and trichloroethanol leading to induction of lipid peroxidation via a free radical mechanism. Drug Metab Dispos 24(1):81–90

Nieuwenhuijsen MJ, Grellier J, Iszatt N, Martinez D, Rahman MB, Villanueva CM (2010) Literature review of meta-analyses and pooled analyses of disinfection by-products in drinking water and cancer and reproductive health outcomes. In: Halden RU (ed) Contaminants of emerging concern in the environment: ecological and human health considerations. American Chemical Society, Washington DC, pp 483–496

Nieuwenhuijsen MJ, Grellier J, Smith R et al (2009) The epidemiology and possible mechanisms of disinfection by-products in drinking water. Philos Trans A Math Phys Eng Sci 367(1904):4043–4076

Nuckols JR, Ashley DL, Lyu C, Gordon SM, Hinckley AF, Singer P (2005) Influence of tap water quality and household water use activities on indoor air and internal dose levels of trihalomethanes. Environ Health Perspect 113(7):863–870

Pandian AMK, Rajamehala M, Singh MVP, Sarojini G, Rajamohan N (2022) Potential risks and approaches to reduce the toxicity of disinfection by-product–a review. Sci Total Environ 822:153323

Pegram RA, Andersen ME, Warren SH et al (1997) Glutathione S-transferase-mediated mutagenicity of trihalomethanes in salmonella typhimurium: contrasting results with bromodichloromethane off chloroform. Toxicol Appl Pharmacol 144(1):183–188

Pereira MA, Kramer PM, Conran PB et al (2001) Effect of chloroform on dichloroacetic acid and trichloroacetic acid-induced hypomethylation and expression of the c-myc gene and on their promotion of liver and kidney tumors in mice. Carcinogenesis 22(9):1511–1519

Rahman MB, Driscoll T, Cowie C et al (2010) Disinfection by-products in drinking water and colorectal cancer: a meta-analysis. Int J Epidemiol 39(3):733–745

Richardson SD, Plewa MJ, Wagner ED et al (2009) Occurrence, genotoxicity, and carcinogenicity of regulated and emerging disinfection by-products in drinking water: a review and roadmap for research. Mutat Res 636(1–3):178–242

Righi E, Bechtold P, Tortorici D et al (2012) Trihalomethanes, chlorite, chlorate in drinking water and risk of congenital anomalies: a population-based case–control study in northern Italy. Environ Res 116:66–73

Robinson D, Mead GC, Barnes KA (1981) Detection of chloroform in the tissues of freshly eviscerated poultry carcasses exposed to water containing added chlorine or chlorine dioxide. Bull Environ Contam Toxicol 27(2):145–150

Ross MK, Pegram RA (2003) Glutathione transferase theta 1-1-dependent metabolism of the water disinfection byproduct bromodichloromethane. Chem Res Toxicol 16(2):216–226

Ross MK, Pegram RA (2004) In vitro biotransformation and genotoxicity of the drinking water disinfection byproduct bromodichloromethane: DNA binding mediated by glutathione transferase theta 1-1. Toxicol Appl Pharmacol 195(2):166–181

Salas LA, Villanueva CM, Tajuddin SM, Amaral AFS, Fernandez AF, Moore LE et al (2014) LINE1 methylation in granulocyte DNA and trihalomethane exposure is associated with bladder cancer risk. Epigenetics 9(11):1532–1539

Saleh MA, Abdel-Rahman FH, Woodard BB, Clark S, Wallace C, Aboaba A et al (2008) Chemical, microbial and physical evaluation of commercial bottled waters in greater Houston area of Texas. J Environ Sci Health A Tox Hazard Subst Environ Eng 43(4):335–347

Savitz DA, Singer PC, Herring AH et al (2006) Exposure to drinking water disinfection byproducts and pregnancy loss. Am J Epidemiol. 164(11):1043–1051

Shimokura GH, Savitz DA, Symanski E (1998) Assessment of water use for estimating exposure to tap water contaminants. Environ Health Perspect 106(2):55–59

Smith MK, George EL, Zenick H, Manson JM, Stober JA (1987) Developmental toxicity of halogenated acetonitriles: drinking water by-products of chlorine disinfection. Toxicology 46(1):83–93

Smith MK, Randall JL, Read EJ, Stober JA (1992) Developmental toxicity of dichloroacetate in the rat. Teratology 46(2):217–223

Smith MK, Randall JL, Read EJ, Stober JA (1989a) Teratogenic activity of trichloroacetic acid in the rat. Teratology 40(5):445–451

Smith MK, Randall JL, Stober JA, Read EJ (1989b) Developmental toxicity of dichloroacetonitrile: a by-product of drinking water disinfection. Fundam Appl Toxicol 12(4):765–772

Smith MK, Randall JL, Tocco DR, York RG, Stober JA, Read EJ (1998) Teratogenic effects of trichloroacetonitrile in the Long-Evans rat. Teratology 38(2):113–120

Smith RB (2011) Exposure to disinfection by- products during pregnancy, in an epidemiological study examining associated risk of adverse fetal growth outcomes, (PhD thesis–Department of Epidemiology and Biostatistics, Imperial College London), 2011

Tao L, Wang W, Li L et al (2004) Effect of dibromoacetic acid on DNA methylation, glycogen accumulation, and peroxisome proliferation in mouse and rat liver. Toxicol Sci 82(1):62–69

Tardiff RG, Carson ML, Ginevan ME (2006) Updated weight of evidence for an association between adverse reproductive and developmental effects and exposure to disinfection by-products. Regul Toxicol Pharmacol 45(2):185–205

Testai E, de Curtis V, Gemma S et al (1996) The role of different cytochrome P450 isoforms in in vitro chloroform metabolism. J Biochem Toxicol 11(6):305–312

Villanueva CM, Cordier S, Font-Ribera L, Salas LA, Levallois P (2015) Overview of disinfection by-products and associated health effects. Current Environmental Health Reports 2(1):107–115

Weinberg HS, Pereira VR, Singer PC, Savitz DA (2006) Considerations for improving the accuracy of exposure to disinfection by-products by ingestion in epidemiologic studies. Sci Total Environ 354(1):35–42

Weisel CP, Chen WJ (1994) Exposure to chlorination by-products from hot water uses. Risk Anal 14(1):101–106

Weisel CP, Jo WK (1996) Ingestion, inhalation, and dermal exposures to chloroform and trichloroethene from tap water. Environ Health Perspect 104(1):48

Whitaker HJ, Nieuwenhuijsen MJ, Best NG (2003) The relationship between water concentrations and individual uptake of chloroform: a simulation study. Environ Health Perspect 111(5):688–694

Williams BL, Florez Y, Pettygrove S (2001) Inter- and intra-ethnic variation in water intake, contact, and source estimates among Tucson residents: implications for exposure analysis. J Expo Anal Environ Epidemiol 11(6):510–521

Wu WW, Benjamin MM, Korshin GV (2001) Effects of thermal treatment on halogenated disinfection by-products in drinking water. Water Res 35(15):3545–3550

Xie SH, Li YF, Tan YF et al (2011) Urinary trichloroacetic acid levels and semen quality: a hospital-based cross-sectional study in Wuhan, China. Environ Res 111(2):295–300

Xu X, Mariano TM, Laskin JD, Weisel CP (2002) Percutaneous absorption of trihalomethanes, haloacetic acids, and haloketones. Toxicol Appl Pharmacol 184(1):19–26

Xu X, Weisel CP (2003) Inhalation exposure to haloacetic acids and haloketones during showering. Environ Sci Technol 37(3):569–576

Xu X, Weisel CP (2005) Human respiratory uptake of chloroform and haloketones during showering. J Expo Anal Environ Epidemiol 15(1):6–16

Yulin T, Long X, Wu M, Yang S, Gao N, Xu B, Dutta S (2020) Bibliometric review of research trends on disinfection by-products in drinking water during 1975–2018. Separation and Purification Technology 241:116741

Zender R, Bachand AM, Reif JS (2001) Exposure to tap water during pregnancy. J Expo Anal Environ Epidemiol 11:224–230

Zeng Q, Wang YX, Xie SH et al (2014) Drinking-water disinfection by- products and semen quality: a cross-sectional study in China. Environ Health Perspect 122(7):741–746

Zhou WS, Xu L, Xie SH et al (2012) Decreased birth weight in relation to maternal urinary trichloroacetic acid levels. Sci Total Environ 416:105–110

Detection and Measurement of Disinfection By-Products in Drinking Water

Vivek Dalvi, Afzal Ansari, Shweta Kalia, Rubia Gaur, Beni Lew, and Abid Ali Khan

Abstract It is very important to identify and precisely quantify disinfection by-products (DBPs) in water to make sure it is safe for drinking and of good quality. During the process of disinfecting water, disinfectants like chlorine combine with the organic and inorganic matter present in the water. Consumers get at the risk for health problems from these DBPs, so it's important to find and measure them correctly. In order to address newly discovered contaminants and fulfil regulatory requirements, efforts are now being made to create and hone methods of detection and testing for DBPs in drinking water.

DBPs present in water can be identified and measured using various chemical methods. However, advanced methods are required for precise measurement of DBPs in water. These methods encompass advanced techniques like gas chromatography, liquid chromatography, mass spectrometry, and spectroscopic methods. Advances in technology have allowed for improved sensitivity, selectivity, and speed of analysis, enabling the identification and quantification of a wide array of DBPs. The present chapter gives an overview of types of disinfectants used for removal of pathogens, their environmental implications due to by-product formation, and different techniques of analysis. The findings indicate continuous advancements in analytical instrumentation, data analysis techniques, and quality assurance protocols that contribute to the comprehensive monitoring and assessment of DBP levels in drinking water.

Keywords DBPs · Disinfectants · Health risks · Public health · Analytical techniques

V. Dalvi · A. Ansari · S. Kalia · A. A. Khan (✉)
Department of Civil Engineering, Jamia Millia Islamia, New Delhi, India
e-mail: akhan23@jmi.ac.in

R. Gaur
Tech-Bulls Research & Engineering Consultants, New Delhi, India

B. Lew
Department of Civil Engineering, Ariel University, Ari'el, Israel

S. Madhav et al. (eds.), *Drinking Water Disinfection By-products*,
https://doi.org/10.1007/978-3-031-49047-7_8

1 Introduction

According to United Nations General Assembly's Sustainable Development Goal 6 targeted for clean water and sanitation states that, access to clean drinking water, sanitation, and good health is one of the most important things for people's health and well-being. Due to increase in population in recent decade, water consumption has increased leading to a worldwide problem of water pollution (Srivastav et al. 2020). About 2.1 billion humans around the world are at risk of water stress. There is an urgency for improvements in the quality of drinking water because many children below 5 years die every year because of diarrhoea and other infections that spread through uncleaned drinking water (Dubey et al. 2020).

Different types of contaminants such as physical (colour, odour, suspended solids, etc.), chemical (metals, dyes, salts, etc.), and biologicals (bacterial, fungus, virus, pathogens etc.) are present in water used for daily purposes, where the reason could vary due to source of water used (Singh et al. 2012, 2015). These contaminants cause acute health problems. Since decades, the need to treat the water to make it fit for drinking has become essential (Srivastav et al. 2015; Chaves et al. 2019). Disinfection is a commonly applied treatment technique to destroy the pathogens present in liquid to help inhibit the spread of water borne infections (Li and Mitch 2018). Different types of disinfectants such as chemicals or UV radiations are used to treat the water at large scale. Chlorine, chloramines, ozone are the examples of most commonly applied disinfectants to get rid of water contaminants (Tram et al. 2014; Tang et al. 2020). The low cost, easy purchase and handling, and the longer persistence nature in water distribution system make chlorine as choice of disinfectant (Sadiq and Rodriguez 2004). However, the use of these disinfectants resulted in the formation of disinfectant by-products (DBPs) (Mazhar et al. 2020), which pose a hazard to human well-being and aquatic habitats.

When free oxidants of chlorine or bromine get in contact with organic matter, the DBP creation takes place (Richardson et al. 2007). Numerous classes of DBPs generated by the use of chlorine are trihalomethanes (THMs), haloacetic acids (HAAs), chlorophenol, etc. (Nieuwenhuijsen et al. 2000). Chloramines are also disinfectants that react to form DBPs such as Haloacetamides (HAAms), Halonitromethanes (HNMs), Haloacetonitrile (HANs), etc. (Li and Mitch 2018). Ozone-based disinfectant releases bromate in presence of bromine ions which is highly injurious. Due to extensive use of chlorine disinfectant, significant levels of THMs and HAA are reported in potable water. Therefore, the release of THMs and HAAs has been controlled by department of public safety along with reaching highest possible disinfection. The level of DBPs is regulated to minimize the harmful effects caused due to them (Chhipi-Shrestha et al. 2018).

Detection, quantification, as well as estimation of disinfection by-products are therefore necessary, as accurate amounts of DBPS in potable water under varying conditions are crucial for controlling its formation and mitigating its toxicity. Discrete analytical methods are used for estimation of DBPs within disinfected liquid. Gas chromatography along with mass spectroscopy has been predominantly practised for the determination of DBPs. Liquid chromatography and ion exchange chromatography has not been explored much for DBPs analysis. However, for HAAs, anions existed in drinking water could be identified by ion exchange chromatography (Paull and Barron 2004).

2 Types of Disinfectants

Water disinfection is an essential step in preserving public health since waterborne infections present serious health hazards. Chlorination, ultraviolet disinfection, ozonolysis, and chloramination are the four main categories of disinfectants.

2.1 *Chlorination*

Chlorination is a method that is used a lot because it works well, is cheap, and is easy to use. Pathogenic bacteria are killed or stopped from spreading disease by adding chlorine to the water. The main ways that chlorination works are through oxidation and chlorination. In the oxidation process, microbes are killed by the highly reactive nature of chlorine, which messes up their cell structures and how their metabolism works. Chlorination is the process of making hypochlorous acid and hypochlorite ions, which then mix with cell parts like proteins and enzymes to kill microorganisms.

A wide range of germs, such as bacteria, viruses, and protozoa, can be killed very well by chlorination. The fact that chlorine can get inside microbial cells and mess up their important processes makes it a good disinfectant for a wide range of things. Some of the things that affect how well chlorine works are the amount of chlorine, how long it stays in contact with the water, the pH of the water, the temperature, and the presence of organic matter. There are many benefits to chlorination, such as its wide use, low cost, and quick ability to kill germs. It leaves a disinfectant in the water, which helps keep the water quality high throughout the transportation system. But there are some pathogenic organisms against which chlorination may not help. It is also reported that chlorination might not work as well against Cryptosporidium and other microorganisms that are resistant to chlorine. Also, chlorine disinfection doesn't get rid of all chemical pollution or deal with things like taste and smell problems.

2.2 Ultraviolet Disinfection

Ultraviolet (UV) disinfection is a new method that uses UV light to kill or stop the growth of harmful bacteria in water. Microorganisms are mostly killed by UV light because it damages their DNA. UV light kills microbes by messing up their DNA, which stops them from making copies of themselves and kills the cells. UV light is most likely to damage a microorganism's DNA. This is because UV energy causes covalent links to form between nearby nucleotides, which breaks up the genetic code.

For most of the time, the wavelength of UV light used is between 200 and 300 nm. Most UV disinfection in water treatment is done at a frequency of about 254 nm, which is in the UV-C range. UV-C light at 254 nm is especially good at killing microbes because it has a strong germicidal effect and can damage their DNA and RNA, which is their genetic material. The nucleic acids in microbial cells receive a lot of this wavelength, which messes up their genetic material and makes them unable to divide or cause an infection. It's important to remember that different types of UV light have different effects on germs. UV-A (315–400 nm) and UV-B (280–315 nm) are less effective at killing germs than UV-C, but they can still disinfect pathogens up to certain level. But UV-C at 254 nm is the most widely used wavelength for disinfecting water because it has strong disinfecting properties. UV disinfection uses UV lamps or LEDs that send out UV light at the right range to clean the water. The water to be treated goes through a UV chamber or reactor, where it is exposed to UV light. This lets the light get into the water and kill any microorganisms that are there.

Overall, the exact wavelength of UV light needed to kill microorganisms in water depends on the microorganisms you want to kill and how much you want to kill them. But UV-C light at about 254 nm is generally known to kill a wide range of microorganisms, such as bacteria, viruses, and protozoa. It gives a quick and effective way to clean water without using chemicals. But UV disinfection may not be as effective against bacteria that are resistant to UV light and may need to be combined with other treatments to fully control water quality.

UV disinfection has many benefits, such as being able to kill a wide range of bacteria, even those that are resistant to chlorine. It doesn't add chemicals or change the way water tastes or smells. UV devices only need a short amount of contact time and don't make any harmful by-products. But there are some problems, such as the need for a good system design, regular upkeep, and a power source. UV disinfection works best when the UV dose, water quality measures (like turbidity and suspended solids), water temperature, and UV system design are all right. To get the best results from UV disinfection, all above things must be taken into account.

2.3 Ozonolysis

Ozonolysis is yet another advanced oxidation method that is used to disinfect water by getting rid of pathogens and organic pollutants. Ozonolysis is the process of making water safe to drink by using ozone (O3). It is a multistep process that starts with the breakdown of ozone, which makes reactive oxygen species (ROS). ROS like hydroxyl radicals (•OH) and peroxone species (O3•) have a high oxidative potential and can combine with microorganisms and organic pollutants to kill or break them down. Ozonolysis is a very effective way to kill germs, like most bacteria, viruses, and protozoa. Because ozone and ROS are oxidizing, they damage the cell walls, proteins, and genetic material of microorganisms, which leads to the loss of important functions and cell death. How well ozonolysis works relies on things like the amount of ozone, how long it stays in contact with the water, the quality of the water, and the temperature.

Ozonolysis has a lot of benefits, like being able to clean a wide range of substances, leaving no residual disinfection by-products, and getting rid of taste and odour compounds. It works against bacteria that are resistant to chlorine and works quickly. But there are some problems, like the need for careful tracking and control of the amount of ozone, the possibility of corrosion, and the high amount of energy needed to make ozone. But many things can affect how well ozonolysis works, such as the amount of ozone, the amount of time the two substances are in touch, the pH of the water, the temperature of the water, and the presence of substances that interfere. For ozonolysis to work well to disinfect water, the system must be designed well, the working parameters must be optimized, and the water quality must be controlled. By thinking about the pros, cons, and best ways to use ozonolysis processes, people who work in water treatment can use this advanced oxidation process to make sure people have safe, high-quality drinking water.

2.4 Chloramination

Chloramination is a way of disinfecting water via introducing chloramine compounds to kill microorganisms and keep the water's quality stable. Chloramination works by letting chlorine and ammonia slowly leak out, making disinfectants like monochloramine. Monochloramine has the ability to kill a wide range of bacteria by messing up their cell structures and metabolic processes.

The primary means by which chloramines work is by entering into the cells of microorganisms and damaging proteins, enzymes, and genetic material. Since disinfectant residue persists in the system for a long time, the water stays clean, and bacteria can't grow back. The chloramination process works optimally when there is a sufficient amount of chloramine, ample contact time, an ideal pH and temperature, and absence of organic matter or other cleaning by-product precursors. Chloramination is used to clean water in places like swimming pools, public water resources, and wastewater treatment facilities. It is especially useful when a sanitizer needs to be kept in the distribution system. Chloramination can stop germs from growing, cut down on cleaning waste, and allow disinfection last for longer.

Chloramination has a number of advantages, such as being able to cleanse water for a long time, making fewer by-products of disinfection than chlorine, and making cleaned water taste and smell better. It gives residual disinfection, which helps stop microorganisms from growing and keeps the water quality in the distribution system high. However, the chloramination has some problems, such as the possibility of nitrification. For chloramination to function effectively, the right quantity of chloramine compounds must be used, residual levels have to be tracked, pH and levels of ammonia need to be monitored, and tools and infrastructure need to be maintained regularly. It is of the utmost significance to follow the rules and guidelines that the government establishes for the amount of leftover chloramine and disinfection by-products.

3 Disinfectant By-Products

The treatment of water with disinfectants led to DBPs formation that are toxic in nature. The generation of DBPs depends upon the dosage of disinfectant during disinfection of water (Carter and Joll 2017). In 1970, presence of DBPs were first observed in portable water (Richardson and Postigo 2015). Thereafter, due to addition of chlorine disinfectant in drinking water, formation of trichloromethane was observed (Bellar et al. 1974). Based on the literature, approximately 800 DBPs have been reported till now. According to Chhipi-Shrestha et al. (2018), most widely reported DBPs are HANs, HKs, and HNMs.

DBPs are found both in organic (chloroform) and inorganic (chlorate, chlorite, bromate) form. However, it was observed that organic DBPs are present in higher in chlorine-treated water as compared to the inorganic ones (Krasner 2009). The chlorinated DBPs such as THMs, HKs, HAA, HANs are more as compared to other DBPs in disinfected water. According to Ding et al. (2019), three main classes of DBPs, i.e. THMs, HAAs, HANs are reported in water. THMs and HAAs class of DBPs are focussed by the researchers and its level in drinking water has been monitored by various countries (Goslan et al. 2006). Among DBPs, HNMs, HAAms, and HANs are highly carcinogenic and their frequency of occurrence in disinfected water is more (Krasner et al. 2006). The entrance of DBPs to human bodies is through different modes such as inhalation, ingestion, and absorption. Due to exposure to DBPs, human faces respiratory diseases such as asthma and wheezing, whereas it also develops a risk of colon cancer, bladder cancer, and reproductive problems (Hariganesh et al. 2020). The DBPs generated from the disinfectants and health effects caused due to it are represented in Table 1. According to literature types of DBPs reported are as follows:

Table 1 Harmful effects of DBPs generated from different sources

Disinfected by-products	Sources	Harmful effects	References
Trihalomethanes	Drinking water, pool water	Causes cancer, effects liver, kidney, lungs, adipose tissues	de Castro Medeiros et al. (2019)
Haloacetic acids	Drinking water	Increasing risk of cancer	Paull and Barron (2004)
Haloacetonitriles	Drinking water	Cancer, clastogenic, mutagenic effects	Dubey et al. (2020)
Haloketones	Swimming pool water	Irritation in eyes, skin, mucous membrane	Zwiener et al. (2007a, b)
Halonitromethane	Drinking water	Cytotoxic and genotoxic	Bond et al. (2011)
Halophenols	Drinking water	Bladder cancer and other reproductive anomalies	Kali et al. (2021)
Haloacetamides	Drinking water	Bladder cancer, colorectal cancer, birth effects	Yang et al. (2019)
Halofuranones	Drinking water	Genotoxicity and mutagenicity	Richardson et al. (2007)

- *Trihalomethanes (THMs)*

THMs and chloroform are the most prevalent and monitored DBPs. Initially, it was discovered in potable water and in swimming pool water. During water disinfection when unbound chlorine reacts with natural organic matter (NOM) in the water, THMs are formed. Significant levels of THMs were present in potable water due to widespread use of chlorine in water disinfection. It is a volatile organic compound having a methyl group and 3 halogen (CH_3X). THMs compounds formed from chlorine and bromine are chloroform, bromodichloromethane, dibromochloromethane, and bromoform. Iodinated THMs have also been discovered in chlorinated water however, they are not widely measured through the compounds are more toxic than brominate and chlorinated THMs (Richardson 2003). Amy et al. (1991) investigated that natural water contains 50% higher bromine level than chlorine. Bromine DBPs are genotoxic compared to chlorine DBPs (Du et al. 2017). Factors such as contact time, temperature, pH, presence of halides, chlorine dosage, etc. affect the formation of THMs (Liang and Singer 2003). However, these factors do not depend upon each other and can affect the disinfection process alone.

- *Haloacetic acids (HAAs)*

HAAs are the most prominent DBPs found in chlorinated water after THMs. They are formed by chlorine and chloramines reaction with (NOM) during disinfection process. HAAs were first investigated in swimming pool water (Martínez et al. 1999). They are the acetic acid substituted with one or more halogen. Different types of HAAs are monochloroacetic acid, dichloroacetic acid, trichloroacetic acid, monobromoacetic acid, dibromoacetic acid, tribromoacetic acid, bromochloroacetic acid, bromodichloroacetic acid, and dibromochloroacetic acid (Li et al. 2013). The factors affecting the formation of HAAs are concentration of disinfectants, high temperature, longer contact time, acidic environment (pH -6.0), and organic or inorganic precursors. It has been studied that swimming pool water and wastewater contain high concentration of HAAs compared to the drinking water due to the presence of anthropogenic contaminants (Li et al. 2022).

HAAs are carcinogenic, mutagenic, and have genotoxic potential. They enter the human body through permeation and cause eye and skin irritation. The permeability of HAAs depends upon its polarity. More polar HAAs have low volatility and can penetrate the skin easily (Xu et al. 2002).

- *Haloacetonitriles (HANs)*

HKs are also one of the prevalent classes of DBPs that have been investigated to be genotoxic in bacterial systems. Exposure to these compounds poses a threat to human health (Xu and Weisel 2005). According to Nieuwenhuijsen et al. (2000), showering for 10 min can result in >10% daily ingestion doses of HKs. Studies reported that pools having chlorine contain prominent concentration of 1,1-dichloropropanone (7.7 μg L^{-1}) and 1,1,1-trichloropropnone (15 μg L^{-1}) (Hang et al. 2016).

- *Haloketones (HKs)*

HKs are also one of the prevalent classes of DBPs that is reported to be genotoxic in bacteria. Exposure to these compounds poses a threat to human health (Xu and Weisel 2005). According to Nieuwenhuijsen et al. (2000)), showering for 10 min can result in >10% daily ingestion doses of HKs. Studies reported that pools having chlorine contain prominent concentration of 1,1-dichloropropanone (7.7 μg L^{-1}) and 1,1,1-trichloropropnone (15 μg L^{-1}) (Hang et al. 2016).

- *Halonitromethanes (HNMs)*

HNMs are the emerging DBPs that are cytotoxic, genotoxic, and have high toxicity than THMs and HAAs (Plewa et al. 2004). In drinking water, 9 bromine and chlorine substituted HNMs are reported such as bromonitromethane (BNM), bromonitromethane (DBNM), tribromonitromethane (TBNM), chloronitromethane (CNM), dichloronitromethane (DCNM), trichloronitromethane (TCNM), bromochloronitromethane (BCNM), bromodichloronitromethane (BDCNM), dibromochloronitromethane (DBCNM) (Song et al. 2010). Study revealed that swimming pool and spa water contain TCNM and HNM at 5 μg L^{-1} concentration (Tardif et al. 2015). HNMs such as bromonitromethane and bromochloronitromethane contain concentration range from 1.2–11 μg L^{-1} (Yeh et al. 2014).

- *Haloacetamides (HAAms)*

HAAms are known to be more genotoxic and cytotoxic than HAAs. Till date five bromine and chlorine substituted HAAms have been identified in drinking water (Liew et al. 2012). Studies showed that chlorinated water contains dibromoacetamide, dichloroacetamide, and trichloroacetamide at a maximum concentration range from 2.0–3.1 μg L^{-1} (Yeh et al. 2014).

- *Halophenols*

Chlorination of natural waters led to the formation of chlorine, bromine substituted phenols. The concentration of halophenols (2,4,6-dichlorophenol) in chlorinate water is lower as compared to the standards for protection of public health still they pose a serious problem to the drinking water suppliers (Agus et al. 2009).

4 Detection and Estimation of DBPs

Disinfection by-products (DBPs) are chemical compounds that form when disinfectants, such as chlorine, react with organic matter in water during the disinfection process. These DBPs can pose potential health risks, so it is important to detect and measure their levels in drinking water to ensure water safety. There are several methods and techniques used for the detection and measurement of DBPs. Gas chromatography (GC) is a widely used technique for analysing DBPs. It involves the separation of DBPs from water samples using a gas chromatograph and detection using a suitable detector, such as mass spectrometry (GC-MS). A variety of DBPs can be identified and quantified using this method. The sophisticated analytical method known as GC-MS combines mass spectrometry detection with gas chromatography. It allows for the identification and quantification of DBPs based on their mass spectra, providing valuable information about the specific compounds present. Another frequently used technique for detecting polar and high molecular weight DBP analysis is liquid chromatography-mass spectrometry (LC-MS). It combines separation by liquid chromatography with detection by mass spectrometry. The sensitivity and selectivity of LC-MS technologies enable for the identification and measurement of diverse DBPs. Techniques such as comprehensive two-dimensional gas chromatography-mass spectrometry (GCxGC-MS) can provide a broad overview of DBP composition. These methods allow for the detection of known and unknown DBPs and can be useful for identifying emerging or unexpected DBPs in drinking water. Disinfection By-Product Formation Potential (DBPFP) Assays: DBPFP assays involve subjecting water samples to controlled disinfection conditions and measuring the formation potential of DBPs. These assays provide an indication of the potential DBP levels that can be formed during disinfection processes. It is vital to highlight that the method of choosing is determined by the individual DBPs of interest, their concentration levels, and the analytical capabilities available. Regulatory agencies and water utilities often employ a combination of these techniques to comprehensively monitor and assess DBP levels in drinking water supplies.

- *Gas chromatography*

In order to detect and quantify disinfection by-products (DBPs) in water samples, gas chromatography (GC) is a common analytical technique. DBPs are chemical compounds that develop during the disinfection process when organic and inorganic materials in the water react with disinfectants like chlorine or chloramines. As potential health hazards, it is crucial to monitor and assess the presence of these DBPs in water supplies. To increase the sensitivity of the analysis, it may be necessary to pre-treat or concentrate the water samples to eliminate interferences or increase their sensitivity, respectively. Samples preparation can be done with methods like solid-phase extraction (SPE) or liquid–liquid extraction (LLE). The components of a gas chromatograph are an injection port, separation column, detector, and data system. A fused silica capillary covered with a stationary phase suitable for DBP analysis typically acts as the separation column. Using an injector, the water sample is injected into the GC system. A suitable injection technique (e.g., liquid injection, headspace injection, or purge and capture) is employed depending on the DBPs of interest. The sample is vaporized before entering the separation column. The column differentiates the DBPs based on their volatility and affinity for the stationary phase. Depending on the DBPs being analysed, various column categories, such as polar and nonpolar, may be utilised. Flame ionization detectors, electron capture detectors, and mass spectrometers are just some of the detectors that can be used in GC analysis (Hariganesh et al. 2020). The choice of detector is determined by the requisite sensitivity and selectivity of the DBPs being analysed. A data system analyses the detector's output to determine the retention periods and peak areas of the separated DBPs. Analysing standards containing known concentrations of the target DBPs is performed to generate a calibration curve. These standards are created by adding known quantities of DBPs to a matrix that resembles the water sample. The prepared water samples are injected into the GC system, where DBPs are separated and detected. By contrasting the peak areas or heights produced from the detector response with the calibration curve, the concentration of DBPs in the water sample is identified. Notably, the specific methodology and parameters may alter based on the DBPs of interest, regulatory requirements, and available instrumentation. Gas chromatography is only one of several analytical techniques used for DBP analysis; liquid chromatography (LC) and mass spectrometry (MS) may also be used, depending on the specific DBPs that are being analysed.

- *Liquid chromatography*

Liquid chromatography is a common method for detecting and analysing polar, high molecular weight, and thermally unstable DBPs (Richardson and Postigo 2018). In liquid chromatography (LC), components of a liquid mixture are separated from one another and analysed based on their interaction with a stationary phase and a mobile phase. We used MS detectors in our analyses. It is possible to identify DBPs using several forms of liquid chromatography, such as high-performance liquid chromatography (HPLC) and ultra-high-performance liquid chromatography (UHPLC).

In comparison with GC/MS procedures (gas chromatography combined with mass spectrometry), LC/MS techniques (liquid chromatography coupled with mass spectrometry) are particularly useful for detecting polar or extremely polar disinfection by-products (DBPs). This is because LC techniques are better suited for analysing compounds with higher polarity, as they utilize a liquid mobile phase and stationary phase separation. Highly polar DBPs, such as haloquinones and dihalo-4 hydroxybenzoic acids, can be effectively detected using LC/MS techniques. Furthermore, LC techniques are also advantageous for measuring thermally unstable DBPs that may decompose or undergo significant changes when subjected to high temperatures. For example, nitrosodiphenylamine, a thermally unstable DBP, can be measured using LC techniques but not with GC-related techniques that involve vaporization and high-temperature conditions. Cl-/Br-benzoquinones (BQs), which were believed to be extremely dangerous as potential DBPs, were the motivation for the invention of LC/MS-MS methods for detection. This was due to the unusual electrochemical reduction of haloquinones on the electrospray ionization (ESI) tip. The production of [M + H]- ions as a result of this reduction process was detected and analysed using LC/MS-MS methods. Researchers investigated at how hydroxylated benzoquinones (HBQs) change when water is exposed to UV light on a bench scale using HPLC/MS-MS analysis with quadrupole time-of-flight (QTOF) mass spectrometry. The goal of the study aimed to understand the processes that convert HBQs into hydroxylated HBQs in drinking water distribution networks. A method for simultaneously quantifying HBQs and hydroxylated HBQs was devised using SPE, UHPLC, and MS-MS coupled with solid-phase extraction (SPE). The relationship between the characteristics of naturally occurring organic matter (NOM) in source waters and the production of HBQ during alum coagulation was also investigated in a laboratory investigation utilizing LC/MS-MS. The purpose of this study was to comprehend how NOM properties influence the development of HBQs during the water treatment process.

- *Ultra-High Resolution Mas Spectroscopy*

Ultra-high resolution mass spectrometry (UHRMS) is a powerful analytical technique for detecting and identifying distinct chemicals in complex samples. UHRMS offers several advantages over conventional mass spectrometry techniques. With its improved mass accuracy and resolving capacity, a wider variety of chemicals may be detected and characterized with greater precision. This is especially crucial when working with very complex combinations such as found in environmental samples or purified water. When it comes to DBP analysis, UHRMS enables the identification of a broader spectrum of compounds compared to traditional analytical methods. It can detect both known and unknown DBPs, which is valuable for comprehensive monitoring and risk assessment (Yang et al. 2019). By utilizing high-resolution mass spectrometry, it becomes possible to differentiate between various DBP isomers and accurately determine their molecular formulas. UHRMS may provide structural information about the discovered DBPs in addition to accurate mass measurements. Tandem mass spectrometry (MS/MS) or high-resolution mass spectrometry with fragmentation (HRMS-MS/MS) can be used to elucidate the fragmentation patterns of the compounds, both of which aid in chemical identification and confirmation. This cutting-edge approach enables the direct estimation of molecular formulas by delivering exact molecular masses. As a result of this research, a substantial number of previously undiscovered DBPs have been uncovered and made public.

HRMS/UHRMS/MS are becoming increasingly desirable for elucidating the chemical structures of unknown DBPs, particularly when working with complex matrices. These techniques excel at precisely determining the mass of analytes. However, it is worth noting that the sensitivity for detecting new trace-level chemicals in time-of-flight (TOF) and QTOF instruments may be limited by the lower sensitivity that comes with better resolution. To address this limitation, ion cyclotron resonance (ICR), orbitrap, and ultrahigh-resolution Fourier-transform (FT) mass spectrometers can be employed to achieve sensitive and precise mass analyses. Significant DBP precursors include various components of NOM. In comparison with the chlorine-containing compounds present prior to chlorination, the DBPs produced after disinfection exhibit lower hydrogen-to-carbon (H/C) ratios, higher oxygen-to-carbon (O/C) ratios, higher average carbon oxidation states, and more double bond equivalents per carbon. These differences suggest that electrophilic substitution reactions, favouring compounds with higher O/C ratios and more double bond equivalents, likely occur during the disinfection process.

- *TOX (Total organic halogen analysis)*

A technique used to identify and measure the presence of unknown and recognized DBPs in water or other samples is total organic halogen (TOX) analysis (Ackerson et al. 2018). The total amount of halogenated organic chemicals in a sample is determined using a TOX analysis. These compounds include halogenated acids, halogenated aldehydes, halogenated ketones, and other halogenated organic compounds that are formed as DBPs. The halogens commonly analysed in TOX analysis include chlorine, bromine, and iodine. By monitoring TOX levels, water treatment facilities can optimize disinfection processes and take necessary steps to minimize the formation of harmful DBPs. An offline ion chromatograph with a conductivity detector can perform halogen-specific TOX analysis instead of online micro-coulometric titration. This approach determines TOCl, TOBr, and TOI individually. Halogen-specific TOX analysis is becoming increasingly popular since iodinated and brominated DBPs are more hazardous than chlorinated compounds. Investigation of chlorination induced brominated DBPs synthesis requires an understanding of bromine electrophilic substitution, electron transfer, and recycling. Two kinetic models were developed using TOX studies and a careful selection of halogen-related processes to accurately predict the generation of halogenated DBPs during chlorination and chloramination, with an emphasis on brominated DBPs (Yang et al. 2019).

- *Ion Exchange Chromatography*

Ion exchange chromatography can be used to find disinfection by-products (DBPs) in water samples. Ion exchange chromatography is a separation technique that uses the differences in the charge of ions to separate and analyse sample compounds. Ion exchange chromatography can be used to separate and quantify specific ions or classes of DBPs in the case of DBPs.

For controlling DBP precursors in the context of water treatment, ion exchange resin adsorption is a common technique. Haloacetic acids (HAAs) and brominated

DBPs may be produced when chlorine-based disinfectants interact with natural organic matter (NOM) and halides in the source water. NOM and halides can be adsorbed and removed using anion exchange resins, which are effective in eliminating negatively charged organic molecules, reducing the possibility of DBPs will form. Anion exchange resin adsorption is particularly effective at eliminating bromide ions, which are normally found in source waters at a concentration of 0.1 mg/L. During chlorination, bromide can be swiftly converted to hypobromous acid (HOBr), and HOBr can react with NOM much more quickly than hypochlorous acid (HOCl), resulting in the production of more hazardous brominated DBPs (Br-DBPs). By eradicating bromide ions through anion exchange, Br-DBP formation can be drastically reduced. Various methodologies can be used to analyse DBPs depending on the specific compounds of interest. Methods such as ion chromatography and electrospray ionization-tandem mass spectrometry (ESI-MS/MS) can be used to identify HAAs. Labs often utilize ion chromatography with a conductivity detector to search for bromated and chlorinated compounds. It is a dependable method for measuring trace levels of bromate in potable water. An alternate approach for assessing bromate and other oxyhalide anions in potable water is ICPCR-UV/Vis (ion chromatography followed by post-column reaction).

5 Conclusion

Disinfection by-products (DBPs) must be identified and quantified in drinking water to guarantee its purity and safety. Accurate identification and quantification of a wide variety of DBPs have been made possible by the invention and optimisation of analytical procedures. The findings of the present book chapter provide an insight about the discrete analytical methods that are used for estimation of DBPs within disinfected liquid. Gas chromatography along with mass spectroscopy has been predominantly practised for the determination of DBPs. Liquid chromatography and ion exchange chromatography has not been explored much for DBPs analysis. However, for HAAs, anions existed in drinking water could be identified by ion exchange chromatography. The sensitivity and selectivity of analysis have been enhanced by the development of cutting-edge methods like gas chromatography, liquid chromatography, mass spectrometry, and spectroscopy. The recommendations detection and monitoring technologies will improve as our knowledge of DBPs and their potential health hazards, leading to smarter approaches to management and a steadily rising standard of treatment for drinking water.

Acknowledgement The authors are very thankful and acknowledge the administration of Jamia Millia Islamia and administration of other institutions to which authors represent.

References

Ackerson NOB, Machek EJ, Killinger AH, Crafton EA, Kumkum P, Liberatore HK, Plewa MJ, Richardson SD, Ternes TA, Duirk SE (2018) Formation of DBPs and halogen-specific TOX in the presence of iopamidol and chlorinated oxidants. Chemosphere 202:349–357

Agus E, Voutchkov N, Sedlak DL (2009) Disinfection by-products and their potential impact on the quality of water produced by desalination systems: a literature review. Desalination 237(1–3):214–237

Amy GL, Tan L, Davis MK (1991) The effects of ozonation and activated carbon adsorption on trihalomethane speciation. Water Res 25:191

Bellar TA, Lichtenberg JJ, Kroner RC (1974) Occurrence of Organohalides in chlorinated drinking waters. J Am Water Works Ass 66(12):703–706

Bond T, Huang J, Templeton MR, Graham N (2011) Occurrence and control of nitrogenous disinfection by-products in drinking water–a review. Water Res 45(15):4341–4354. Available at:. https://doi.org/10.1016/j.watres.2011.05.034

Carter RAA, Joll CA (2017) Occurrence and formation of disinfection by-products in the swimming pool environment: a critical review. J Environ Sci (China) 58:19

de Castro Medeiros L, de Alencar FLS, Navoni JA, de Araujo ALC, do Amaral VS (2019) Toxicological aspects of trihalomethanes: a systematic review. Environ Sci Pollut Res 26(6):5316–5332

Chaves RS, Guerreiro CS, Cardoso VV, Benoliel MJ, Santos MM (2019) Hazard and mode of action of disinfection by-products (DBPs) in water for human consumption: evidences and research priorities. Comp Biochem Physiol C Toxicol Pharmacol 223:53

Chhipi-Shrestha G, Rodriguez M, Sadiq R (2018) Unregulated disinfection by-products in drinking water in Quebec: a meta-analysis. J Environ Manag 223:984

Ding S et al (2019) Disinfection byproduct formation during drinking water treatment and distribution: a review of unintended effects of engineering agents and materials, vol 160. Water Res, p 313

Du Y et al (2017) Formation and control of disinfection byproducts and toxicity during reclaimed water chlorination: a review. J Environ Sci (China) 58:51–63

Dubey S, Gusain D, Sharma YC, Bux F (2020) The occurrence of various types of disinfectant by-products (trihalomethanes, haloacetic acids, haloacetonitrile) in drinking water. In: Disinfection by-products in drinking water: detection and treatment

Goslan EH, Gurses F, Banks J, Parsons SA (2006) An investigation into reservoir NOM reduction by UV photolysis and advanced oxidation processes. Chemosphere 65:1113

Hang C, Zhang B, Gong T, Xian Q (2016) Occurrence and health risk assessment of halogenated disinfection byproducts in indoor swimming pool water. Sci Total Environ 543:425–431

Hariganesh S, Vadivel S, Maruthamani D, Rangabhashiyam S (2020) Disinfection by-products in drinking water: detection and treatment methods. In: Disinfection by-products in drinking water: detection and treatment

Kali S et al (2021) Occurrence, influencing factors, toxicity, regulations, and abatement approaches for disinfection by-products in chlorinated drinking water: a comprehensive review. Environ Pollut 281:116950. Available at:. https://doi.org/10.1016/j.envpol.2021.116950

Krasner SW et al (2006) Occurrence of a new generation of disinfection byproducts. Environ Sci Technol 40(23):7175–7185

Krasner SW (2009) The formation and control of emerging disinfection by-products of health concern. Philos Trans R Soc A Math Phys Eng Sci 367(1904):4077–4095

Li J et al (2022) Haloacetic acids in the aquatic environment. Sep Purif Rev:1

Li W, Liu Y, Duan J, Mulcahy D (2013) Determination of ten haloacetic acids in water using gas chromatography-triple quadrupole mass spectrometry. Anal Methods 5:2258

Li XF, Mitch WA (2018) Drinking water disinfection byproducts (DBPs) and human health effects: multidisciplinary challenges and opportunities. Environ Sci Technol 52(4):1681–1689

Liang L, Singer PC (2003) Factors influencing the formation and relative distribution of haloacetic acids and trihalomethanes in drinking water. Environ Sci Technol 37:2920

Liew D, Linge KL, Joll CA, Heitz A, Charrois JWA (2012) Determination of halonitromethanes and haloacetamides: an evaluation of sample preservation and analyte stability in drinking water. J Chromatogr A 1241:117

Martínez D, Borrull F, Calull M (1999) Evaluation of different electrolyte systems and on-line preconcentrations for the analysis of haloacetic acids by capillary zone electrophoresis. J Chromatogr A 835:187

Mazhar MA et al (2020) Chlorination disinfection by-products in municipal drinking water–a review. J Clean Prod 273. Available at: http://www.scopus.com/inward/record.url?eid=2-s2.0-85088146297&partnerID=MN8TOARS:123159

Nieuwenhuijsen MJ, Toledano MB, Elliott P (2000) Uptake of chlorination disinfection by-products; a review and a discussion of its implications for exposure assessment in epidemiological studies. J Expo Anal Environ Epidemiol 10:586

Paull B, Barron L (2004) Using ion chromatography to monitor haloacetic acids in drinking water: a review of current technologies. J Chromatogr A 1060:1

Plewa MJ et al (2004) Halonitromethane drinking water disinfection byproducts: chemical characterization and mammalian cell cytotoxicity and genotoxicity, vol 38. Environ Sci Technol, p 62

Richardson SD (2003) Disinfection by-products and other emerging contaminants in drinking water. TrAC–Trends Analyt Chem

Richardson SD, Plewa MJ, Wagner ED, Schoeny R, DeMarini DM (2007) Occurrence, genotoxicity, and carcinogenicity of regulated and emerging disinfection by-products in drinking water: a review and roadmap for research. Mutat Res Rev Mutat Res 636(1–3):178–242

Richardson SD, Postigo C (2015) Formation of DBPs: state of the science. ACS Symp Ser 1190:189–214

Sadiq R, Rodriguez MJ (2004) Disinfection by-products (DBPs) in drinking water and predictive models for their occurrence: a review. Sci Total Environ 321(1–3):21–46

Singh PK, Banerjee S, Srivastava AL, Sharma YC (2015) Kinetic and equilibrium modeling for removal of nitrate from aqueous solutions and drinking water by a potential adsorbent, hydrous bismuth oxide. RSC Adv 5:35365

Singh PK, Srivastav AL, Ghosh DK, Sharma YC (2012) Preparation and properties of hydrous bismuth oxides for nitrate removal from aqueous solutions. Desalin Water Treat 40:144

Song H, Addison JW, Hu J, Karanfil T (2010) Halonitromethanes formation in wastewater treatment plant effluents. Chemosphere 79(2):174–179

Srivastav AL, Patel N, Chaudhary VK (2020) Disinfection by-products in drinking water: occurrence, toxicity and abatement. Environ Pollut 267:115474

Srivastav AL, Singh PK, Sharma YC (2015) Synthesis of a novel adsorbent, hydrous bismuth oxide (HBO2) for the removal of fluoride from aqueous solutions. Desalin Water Treat 55:604

Tang Y et al (2020) Bibliometric review of research trends on disinfection by-products in drinking water during 1975–2018. Sep Purif Technol 241:116741

Tardif MR et al (2015) Secretion of S100A8, S100A9, and S100A12 by neutrophils involves reactive oxygen species and potassium efflux. J Immunol Res 2015:1

Tram VO P et al (2014) A mini-review on the impacts of climate change on wastewater reclamation and reuse. Sci Total Environ 494–495:9–17

Xu X, Mariano TM, Laskin JD, Weisel CP (2002) Percutaneous absorption of Trihalomethanes, Haloacetic acids, and Haloketones. Toxicol Appl Pharmacol 184(1):19–26

Xu X, Weisel CP (2005) Human respiratory uptake of chloroform and haloketones during showering. J Expo Anal Environ Epidemiol 15(1):6–16

Yang M, Liberatore HK, Zhang X (2019) Current methods for analyzing drinking water disinfection byproducts. Curr Opin Environ Sci Health 7:98–107

Yeh RYL et al (2014) Bioanalytical and chemical evaluation of disinfection by-products in swimming pool water. Water Res 59:172

Zwiener C et al (2007a) Drowning in disinfection byproducts? Swimming pool water qualtiy. Water 34(7):42–46

Zwiener C, Richardson SD, DeMarini DM, Grummt T, Glauner T, Frimmel FH (2007b) Drowning in disinfection byproducts? Swimming Pool water quality: THMs and other toxics exist in pool water. Water-Melbourne Then Artarmon 34(7):42

Recent Trends in Controlling the Disinfection By-Products Before their Formation in Drinking Water: A Review

Mehreen Shah, Mohd. Aamir Mazhar, Sirajuddin Ahmed, Beni Lew, and Nadeem Khalil

Abstract Disinfection is a technique to achieve clean and safe drinking water. But often, disinfection in itself leads to the formation of disinfection by-products (DBPs) that are formed by the reaction of disinfectants, such as chlorine, ozone, chloramines, chlorine dioxide with natural organic matter present in the water. In order to prevent the formation of DBPs in water one must understand the conditions under which DBPs are formed and not allow these favourable conditions to occur. DBPs form once there is a chemical reaction between the organic precursors such as humic acid present in water with the disinfectant. Removing the organic precursors present in water through effective removal strategies is an effective strategy to prevent the formation of DBPs. This review paper contains various scientific findings that help us understand about the various ways DBPs are formed and how to prevent formation by timely removal of precursors and antecedent conditions.

Keywords Drinking Water · Disinfection By-products · DBPs control

M. Shah
Department of Environmental Science, Faculty of Engg. & Tech., Jamia Millia Islamia (Central University), New Delhi, India

S. Ahmed (✉)
Department of Environmental Science, Jamia Millia Islamia, New Delhi, India
e-mail: suahmed@jmi.ac.in

M. A. Mazhar
Department of Civil Engineering, Jamia Millia Islamia (Central University), New Delhi, India

B. Lew (✉)
Department of Civil Engineering, Ariel University, Ariel, Israel
e-mail: benilew@ariel.ac.il

N. Khalil
Department of Civil Engineering, Aligarh Muslim University, Aligarh, U.P., India

S. Madhav et al. (eds.), *Drinking Water Disinfection By-products*,
https://doi.org/10.1007/978-3-031-49047-7_9

1 Introduction

Water is an important factor that contributes to the socio-economic security of a nation and is an important need for the growth of a nation and support of the growing needs of a population (Katko and Hukka 2015). Water demand rises exponentially as the population increases, hence the water crisis may rise if water is not effectively cleaned and supplied (Balaji et al. 2021). Urban areas pose a further challenge towards clean water demands as increasing amounts of pollutants find ways to contaminate water (Madhav et al. 2020). Clean drinking water is an important goal towards which various scientific endeavours have been invested which primarily involves disinfection by chlorination. This step attenuates harmful pathogens and helps remove large impurities in earlier stages such as screening (Golea et al. 2017). But disinfection in itself can lead to chemical reactions between the disinfectant chlorine and the organic molecules present in the water leading to formation of undesirable entities called disinfection by-products which are harmful in nature ranging from hepatotoxicity to carcinogenicity in nature (Evan et al. 2020; Mazhar et al. 2020; Gilca et al. 2020). The best way to reduce the presence of disinfection by-products in water is to prevent their formation in the first place (Liu 2013). Various steps can be taken to achieve this by becoming aware of processes and conditions that are favourable for DBPs formation such as elevated temperature post-filtration, pH ranges, disinfectant dosage administered, the contact duration, etc. To reduce disinfection by-products formation the best preventive strategy remains to remove the organic precursors that react with the disinfectant to give rise to DBPs, i.e., the NOMs (natural organic matter) present in water such as humic acids, amino acids, etc. (Abu et al. 2020). Various procedures such as coagulation-filtration, activated carbon adsorption, ozonation, activated carbon adsorption, membrane filtration, reverse osmosis can be adopted in order to remove the precursors of disinfection by-products hence preventing effectively their formation in water upon contact with the disinfectant (Murray and Parsons 2004; Cheng et al. 2005; Luukkonen et al. 2014).

2 Disinfection By-Products (DBPs)

Drinking water undergoes various steps in order to remove impurities to be considered safe enough for drinking but these steps such as chlorination for achieving disinfection lead to formation of DBPs or disinfection by-products. These disinfectants may react with organic precursors present in water in order to lead to formation of DBPs. Some DBPs are trihalomethanes (THMs), haloacetic acids (HAAs), bromate, chlorite, etc. DBPs found in drinking water have been linked to cancer (Mazhar et al. 2022; Evan et al. 2020). Chlorination is the most common disinfection practice, but this leads to formation of trichloromethane (TCM). DBPs can be prevented from formation by two strategies: (i) Avoiding favourable ranges of pH, temperature which support DBPs formation (ii) Removing the natural organic precursors of DBPs from water from favourable treatments (Liu 2013; Gilca et al. 2020).

3 Parameters that Influence the Formation of DBPs in Water

3.1 *Temperature*

Temperature of the water has a significant impact on DBPs formation in the drinking water. If the water was incubated at a higher temperature prior to the disinfection step, then it can lead to lesser formation of DBPs because it leads to higher microbial activity which causes the microbes to consume the organic precursors of disinfection by-products in the water as a substrate hence lowering the total dissolved carbon (TOC). But post-chlorination elevated temperatures of 35 °C lead to higher formation of disinfection by-products particularly trihalomethanes (THM) when water is abundant with bromides leading to formation of Br-DBPs (Liu 2013). These can be reduced by firstly reducing the temperature post the disinfection step and also avoiding conventional chlorination disinfection (Doederer et al. 2014).

3.2 *pH*

The effect of pH cannot be ignored when studying the effect of formation of undesirable toxic DBPs and chloroforms in water after disinfection. The reactions during chlorination are pH sensitive hence pH is an important parameter in the formation of DBPs during chlorination. As pH decreases, trihalomethanes (THMs) formation also decreases but HAA (haloacetic acid) increases. As pH is acidic there is an increased formation of chloroform in the water (Hansen et al. 2012).

3.3 *Contact Time*

Increased contact time with the chlorine disinfectant led to an increase in the formation of chlorine-based disinfectant by-products or C-DBPs (such as THM or trihalomethanes and HAA or haloacetic acids (with some excluding dichloroacetamide chloroform, 4-Chlorophenol, etc.) (Hong et al. 2013).

3.4 *Disinfectant Dosage*

The most common disinfectant is chlorine for drinking water and there is a positive correlation between the dosage of chlorine administered and the amount of DBPs formed in water. The levels of THMs and HAAs increased with the increase of chlorine dosage. The organic precursors which are hydrophobic are important for production of TCAA and THM (trichloroacetic acid and trihalomethane) during the chlorination process (Golea et al. 2017).

3.5 Effect of Retention Time in Disinfection Tank

The effect of retention time is an important factor as there is a prolonged exposure to the disinfectant to the natural organic matter in the water as there is increased contact there is a continuity of a chemical reaction which leads to more generation of DBPs such as haloacetic acid (Sun et al. 2009).

3.6 Effect of Pipes Used in Disinfection Tank

The pipe material used also has an influence on the formation of the DBPs as the pipes walls may lead to humification and due to accumulation of microorganisms. The steel and iron pipes corrode in prolonged exposure to water and bacterial bio-film also contributes to the NOMs which are precursors to DBPs. The PE pipes are the most inert piping material which led to the least contribution of DBPs in water (Zhang et al. 2017).

4 Pre-Treatment Techniques to Prevent DBPs Formation

4.1 Coagulation-Flocculation

Coagulation is a popular procedure for removal of precursors of DBPs from water by addition of a dosage of a coagulant in water which act to neutralise the charged moieties in the water, and the counteracted charges react together in a neutralisation process leading the suspended colloids to settle down (Dayarathne et al. 2021). The most popular coagulants are metal ion-based charged moieties such as the most common one called ferric chloride ($FeCl_3{\cdot}6H_2O$). The reaction may be coupled with flocculation (Matilainen et al. 2010; Butler et al. 2011). There is aggregation of the particles in the form of flocs or masses that soon join the microscopic small masses into visible large colloids that eventually become heavy and are settled down by gravity. These are then subsequently separated by sedimentation, floatation, or filtration (Abu et al. 2020). The process of coagulation is dependent on various factors that may hamper or enhance the efficiency such as temperature, coagulant dosage, pH/alkalinity, contact time, etc. (Karanfil et al. 2006; Newcombe 1999). Humic acid removal is essential for lowering the DBPs formed but its removal is dependent on the pH conditions which are most conducive to acidic and better removed by processes such as adsorption (Gilca et al. 2020; Jiang 2015). The efficiency of coagulation is highly improved when done followed by processes such as membrane-based removal, activated carbon uptake, ion exchange, oxidation removal based on photo-oxidation/Fenton processes (Matilainen et al. 2011; Capodaglio et al. 2017; Szlachta and Adamski 2009; Murray and Parsons 2004).

Electrocoagulation is an advancement where electricity is passed through the solution and since electric charge has become the source of charged moiety there is a complete elimination of the need to chemically add a dosage of coagulant to neutralise the charge present on the NOMs present in the water to be treated (Butler et al. 2011; Shammas et al. 2010; Vasudevan et al. 2010; Das et al. 2022). Electrocoagulation offers a newer alternative to the conventional coagulation. When the charge is passed there is a visible agglomeration of the colloids being accumulated together as flocs hence can be easily removed and this reduces the burden of adding additional coagulants which too pose a task for its subsequent removal from the water (Ebba et al. 2022).

4.2 *Ozonation*

Ozonation helps in removal of organic precursors of DBPs in water by various mechanisms such as volatilization, stripping, or mineralization. This procedure also aids in the stability of the flocs formed hence improving the efficiency of coagulation techniques for effective removal of natural organic matter in water or NOMs (Loganathan et al. 2022). Using pre-ozonation as a technique helped in reducing the DBPs formation in water by subsequent removal of NOMs (Ryan et al. 2013; Ryan and Singer 2013). It also improved the efficiency with which improved the rates of removal by coagulation-flocculation processes (Pei et al. 2007). It helped in removal of organic matter in water which helped in improved aesthetic quality, odour reduction, turbidity levels and showed lesser NOMs in water when tested for BOD, COD, TOC, DO, etc. (Wildhaber et al. 2015). Usually there is a risk of bromate formation in water as an undesirable side effect in various techniques of NOM removals but using pre-ozonation helps in reducing the risk of bromate formation (Myllykangas et al. 2000).

4.3 *Ion Exchange*

Ion exchange is a popular method for NOM removal in water particularly humic acid. The NOMs have a chain of phenolic and carboxylic groups which leads to humic acid and other NOMs in water to have an anionic charge in them (Adusei-Gyamfi et al. 2019; Kennedy et al. 2008). The concentration of natural organic matter in water can be up to 10 ppm in water which when reacted with a disinfectant like chlorine will lead to disinfectant by-product formation hence to provide safe drinking water the natural organic matter in water must be removed priorly and ion exchange as a strategy helps in exploiting the charge polarity between the anionic natural organic matter and the resin bed which acts as an exchange media (Irina et al. 2018).

The resin bed can be regenerated helping achieve economically feasible ways to remove natural organic matter from water over multiple cycles. There is also a knowledge required of hydrophobic and hydrophilic natural organic matter. Hydrophilic resins are better attracted to the weakly basic natural organic matter present in water compared to the conventional quaternary ammonium resin beds (Bolto et al. 2003). Ion-exchange taking place prior to other purification and filtration steps is beneficial as one challenge that is commonly faced is membrane biofouling, which means excess biofilm formation in the nanofiltration membrane due to excess load at the membrane (Kim et al. 2007; Kim and Dempsy 2013; Kennedy et al. 2008).

5 Removing Precursors of DBPs

Precursors of DBPs are natural organic matters found dissolved or suspended in water hence it is necessary to remove them from water in order to deal with the problem of DBPs. NOMs or natural organic matter can be chained with amino acids and have several functional group chains, tannins, humic acids, phenols, etc. (Marc et al. 1993). NOMs are often complex compounds and can be assessed by the carbon-content in water. Krasner (2009) used activated carbon to remove NOMs from water and found significant decline in the DBPs in water. Watson et al. (2012) used nanofiltration to remove NOMs from water, biofiltration techniques also removed NOMs from water. Several physical techniques can be used to assist chemical methods to effectively remove NOMs from water in order to deal with the production of DBPs in water (Evan et al. 2020).

5.1 Natural Organic Matter (NOMs)

NOMs (natural organic matters) in water is a complex mixture of heterogeneous compounds that are derived from sources of various origins which may be bacterial in origin or may be derived from soil organic matter (Hu et al. 2013). They are further classified as hydrophobic or hydrophilic in nature. Hydrophobic (water-repellent in nature) NOMs are anionic in nature or neutral (based on charge) (Copetti et al. 2017; Crouè et al. 2000). Anionic (having negative charge) are derivatives of humic, fulvic, and tannic acid (Aiken et al. 1985). Neutral NOMs (which are hydrophobic in nature) are hydrocarbons (both long- and short-chain carbon chains), tannins (which derive a characteristic colour to compounds in nature), aromatic amines (having a cyclic-structure with a benzene like ring which follows the Kekulé rule for resonance) (Krygowski and Szatylowicz 2015) and resorcinol (bromophenols on reaction with KOH/Potassium hydroxide yield resorcinol, an aromatic compound) (Al-Muhtaseb and Ritter 2003). NOMs are then classified, on the other hand, into anionic (having negative charge) and neutral (having neither a positive nor negative charge) hydrophilic compounds (water-attractive). Hydrophilic anionic NOMs are classified as carboxylic acids, polyuronic acids and amines like aspartic and glutamic acid. Neutral hydrophilic NOMs are classified further into amino acids, peptides (long-chains of amino acids), carbohydrates. NOMs in water react under favourable conditions such as a contact with disinfectant such as chlorine and lead to formation of DBPs (McCreary and Snoeyink 1980).

NOMs in water affect its aesthetic quality and render it unfit for consumption by causing changes in taste, turbidity, colour, odour, etc. and lower treatment efficiency by causing blockages in pores of membrane-reactors, corrosion to pipes, increased need for more dosage of coagulant for flocculation/coagulation-based removal, overload on activated carbon membranes (Eikebrokk et al. 2004). Presence of NOMs in water acts as an electron donor which leads to proliferation of bacterial colonies in water after its successful removal hence rendering the disinfection procedure ineffective (Eikebrokk et al. 2006).

5.2 *Activated Carbon (AC) Adsorption*

Activated carbon adsorption is a brilliant technique for uptake of pollutants in water that are hazardous and negatively affect the quality of the water. It improves the optical clarity, removes characteristic colour, odour, suspended particles and improves the aesthetic quality of the water (Newcombe et al. 2002). There are factors such as the size of the impurity and the pore size of the activated carbon that will be adsorbing the pollutants (Jjagwe et al. 2021). Removing a majority of the NOMs such as humic acid with a prior coagulation technique is highly advisable as it improves the adsorption capacity of the activated carbon. The NOMs such as humic acid if not priorly reduced from the water using coagulation may block adsorption sites in the activated carbon. The activated carbon in itself may be susceptible to biological colonisation (Sabzehmeidani et al. 2021). The AC filters can be treated with heat such as thermal treatment which too under intense temperatures may lead to damage to filters (Swietlik et al. 2002). The treatment of filters with ozone is suitable. The activated carbon media might get saturated with the NOMs present in water hence it is important to remove as many NOMs as possible from the water by pre-treatment with coagulation so as to prevent blocking of the pores which act as sites of adsorption (Korotta-Gamage and Sathasivan 2017; Quinlivan et al. 2005).

5.3 *Iron Oxide-Based Adsorption*

Iron oxide-based adsorption is a novel solution for aiding in the control of DBPs formation in drinking water by effectively removing NOMs. It is a feasible technique that allows the uptake of NOMs present in the water (Cui and Choo 2013).

The uptake takes place via a complementary process where both active adsorption takes place through the surface adsorption and by the attraction that arises through attraction between the charged moieties. There is an exchange with the humic acid exchanging its ligands with the anionic moieties of H2O and OH− of the Fe-$OH^{2+}$and Fe-OH groups which increases as there is an increase significantly in the degree of protonation of humic acid as it is adsorbed at constant pH (Harman et al. 2010). The uptake can be confirmed with a linear correlation with uniformly increasing Freundlich affinity parameters (KF) that acts to affirm the confirmation of NOM uptake. Iron oxide particles coated with pumice are an effective way (Baalousha et al. 2008).

5.4 Oxidation Processes

Oxidation techniques are an important class of water treatment for the effective removal of NOMs by their conversion into lesser harmful forms. This is brought to affect by reactive radical intermediates such as hydroxyl radicals. The oxidation of NOMs take place via OH generation which may be facilitated by the usage of catalytic oxidants like ozone-ultraviolet (O_3-UV) in combination with peroxide (H_2O_2-UV) (Gowland et al. 2021; Lamsal et al. 2011). Photooxidation processes involving the Fenton process are also used with great success. There is involvement of a radicle of Fe^{+3} which aids the process. Sunlight provides energy to the electron which has enough energy to jump to the higher energy level and bypass the bandgap. Oxidation helps in the reduction of the NOMs in the water into harmless by-products (Capodaglio et al. 2016). Hydrogen peroxide is a strong oxidant but alone cannot function that effectively in presence of refractory materials in water. When UV light or sunlight or salts of metals such as Fe^{+3} are used in conjunction it allows the peroxide to form hydroxyl radicals which are stronger oxidants (Matilainen and Sillanpää 2010; Sillanpää et al. 2018).

Photo-Fenton's process is an advanced oxidation process in which hydroxyl radicals are generated to degrade via oxidation of the hazardous and recalcitrant compounds that are present in wastewater. H_2O_2 or peroxide added is decomposed catalytically by Fe^{+2} and the generated radicle of ·OH oxidizes the contaminant in water. Bello et al. (2019) showed a complex formation between SiO_2 and H_2O_2 via a siloxane bridge (Si–O–Si) which enabled dye degradation from wastewater. The coupling with SiO2 enabled a higher COD removal (20%) than without the semiconductor. In the ferrous cycle, Fe^{3+} is converted to Fe^{2+} by the reduction of H_2O_2 and the Fenton cycle is repeated multiple times (Lin et al. 2022). Fe^{+3} ion is a powerful coagulating agent that helps in the sedimentation through coagulation of the pollutants present in the water (Sinha and Dahiya 2022). The radicle that is generated in water operates via electron transfer, hydroxyl and hydrogen addition hence reacts with compounds in water for their effective oxidation in harmless compounds (Kalajdžić et al. 2013).

5.5 *Photocatalysis*

Photocatalysis is an important process that is in motion to remediate NOMs or natural organic matter in water. It refers to the degradation of organic wastes and compounds present in the wastewater by the action of UV or light from a source such as the sun. The photons of light act as photocatalysts hence helping speed up the rate of the biodegradation reaction. Photocatalysis is an oxidative reaction where oxidative degradation of wastewater contaminants occurs. TiO_2 is a popular choice for this reaction which takes place via oxidative-reduction as particles are irradiated with light of 200–400 nm wavelength causing excitation of the electron and leading to a cascade of oxidative-reduction steps in which wastewater pollutants are biodegradable into harmless compounds such as water and CO_2 (Murray et al., 2006). Photocatalysis can be exploited for a higher efficiency rate if we use nanoparticles (Khan et al. 2019). Due to large surface area/volume ratio nanoparticles are more effective over their macro counterparts (Chen et al. 2019). The UV irradiation helps in overcoming the energy gap and doping with gas such as nitrogen further reduces the band width. The electron jump to a higher level will cause reduction of oxygen (Gomes et al. 2019).

5.6 *Ballasted Flocculation (BSF)*

Ballasted flocculation is a common method for the physico-chemical method of water treatment that involves the flocs formed to be heavier than the conventionally formed flocs due to addition of chemical additives that increases the density of the aggregate formed which leads to settlement (Kumar et al. 2016). The ballasted flocculation process involves flocculants such as aluminium sulphate, sodium aluminate, ferric chloride, anionic and cationic agents, etc. Coagulants such as ferric chloride/alum (Lapointe et al. 2021). Microsand of the dimensions (40–150 μm) can reduce the overflow rate and the settling velocity and make the process better compared to conventional processes (Zafisah et al. 2020). Magnetic powder (MP) is an excellent ballast agent compared to the microsand and helps in creation of larger flocs that helps in efficiency in removal. There are 3 times larger flocs compared to micro sand flocs and 6 times larger flocs compared to the general coagulation techniques (Su et al. 2014).

5.7 *Membrane Filtration*

Membrane filtration is a popular and oft-employed practice for water filtration, mostly drinking water. It employs nanofiltration (NF), reverse osmosis (RO), ultrafiltration (UF), and microfiltration (MF) processes (Shen and Schafer 2015; Zhai et al. 2022; Shan et al. 2016). These systems employ a low-pressure gradient that leads to molecules to pass through the membrane hence are filtered out. Membrane filtration is very efficient for NOMs removal and since it removes NOMs from water it can lead to reduction in the formation of DBPs in water that is made for drinking (Zhou et al. 2016). Biofouling of the membrane is a common limitation in membrane-based filtration which leads to reduction in filter longevity hence it is recommended that water be removed from NOMs via coagulation-related pre-treatments (Ødegaard et al. 2000). Membrane biofouling leads to build-up of suspended and colloidal particles in the water and also complicates the filtration processes. It increases the contact time required with the chemicals and the coagulants and also leads to increased costs for disinfectants as filters are rendered useless over time (Acero et al. 2016).

Reverse osmosis is an extremely successful technique to remove contaminants from wastewater having a success rate of up to 99% which leads to remarkably low risk of formations of DBPs (Owusu-Agyeman et al. 2019).

5.8 Membrane-Specific Flux Characterisation

The membrane-specific flux characterisation is a way of assessing membrane fouling. Membrane fouling is an indication of biofilm formation (growth of bacterial mass on the membrane) which leads to health hazard issues. The formula for membrane flux characterization is (AlSawaftah et al. 2021):

$$j = \frac{V}{S.T}$$

where J is the membrane flux ($m^3/(m^2{\cdot}h)$); V is the water sample volume (m^3); S is the membrane effective filtration area (m^2); T is the time required to filter water sample (h) (m = metre h = hours).

5.9 Magnetic Ion Exchange (MIEX)

Magnetic ion exchange or MIEX is a method which helps in the removal of NOMs present in water using strong anionic resins for ionic exchange with the macromolecular organic compounds present in the water. This process takes place under a magnetic field. MIEX treatments can be done in pre-coagulation stages in order to improve the efficiencies of the subsequent processes and also improve the density of the flocs formed (Clayton and Singer 2004). MIEX resins are regenerable in nature and can be reused for several cycles. Using MIEX-coagulation in combination leads to lesser needs of dosage of coagulants like alum (Tien et al. 2011).

5.10 Adsorption

Majority of nanoparticles are able to effect water remediation by their ability to adsorb the contaminants onto their surface. The high surface/volume ratio of nanoparticles due to their extremely small size aids them in this process. An adsorption isotherm is a graphical representation of the relationship between the adsorbate that is remaining in liquid phase and the amount absorbed by the adsorbent in equilibrium. Freundlich and Langmuir isotherm are the most studied models to explain the mechanism of metal ion uptake by the adsorbent system. Langmuir isotherm is based on assumptions such as there is a monolayer at surface, each site only holds one single molecule and there are no intermolecular interactions (Langmuir 1918). Freundlich isotherm involves multilayer absorption for heterogeneous molecular systems and as cations and anions both are adsorbed simultaneously. There exists an attractive force between them. There is equilibrium in the solid and the liquid phase (Freundlich et al. 1906).

6 Conclusions

In order to prevent the formation of DBPs in water one must understand the conditions under which DBPs are formed, i.e., temperature, pH ranges, piping materials, etc. and not allow these conditions to occur. DBPs form once there is a chemical reaction between the organic precursors such as humic acid present in water with the disinfectant such as chlorine. To reduce disinfection by-products formation the best preventive strategy remains to remove the organic precursors that react with the disinfectant to give rise to DBPs, i.e., the NOMs (natural organic matter) present in water such as humic acids, amino acids, etc. Various procedures such as coagulation-filtration, activated carbon adsorption, ozonation, activated carbon adsorption, membrane filtration, reverse osmosis can be adopted in order to remove the precursors of disinfection by-products hence preventing effectively their formation in water upon contact with the disinfectant.

References

Abu HH, Muhammad MH, Ismail NI (2020) A review of biological drinking water treatment technologies for contaminants removal from polluted water resources. J Water Process Eng 33:101035

Acero JL, Javier Benitez F, Real FJ, Teva F (2016) Micropollutants removal from retentates generated in ultrafiltration and nanofiltration treatments of municipal secondary effluents by means of coagulation, oxidation, and adsorption processes. Chem Eng J 289:48–58

Adusei-Gyamfi J, Ouddane B, Rietveld L, Cornard J-P, Criquet J (2019) Natural organic matter-cations complexation and its impact on water treatment: a critical review. Water Res 160:130–147

Aiken G, McKnight D, Wershaw R, McCarthy P (1985) Humic substances in soil, sediment and water. John Wiley & Sons, New York, NY, USA

AlSawaftah N, Abuwatfa W, Darwish N, Husseini G (2021) A comprehensive review on membrane fouling: mathematical modelling, prediction, diagnosis, and mitigation. Water 13, no. 9:1327. https://doi.org/10.3390/w13091327

Al-Muhtaseb SA, Ritter JA (2003) Preparation and properties of resorcinol–formaldehyde organic and carbon gels. Adv Mater 15(2):101–114

Baalousha M, Manciulea A, Cumberland S, Kendall K, Lead JR (2008) Aggregation and surface properties of iron oxide nanoparticles: influence of pH and natural organic matter. Environ Toxicol Chem: Int J 27(9):1875–1882

Balaji E, Narsimha A, Madhav S, Srinivasa Gowd S, Keshava Kiran Kumar PL (2021) Salinity problems in groundwater and management strategies in arid and semi-arid regions. In: Groundwater geochemistry: pollution and remediation methods, pp 42–56

Bello MM, Raman AAA, Asghar A (2019) A review on approaches for addressing the limitations of Fenton oxidation for recalcitrant wastewater treatment. Process Saf Environ Prot 126:119–140

Bolto B, Dixon D, Eldridge R, King S, Linge K (2003) Removal of natural organic matter by ion exchange. Water Res 36:5057–5065. https://doi.org/10.1016/S0043-1354(02)00231-2

Butler E, Hung Y-T, Yeh RY-L, Al Ahmad MS (2011) Electrocoagulation in wastewater treatment. Water 3(2):495–525. https://doi.org/10.3390/w3020495

Capodaglio AG (2017) High-energy oxidation process: an efficient alternative for wastewater organic contaminants removal. Clean Techn Environ Policy 19:1995–2006

Capodaglio AG, Callegari A, Molognoni D (2016) Online monitoring of priority and dangerous pollutants in natural and urban waters: a state-of-the-art review. Manag Environ Qual 27:507–536

Chen H, Lin T, Chen W, Tao H, Xu H (2019) Removal of disinfection byproduct precursors and reduction in additive toxicity of chlorinated and chloraminated waters by ozonation and up-flow biological activated carbon process. Chemosphere 216:624–632

Cheng W, Dastgheib SA, Karanfil T (2005) Adsorption of dissolved natural organic matter by modified activated carbons. Water Res 39(11):2281–2290

Clayton JJ, Singer PC (2004) Impact of a magnetic ion exchange resin on ozone demand and bromate formation during drinking water treatment. Water Res 38(17):3738–3750. https://doi.org/10.1016/j.watres.2004.06.021

Copetti D, Valsecchi L, Capodaglio AG, Tartari G (2017) Direct measurement of nutrient concentrations in freshwaters with miniaturised analytical probe: evaluation and validation. Environ Monit Assess 189:144

Crouè J, Korshin GV, Benjamin M (2000) Characterization of natural organic matter in drinking water. American Water Works Association, Denver, CO, USA

Cui X, Choo K-H (2013) Granular iron oxide adsorbents to control natural organic matter and membrane fouling in ultrafiltration water treatment. Water Res 47(13):4227–4237

Das PP, Sharma M, Purkait MK (2022) Recent progress on electrocoagulation process for wastewater treatment: a review. Sep Purif Technol 121058:121058

Dayarathne HNP, Angove MJ, Aryal R, Abuel-Naga H, Mainali B (2021) Removal of natural organic matter from source water: review on coagulants, dual coagulation, alternative coagulants, and mechanisms. J Water Process Eng 40:101820

Doederer W, Gernjak HS, Weinberg MJ (2014) Farré, factors affecting the formation of disinfection by-products during chlorination and chloramination of secondary effluent for the production of high quality recycled water. Water Res 48:218–228

Ebba M, Asaithambi P, Alemayehu E (2022) Development of electrocoagulation process for wastewater treatment: optimization by response surface methodology. Heliyon 8(5):e09383

Eikebrokk B, Juhna T, Melin E, Østerhus SW (2006) Water treatment by enhanced coagulation and ozonation-biofiltration

Eikebrokk B, Vogt RD, Liltved H (2004) NOM increase in northern European source waters: discussion of possible causes and impacts on coagulation/contact filtration processes. Water Sci Technol 4:47

Evans S, Campbell C, Naidenko OV (2020) Analysis of cumulative cancer risk associated with disinfection byproducts in united states drinking water. Int J Environ Res Public Health 17(6):2149. https://doi.org/10.3390/ijerph17062149

Freundlich HMF (1906) Over the adsorption in solution. J Phys Chem 57(385471):1100–1107

Gilca A, Teodosiu C, Fiore S, Musteret CP (2020) Emerging disinfection by-products: a review on their occurrence and control in drinking water treatment processes. Chemosphere 259:127476

Golea A, Upton P, Jarvis G, Moore S, Sutherland SA, Parsons SJ (2017) Judd, THM and HAA formation from NOM in raw and treated surface waters. Water Res 112:226

Gomes J, Matos A, Gmurek M, Quinta-Ferreira RM, Martins RC (2019) Ozone and photocatalytic processes for pathogens removal from water: a review. Catalysts 9(1):46

Gowland DCA, Robertson N, Chatzisymeon E (2021) Photocatalytic oxidation of natural organic matter in water. Water 13(3):288

Hansen S, Willach MG, Antoniou H, Mosbæk H, Albrechtsen HR (2012) Andersen, effect of pH on the formation of disinfection by-products in swimming pool water–is less THM better. Water Res 46:6399

Harman BI, Koseoglu H, Yigit NO, Beyhan M, Kitis M (2010) The use of iron oxide-coated ceramic membranes in removing natural organic matter and phenol from waters. Desalination 261(1–2):27–33

Hong F, Huang F, Wang L, Ding H, Lin Y (2013) Liang, properties of sediment NOM collected from a drinking water reservoir in South China, and its association with THMs and HAAs formation. J Hydrol 467:274

Hu H, Liu S, Jiang F, Shi X, Li M, Xu X (2013) Improvement of the assimilable organic carbon (AOC) analytical method for reclaimed water. Front Environ Sci Eng 7:483–491

Irina L, Márquez JJR, Sillanpää M (2018) Removal of natural organic matter (NOM) from water by ion exchange. Chemosphere 192:90–104. https://doi.org/10.1016/j.chemosphere.2017.10.101

Jiang JQ (2015) The role of coagulation in water treatment. Chem Eng 8:36–44

Jjagwe J, Olupot PW, Menya E, Kalibbala HM (2021) Synthesis and application of granular activated carbon from biomass waste materials for water treatment: a review. J Bioresour Bioprod 6(4):292–322

Khan NA, Khan SU, Ahmed S, Farooqi IH, Dhingra A, Hussain A, Changani F (2019) Applications of nanotechnology in water and wastewater treatment: a review. Asian J Water Environ Pollut 16(4):81–86

Kalajdžić B, Habuda-Stanić M, Željka ROMIĆ, Kuleš M (2013) Removal of natural organic matter from groundwater using Fenton's process. Global NEST J 15(1):13

Karanfil T, Yadav A, Zhang C, Ghosh S, Ahmed S, Jeffcoat SB, Ahmad R (2006) Physico-chemical processes

Katko TS, Hukka JJ (2015) Social and economic importance of water services in the built environment: need for more structured thinking. Procedia Econ Fin 21:217–223

Kennedy M, Kamanyi J, Heijman BGJ, Amy G (2008) Colloidal organic matter fouling of UF membranes: role of NOM composition & size. Desalination 220(1–3):200–213. https://doi.org/10.1016/j.desal.2007.05.025

Kim H-C, Dempsey BA (2013) Membrane fouling due to alginate, SMP, EfOM, humic acid, and NOM. J Membr Sci 428:190–197. https://doi.org/10.1016/j.memsci.2012.11.004

Kim J, Shi W, Yuan Y, Benjamin MM (2007) A serial filtration investigation of membrane fouling by natural organic matter. J Membr Sci 294:115–126

Korotta-Gamage SM, Sathasivan A (2017) A review: potential and challenges of biologically activated carbon to remove natural organic matter in drinking water purification process. Chemosphere 167:120–138

Krasner SW (2009) The formation and control of emerging disinfection by-products of health concern. Philos Trans R Soc A Math Phys Eng Sci 367(1904):4077–4095

Krygowski TM, Szatylowicz H (2015) Aromaticity: what does it mean? Chem Texts 1(3):12. https://doi.org/10.1007/s40828-015-0012-2

Kumar S, Ghosh NC, Kazmi AA (2016) Ballasted sand flocculation for water, wastewater and CSO treatment, vol 5. Environ Technol, Rev, p 57

Lamsal R, Walsh ME, Gagnon GA (2011) Comparison of advanced oxidation processes for the removal of natural organic matter. Water Res 45(10):3263–3269

Langmuir I (1918) The adsorption of gases on plane surfaces of glass, mica and platinum. J Am Chem Soc 40(9):1361–1403

Lapointe M, Papineau I, Peldszus S, Peleato N, Barbeau B (2021) Identifying the best coagulant for simultaneous water treatment objectives: interactions of mononuclear and polynuclear aluminum species with different natural organic matter fractions. J Water Process Eng 40:101829

Lin JL, Nugrayanti MS, Karangan A (2022) Effect of Al hydrates on minimization of disinfection-by-products precursors by coagulation with intensified pre-oxidation towards cyanobacteria-laden water. Sci Total Environ 810:152251

Liu DA (2013) Reckhow, DBP formation in hot and cold water across a simulated distribution system: effect of incubation time, heating time, pH, chlorine dose, and incubation temperature. Environ Sci Technol 47:11584

Loganathan P, Kandasamy J, Jamil S, Ratnaweera H, Vigneswaran S (2022) Ozonation/adsorption hybrid treatment system for improved removal of natural organic matter and organic micropollutants from water–a mini review and future perspectives. Chemosphere 133961:133961

Luukkonen T, Tolonen ET, Runtti H, Pellinen J, Tao H, Rämő J, Lassi U (2014) Removal of total organic carbon (TOC) residue from power plant make-up water by activated carbon. J Water Process Eng 3:46–52

Madhav S, Ahamad A, Singh AK, Kushawaha J, Chauhan JS, Sharma S, Singh P (2020) Water pollutants: sources and impact on the environment and human health. In: Pooja D, Kumar P, Singh P, Patil S (eds) Sensors in water pollutants monitoring: role of material. Advanced Functional Materials and Sensors. Springer, Singapore, pp 43–62

Marc E, Boller M, Benjamin MM (1993) Effect of pre-ozonation on removal of organic matter during water treatment plant operations. Water Sci Technol 27(11):37–45. https://doi.org/10.2166/wst.1993.0262

Matilainen A, Sillanpää M (2010) Removal of natural organic matter from drinking water by advanced oxidation processes. Chemosphere 80:351–365

Matilainen A, Vepsäläinen M, Sillanpää M (2010) Natural organic matter removal by coagulation during drinking water treatment: a review. Adv Colloid Interf Sci 159:189–197

Matilainen A, Gjessing EG, Lahtinen T, Hed L, Bhatnagar A, Sillanpää A (2011) An overview of the methods used in the characterisation of natural organic matter (NOM) in relation to drinking water treatment. Chemosphere 83:1431–1442

Mazhar M, Ahmed S, Husain A, Uddin R (2022) Monitoring of trihalomethanes and its cancer risk assessment in drinking water of Delhi City, India. Pollution 8(3):830–843

Mazhar MA, Khan NA, Ahmed S, Khan AH, Hussain A, Rahisuddin, Changani F, Yousefi M, Ahmadi S, Vambol V (2020) Chlorination disinfection by-products in municipal drinking water–a review. J Clean Prod 273:123159

McCreary JJ, Snoeyink VL (1980) Characterization and activated carbon adsorption of several humic substances. Water Res 14:151–160

Murray CA, Parsons SA (2004) Removal of NOM from drinking water: Fenton's and photo-Fenton's processes. Chemosphere 54(7):1017–1023.59

Murray CA, Parsons SA (2006) Preliminary laboratory investigation of disinfection by-product precursor removal using an advanced oxidation process. Water Environ J 20(3):123–129

Myllykangas T, Nissinen T, Vartiainen T (2000) Bromate formation during ozonation of bromide containing drinking water-a pilot scale study. Ozone: science & engineering, 22, with ozone. Water Res 75:324–335

Newcombe G (1999) Charge vs. porosity—some influences on the adsorption of natural organic matter (NOM) by activated carbon. Water Sci Technol 40(9):191–198

Newcombe G, Morrison J, Hepplewhite C (2002) Simultaneous adsorption of MIB and NOM onto activated carbon: I. Characterization of system and NOM adsorption. Carbon 40:2135–2146

Owusu-Agyeman I, Reinwald M, Jeihanipour A, Schäfer AI (2019) Removal of fluoride and natural organic matter from natural tropical brackish waters by nanofiltration/reverse osmosis with varying water chemistry. Chemosphere 217:47–58

Ødegaard H, Thorsen T, Melin E (2000) Practical experiences from membrane filtration plants for humic substance removal. Water Sci Technol 41:33–41

Pei Y, Yu J, Guo Z, Zhang Y, Yang M, Zhang J, Junji H (2007) Pilot study on pre-ozonation enhanced drinking water treatment process. Ozone Sci Eng 29:317–323

Quinlivan PA, Li L, Knappe DRU (2005) Effects of activated carbon characteristics on the simultaneous adsorption of aqueous organic micropollutants and natural organic matter. Water Res 39(8):1663–1673

Ryan SK, Singer PC (2013) Effect of magnetic ion exchange and ozonation on disinfection by-product formation. Water Research 47(3):1060–1072. https://doi.org/10.1016/j.watres.2012.11.015

Ryan S, Gleeson D, Jordan K, Furey A, O'Sullivan K, O'Brien B (2013) Strategy for the reduction of Trichloromethane residue levels in farm bulk milk. J Dairy Res 80(2):184–189

Sabzehmeidani MM, Mahnaee S, Ghaedi M, Heidari H, Roy VAL (2021) Carbon based materials: a review of adsorbents for inorganic and organic compounds. Material Adv 2(2):598–627

Shammas NK, Pouet M, Grasmick A (2010) Wastewater Treatment by Electrocoagulation–Flotation. In: Wang L (ed) Flotation technology. Springer, New York, NY, USA, pp 99–124

Shan L, Fan H, Guo H, Ji S, Zhang G (2016) Natural organic matter fouling behaviors on superwetting nanofiltration membranes. Water Res 93:121

Shen J, Schäfer AI (2015) Factors affecting fluoride and natural organic matter (NOM) removal from natural waters in Tanzania by nanofiltration/reverse osmosis. Sci Total Environ 527–528:520–529

Sillanpää M, Ncibi MC, Matilainen A (2018) Advanced oxidation processes for the removal of natural organic matter from drinking water sources: a comprehensive review. J Environ Manag 208:56–76

Sinha N, Dahiya P (2022) Removal of emerging contaminants from pharmaceutical wastewater through application of bionanotechnology. In: Development in wastewater treatment research and processes, pp 247–264

Su ZY, Li X, Yang Y, Zhou Z, Song W, Sun S (2014) Treatment of micro-polluted surface water by ballasted flocculation and its floc structural characteristics. Appl Mech. Mater 448-453:1147

Sun XY, Wu QY, Hu HY, Tian J (2009) Effects of operating conditions on THMs and HAAs formation during wastewater chlorination. J Hazard Mater 168:1290–1295

Swietlik J, Raczyk-Stanislawiak U, Bilozor S, Ilecki W, Nawrocki J (2002) Adsorption of natural organic matter oxidized with ClO2 on granular activated carbon. Water Res 36:2328–2336

Szlachta M, Adamski W (2009) Effects of natural organic matter removal by integrated processes: alum coagulation and PAC-adsorption. Water Sci Technol 59:1951–1957

Tien VN, Zhang R, Vigneswaran S, Ngo HH, Kandasamy J (2011) Pram Mathes, removal of organic matter from effluents by magnetic ion exchange (MIEX®). Desalination 276(1–3):96–102. https://doi.org/10.1016/j.desal.2011.03.028

Vasudevan S, Lakshmi J, Sozhan G (2010) Studies relating to removal of arsenate by electrochemical coagulation: optimization, kinetics, coagulant characterization. Separ Sci Tech 45:1313–1325

Watson K, Farré MJ, Knight N (2012) Strategies for the removal of halides from drinking water sources, and their applicability in disinfection by-product minimisation: a critical review. J Environ Manag 110:276–298

Wildhaber Y, Mestankova H, Schärer M, Schirmer K, Salhi E, von Gunten U (2015) Novel test procedure to evaluate the treatability of wastewater with ozone. Water Res 15(75):324–335. https://doi.org/10.1016/j.watres.2015.02.030

Zafisah NS, Ang WL, Mohammad AW, Hilal N, Johnson DJ (2020) Interaction between ballasting agent and flocs in ballasted flocculation for the removal of suspended solids in water. J Water Process Eng 33:101028

Zhai Y, Liu G, van der Meer WGJ (2022) One-step reverse osmosis based on riverbank filtration for future drinking water purification. Engineering 9:27–34

Zhang C, Li X, Zheng J, Zhao G, He T (2017) Zhang, effect of pipe materials on chlorine decay, trihalomethanes formation, and bacterial communities in pilot-scale water distribution systems. Int J Environ Sci Technol 14:85

Zhou Z, Yang Y, Li X, Ji S, Zhang H, Wang S, Zeng Q, Han X (2016) The removal characteristics of natural organic matter in the recycling of drinking water treatment sludge: role of solubilized organics. Ultrason Sonochem 28:259–268

Removal of Disinfection By-Products by Physico-Chemical Treatment Methods

Deeksha Ranjan

Abstract The presence of disinfection by-products (DBPs), such as trihalomethanes (THMs), haloacetic acid (HAA) and N-nitrosodimethylamine (NDMA), in treated drinking water needs to be paid attention. Its prevention and removal are critical for wastewater recycling. The treatment technology should be able to remove DBPs without compromising with the quality of final drinking water. There are various physico-chemical and biological methods to remove DBPs from drinking water. This chapter represents detailed aspects of conventional and emerging physical and chemical methods for the removal of DBPs from drinking water. It was found during the study that out of all the methods available, removal using granular activated carbon, air stripping and membrane filtration are found to be the most effective ones.

Keywords Disinfection by-products · Chlorination · DBPs removal

1 Introduction

Treatment of water is being done since ancient times using various conventional procedures. However, it was only after the ninteenth century, it was discovered and a connection been established that the treated water still might contain certain pathogens causing many life-threatening epidemics like cholera, typhoid, typhus, dysentery, etc. in those times. The need of a final disinfection step then arose, and various techniques were since then discovered to remove pathogens from treated water by annihilating and inactivating them (Namiesnik et al. 2003; Skotnicka-Pitak 2006). Around 1400 pathogenic species have been discovered from surface water, including bacteria and its spores, parasites, viruses and other higher organisms which are dangerous to human and others lives as well. The routine application of disinfection of potable water, as a mandatory step, worldwide, has helped in almost

D. Ranjan (✉)
Department of Applied Sciences and Humanities, Faculty of Engineering and Technology, Rama University, Kanpur, India

S. Madhav et al. (eds.), *Drinking Water Disinfection By-products*,
https://doi.org/10.1007/978-3-031-49047-7_10

eradicating water-borne diseases and the spread of various epidemics (Bond et al. 2012; Weiner 2012; Ram'ırez-Castillo et al. 2015; Dembek and Anderson 2018). Disinfection of water is done for destroying living pathogenic organism and their spores so as to avoid their early development and protect the water distribution system. But during the process of disinfection various by-products are formed which are very hazardous in nature and need to be removed from treated water (Mazhar et al. 2020). The present chapter deals with getting an insight into various physical and chemical methods used for the removal of disinfection by-products from treated water.

Various physical as well as chemical disinfection methods evolved with time are as follows (Solecka 2009; Ishaq et al. 2018):

1.1 Physical Methods for Disinfection

This actually involves separating the solid phase from water with the help of techniques like pasteurization, filtration, microwaves, UV rays, gamma rays, photocatalysis and adsorption.

1.2 Chemical Methods for Disinfection

This involves addition of chemical oxidizing agents to water to be disinfected. The most common is disinfection using chlorination with the help of chlorine gas, sodium hypochlorite solution, solid calcium hypochlorite, chloramines. It also includes use of ozone, potassium permanganate, other halogens like iodine and bromine.

Disinfectants have varying capabilities of inactivating or killing the pathogens. The efficiency depends upon various operating factors like pH, temperature of water, disinfectant dose, contact time, and also on the types and nature of organisms (Wells et al. 2001). An effective disinfection method should destroy not only pathogen but also affect the structure of organic matter present in water to be treated. Each disinfectant method has its own pros and cons. Chlorine is the most commonly and traditionally used disinfectant as it is of low cost and persists in water until not consumed. However, on reacting with the natural organic matter (NOM) present in water, chlorine forms various toxic by-products. Chloramines, on the other hand, form lesser regulated by-products in comparison with chlorine but are associated with formation nitrates and corrosion in the water distribution system (Karthik et al. 2022). Ozone comes as a solution, as it is tasteless and does not form toxic by-products, however, it does not provide long-term protection against pathogens in the distribution system (MDH 2022). UV light is quite effective disinfectant, but it also does not provide residual disinfectant in system during distribution of water, needing a secondary disinfection method. Thus, it becomes necessity to use chemicals as halogens including chlorine and its compounds for the disinfection purpose (MDH 2022; WesTech 2020). Thus, the process of disinfection is associated with few limitations which include its highly diversified disinfection resistance for an individual group of pathogens like bacteria, viruses, etc., the other microbial forms like vegetative and spores and forms which they may take like formation of biofilms or flocculate suspension, etc. Another limitation of using disinfection process is that it activates and reacts with various chemicals like NOM present in water and forms toxic by-products, known as disinfection by-products (DBPs). Here arise the need of regular monitoring, setting regulatory standards and carefully controlling the use of disinfectants and the produced DBPs (Dymaczewski et al. 2019; WesTech 2020).

NOM, which act as precursors of DBPs, consist of a complex mixture of heterogenous chemical fractions of organic compounds having varied chemical composition, molecular weight, charge and polarity. These remain in dissolved and suspended form in water and are produced by the death and decay of plants, animals and microbes. NOM are basically categorized into hydrophilic (like carboxylic acids, amino acids and carbohydrates) and hydrophobic (like humic acid and fulvic acid) compounds (Orecki et al. 2006). Other chemicals which may act as forerunners for the formation of DBPs include industrial chemicals like pharmaceutical and personal care products (PPCPs), dyes pesticides, sweat and urine (Hua et al. 2019; Ghernaout and Elboughdiri 2020a, b).

Above 500 DBPs are reported, however, the nature and properties of few have been accurately known (Weinberg et al. 2002). Most of the DBPs do not biodegrade easily but also show bioaccumulation. As per the Stockholm Convention, 12 of these DBPs have been kept under the category of persistent organic pollutants (POPs) (Durmish et al. 2015; Zainudin et al. 2018). Various disinfectants react with chemicals present in treated water and form a range of DBPs such as trihalomethanes (THMs), haloacetic acids (HAAs), haloaldehydes, haloketones, halogenated hydroxyfuranone derivatives, chlorite, chlorate, bromate, and few emerging DBPs like haloacetonitriles, haloamides, halofuranoes, nitrosamines, halonitomethanes, iodo-acids and iodo-THMs, etc. (WHO 2000; Richardson et al. 2007). DBPs are reported to cause serious health hazards not only in human but also to other forms of life as well. Most of the DBPs are reported to be carcinogenic and mutagenic. Mutations have been reported in bacterial DNA as well as mammalian cell. These also affect the nervous system, reproductive system, liver and kidney. DBPs, specially THMs are reported to cause cancer of bladder, rectal and colorectal. Few DBPs have also shown to be related to reproductive and developmental toxicity like spontaneous miscarriages, stillbirths, low birth weight, etc. (Regulations.gov 2006). The cytotoxic and genotoxic nature of DBPs is found to be related to the total organic bromide and iodide rather than the total organic chloride. Iodinated DBPs are reported to be most toxic, whereas chlorinated DBPs are reported as least toxic. Also, amongst the other groups of DBPs, HANs reportedly cause more toxicity than HAAs which in turn cause more toxicity than THMs (Richardson et al. 2008; Yang et al. 2014; WHO 2015; Bo'zym et al. 2017; Zazouli and Kalankesh 2017).

In order to limit and control the concentration of DBPs and its precursors in treated water, various authorities have given regulations to set the maximum contaminant level of DBPs. The US EPA has set the maximum contaminant level for total THMs of 0.080 mg/L and for five HAAs 0.060 mg/L. On the other hand, according to WHO guidelines, the maximum contaminant level for chloroform is 200 μg/L, for bromoform 100 μg/L, for bromodicholoromethane is 60 μg/L, for dibromochloromethane is 100 μg/L, for dichloroacetonitrile it is 90 μg/L, and for dibromoacetonitrile it is 100 μg/L. For HAAs, maximum contaminant level given by WHO is 50 μg/L for dichloroacetic acid and 100 μg/L for trichloroacetic acid. It is 10 μg/L for chloral (WHO 2006; Gopal et al. 2007; Uyak et al. 2008).

2 Methods of Prevention for DBPs

In drinking water, the DBPs can be prevented and controlled by various methods. First strategy is based on controlling DBPs even before their formation by the removal of precursors with the help of techniques like enhanced coagulation. The organic matter after flocculation and coagulation is removed by filtering. This technique sometimes removes precursors to a level that DBPs concentrations remain below the limit. However, this requires large quantity of coagulant chemicals, increasing the overall processing cost and also disposal issues. Other techniques which remove the precursors are adsorption using activated carbon, ion exchange, nanofiltration and biological treatment of precursors. The second strategy involves the use of disinfection techniques (like ozone and UV rays) that don't involve chemicals. The third strategy is based on using chemicals like chlorine dioxide, which does not form DBPs at all or forms DBPs at very lower concentrations and within the maximum permissible limits. However, these techniques are expensive and do not provide good solution, as they might produce other toxic by-products. As per fourth strategy, DBPs formation can be controlled by limiting the time and concentration of the disinfectants just in order to provide sufficient disinfection killing all the pathogens. Also, it is necessary to frequently turn over the water stored in large tanks to provide good mixing in order to minimize the formation of DBP (Watson et al. 2012; Ghernaout and Boucherit 2015; Kristiana et al. 2017).

The above discussed methods and techniques are helpful in controlling DBPs before their formation and distribution in water system. However, it is sometimes better to remove the DBPs after their formation with the help of various methods.

3 Methods for Removing DBPs from Treated Water

Removing DBPs after their formation is the most adopted strategy and it provides a good solution to prevent the health hazards associated with it. Various physical, chemical and biological methods are used in order to remove the DBPs from treated supply water. The various physical and chemical methods are discussed in detail in the following sections.

3.1 *Removal of DBPs Using Membrane Filtration*

Water and wastewater treatment using membrane filtration is one of the oldest techniques that is being used even now with many modernizations. Jean Antoine Nollet first observed this technique in 1748 and used it in water and wastewater treatment plants (Olszak 2020). It has also been shown to remove various DBPs specially THMs from water. The technique is based on separating particles and molecules via the size exclusion and diffusion mechanism, where the membrane behaves as a thin layer barrier that selectively or preferentially transports the mixture. Separation of the components takes place by purely physical action, and the separated parts are not chemically or biologically modified. It has been noted to show much greater quality of permeate than the regulatory requirement for drinking water consumption (Waniek et al. 2002; Ghernaout and Elboughdiri 2020a, b). During membrane filtration, water stream that is supplied to surface of the membrane from inlet side is referred to as feed. While water stream that passes out through the membrane from outlet side is permeate, the residue which is retained on membrane is referred to as retentate. Permeate is the product of membrane filtration and retentate is the waste. The membranes can be synthesized or natural and depending upon the material used to prepare the membrane, it can be organic or inorganic. Natural membrane includes rubber, wool, cellulose, while few examples of synthetic membranes are polytetrafluoroethylene (Teflon PTFE), polyamide-imide (PAI) and polyvinylidenedifluoride (PVDF). Organic membranes make a thin layer in between the supporting layers of load bearing fabric or nonwoven fabric. Organic membrane includes polysulfone, polyethersulfone, polyimide, polymethylpentene, polyetherimide, polydimethylsiloxane, cellulose acetate, polycarbonate, polyphenyleneoxide etc. Various inorganic membrane examples are for example amorphous silica, zeolites, nanoporous carbon, carbon molecular sieves, mixed conducting perovskites, palladium alloys, etc. (Mohshim et al. 2013). Organic membranes hold largest share of market of membrane filtration and are associated with advantages of being cost effective, good selectivity and easy processing. The main disadvantages, however, are its fouling, non-resistance towards chemicals, limited operating pH, pressure and temperature and short lifetime (Kochkodan and Hilal 2015; Visser and Wessling 2007). Inorganic membranes, on the other hand, hold small market share. However, it has drawn global attention as they have been reported to show higher permeability as well as selectivity. This phenomenon was introduced by Robeson and is known as 'upper bound trade-off curve' (Robeson et al. 2015).

The membranes are used for various purposes on the basis of their characteristics. Various techniques which are membrane based are microfiltration (MF), ultrafiltration (UF), nanofiltration (NF) and reverse osmosis (RO). Selection of a membrane technique is done on various factors like feed solution composition, operating variables, application and goals of separation. For example, if the contaminant is of particle size less than 0.2 micron and it can be precipitated, dissolved and coagulated, then MF or UF can be used. If dissolved organic and inorganic removal is needed, then RO technique can be used. If the ions to be removed are multivalent, then NF technique can be used (Zazouli and Kalankesh 2017).

3.1.1 Microfiltration and Ultrafiltration

The pore size of MF membrane ranges from 0.1 to 10 μm and is made of organic polymers or inorganic ceramics and metals, etc. A very low pressure of around 1–2 bar is used for the separation purpose (Zazouli and Kalankesh 2017). Pore size of UF membranes is from 1 to 100 nm and utilizes low transmembrane pressure of around 1–5 bar. Both MF and UF can be used as stand-alone method or as pre-treatment method before NF or RO process for turbidity removal. Because of the larger pore size MF is not able to remove NOM from water. It has also been noted that UF effectively removes colloids, turbidity, algae, fungi and precipitates but is not very effective in separating THM and HAA precursor humic acid. Rather NOM causes membrane blockage which can be lessened by filtering NOM before feeding through coagulation-flocculation or adsorbing on activated carbon (Zazouli and Kalankesh 2017). UF and NF have been tested for removal of dissolved oxygen carbon (DOC) from water and it was noted that high molecular weight DOC were rejected successfully more than 90% while low molecular weight DOC were only removed 15 to 30% with the help of two UF and two NF (Ates et al. 2009).

3.1.2 Nanofiltration

It is classified under pressure-driven membrane separation process, features of which lie between ultrafiltration and reverse osmosis. NF membrane has pore diameter in range of 1 nm and a transmembrane pressure of about 5–8 bar is applied. Mostly polyamide thin film composite membranes in spiral configurations are utilized for water treatment. NF is effective in removing water hardness, colour, viruses, and also organic compounds above molecular weight 200 g/mol which include THM, HAA, VOC, etc. It has been reported that DBPs along with their precursors NOM can be effectively removed using NF. It is advantageous in being an easily operated, low-cost technique with improved quality of water satisfactory to consumers (Bodzek and Konieczny 2005). Mechanism of rejection and thus separation of solute ions through NF involves charge interaction, size exclusion (via hydrated ions) and also dielectric interaction. The charge and hydration mechanism dominates amongst all the mechanisms. Co-ions have the same charge as on membrane ions while counter ions are the ones having different charges. The co-ions having low charge and mono-valency may increase the rejection while the counter ions may decrease rejection if they have multiple valency (Gao et al. 2016; Thabo et al. 2021). Studies have been done to remove DBPs with the help of NF (Itoh et al. 2001; Samadi et al. 2006; Uyak et al. 2008; Chalatip et al. 2009; De Munari and Schäfer 2010; Bodzek 2013; Yang et al. 2017a, b). NF membranes NF200 and DS5 have been utilized to separate THMs. NF membrane NF200 showed high removal efficiency for THM than DS5 membrane. It also showed better efficiency for dibromochloromethane (DBCM) (Uyak et al. 2008). A report showed removal efficiency up to 99% for THMs (Itoh et al. 2001). Removal of bromate was also studied employing NF membrane and it was found to be 75–100% efficient (Bodzek 2013). A study carried out to remove five HAAs from water using compact, negatively charged NF membrane showed good removal efficiency even at low pressure for all five HAAs as compared to negatively and neutrally charged open membranes (Chalatip et al. 2009). NF membranes have also been tested for the filtration of DBP precursor of THMs, halonitromethane, N-nitrosodimethylamine (NDMA), which is formed by reaction of DMA and monochloramine (Miyashita et al. 2009; Ersan et al. 2016). NF-270 and NF-90 membranes along with alum and ferrous coagulants have been utilized for bromide and bromate removal, and the techniques NF, coagulation and hybrid NF-coagulation showed better efficiency in removing bromate than bromide (Mazloomi et al. 2009).

NF is a compact, modular and scalable technique which is widely used on large scale and requires no chemicals and additives. There is no sludge generation and the monovalent and multivalent ions are selectively removed. The main limitations are relatively higher cost, low removal efficiency for monovalent ions, requirement of a pre-treatment technique, extra cost required for the concentrate treatment and membrane fouling.

3.1.3 Reverse Osmosis

This technique employs semipermeable membrane of pore size around 10^{-4} μm. High hydrostatic pressure of around 20–70 bar is applied at the feed side which allows water to forcibly enter the permeate side, while the impurities are left behind at the feed side in the concentrated form. If required, filtration is done before RO which prevents the semipermeable membrane from clogging and thus fouling. The applied hydrostatic pressure is greater than osmotic pressure and is able to revert the phenomenon of natural osmotic diffusion (Azmi et al. 2018).

$$\Pi = \Delta C^{*} R^{*} T \tag{1}$$

where Π = osmotic pressure (Pa), ΔC = concentration difference in mol/m^3, R = ideal gas constant (8.314 joule/mol/K), T = temperature in Kelvin.

The above equation shows that for same concentration difference, higher osmotic pressure needs to be applied for smaller molecules (Azmi et al. 2018). RO has been applicable in simultaneously filtering organic as well as inorganic DBP precursors. RO provides a good alternative to remove various impurities and DBPs from water specially at lower concentrations. According to some reports RO is the most effective technique in removing bromides and iodides from water. RO membranes have been employed and studies have been carried out for removing THMs, HAAs, NDMA and bromides from water and good removal efficiencies have been obtained (Miyashita 2007; Plumlee et al. 2008; Mazloomi et al. 2009; Valero and Arbós 2010; Rajamohan et al. 2014; Yang et al. 2015, 2017a, b). RO has also been employed in the treatment of bromate from water and a removal efficiency above 97% has been observed (Bodzek 2013). A new technique Flexible RO (FLERO) which is deployed in removing of twenty DBPs and the result showed that around 80.8% removal was achieved for most of the DBPs (Chen et al. 2021).

RO is advantageous in having compact equipment. However, the requirement of high energy, associated high operational and capital cost, and fouling of membrane are few of its drawbacks (Azmi et al. 2018).

3.1.4 Ceramic Membranes

Ceramic-based membranes have attracted attention now-a-days over the polymeric membranes. They show excellent resistance towards acids/alkalis, oxidation, solvents, mechanical abrasion and fouling and have high thermal stability. Membranes have narrow distribution of pore size, maintain constant flux producing high permeate even at low pressure, have extremely long working period and have proven life span of up to 20 years and more. These can be cleaned easily with aggressive chemicals and high-pressure short back-wash, thus reducing loss of water (Abeynayaka et al. 2012; AMTA 2018). Ceramic membranes are composed of support, made up of porous material like alpha alumina or silicon carbide, which is coated with a layer of inorganic porous membrane of aluminium oxide, zirconium oxide, titanium oxide, nitrides or carbides of similar metals, etc. Ceramic membranes are available in flat sheet, monolithic, hollow fibre and multichannel tubular elements. Typically, flat sheet membranes have outside coating while tubular membranes having flow channels are coated from inside. The combination of support and coated membrane layer is referred to as 'membrane element' or 'ceramic element' (AMTA 2018). Size of tubular types may be 5 ft. long and diameters range from 20 to 200 nm. Surface area of flat types may be around 6 ft^2 per sheet. Ceramic membranes had restricted implementations because of associated high cost, nevertheless, the production of better water quality and its global life cost have encouraged their employment in removal of various water impurities including NOMs as well as DBPs. A report (Moslemi et al. 2012) shows the impact of pH and $CaCl_2$ addition on removal of bromate and bromide using ceramic membrane. Removal efficiencies were 68% and 63% for bromate and bromide, respectively, at pH 8.

3.2 *Removal of DBPs Using Air Stripping*

Air stripping involves immediate contact of wastewater and air allowing the undesirable volatile substances present in liquid phase, i.e., wastewater to be carried away by air. Term air stripping encompasses processes like mechanical surface aeration, diffused aeration, spray fountains, spray or tray towers and counter-current packed towers. In air stripping, conditions are such that a large surface area of liquid phase is exposed to the gaseous phase, promoting transferring volatile impurities from liquid to gaseous phase. Further the treatment of volatile impurities containing air is done with the help of systems like active carbon installation and bio-filter (Srinivasan et al. 2009). The air stripping devices are of two types: packed tower and sieve tray. Low profile sieve tray systems based on cross-flow principle, in which perforated plates are used to intensively aerate the flow of liquid. Sieve trays are less common (EPA 2012).

3.2.1 Packed Column Air Strippers

These comprise of vertical column filled with packing material and work on the principle of counter-flow. Water is sprayed from the top over the porous packing in the column and air is blown from the bottom side. As the volatile impurities transfer from water into air, treated water leaves the column from the bottom (Fig. 1). Columns are reported to show good removal efficiencies even at low pressures, is cost effective, can operate at wide range of air flow rates, requires lower ratio of air to water. On the other hand, it requires cleaning of column bed by removing and washing with acid, fouling of the bed occurs, sometimes to achieve desired higher efficiencies taller columns are required which increases the overall cost (Srinivasan et al. 2009; EPA 2012).

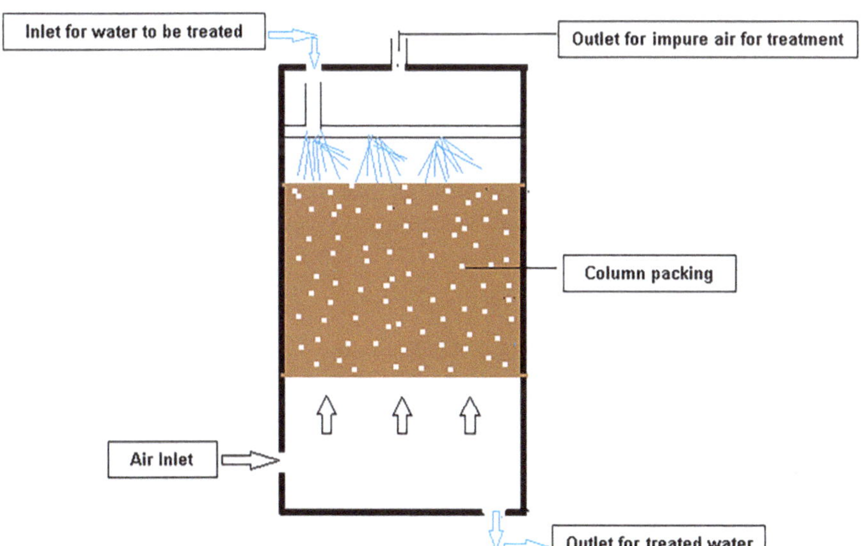

Fig. 1 Packed Column Air Strippers

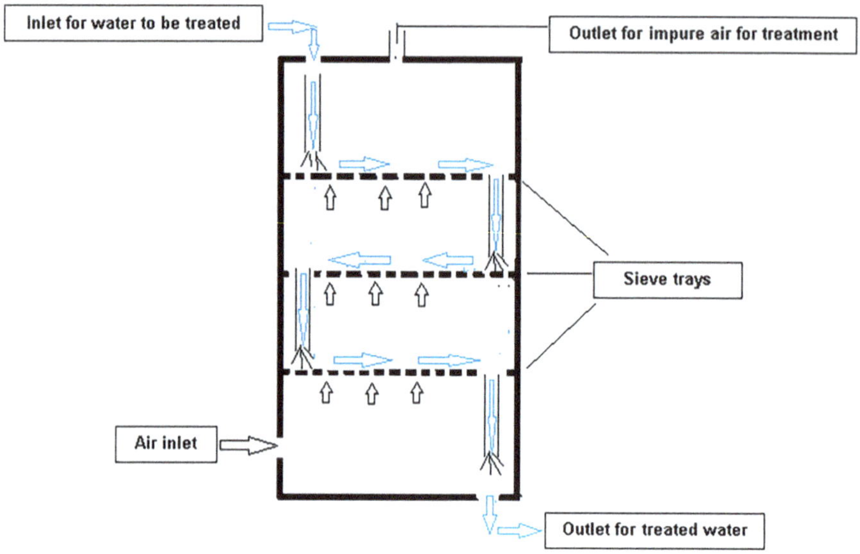

Fig. 2 Sieve-Tray Air Strippers

3.2.2 Sieve-Tray Air Strippers

These have multiple perforated trays enclosed in stacks and water trickles down from top over these trays progressing from tray by tray. Air is bubbled through these trays and volatile impurities are transferred from water to air. Treated water then exists out after passing through last tray (Fig. 2). Trays can be sieve, valve and bubble cap. Sieve trays are able to achieve around 100% efficiency, trays can be easily taken out and cleaned thus preventing fouling, in order to achieve higher treatment efficiencies additional trays can be added, has more compact design for smaller areas. However, the trays need to be placed specifically and in geometry which is designed by manufacturer requires narrow range of airflow rates for volatizing without interfering the flow of water (Srinivasan et al. 2009; EPA 2012).

3.2.3 Diffused-Aeration Tanks for Air Stripping

Here the air directly moves through the water as it is bubbled into the water aeration tank. Water to be treated is pooled into the tank from the top inlet. Air is then bubbled from bottom air inlet and is allowed to diffuse through water, taking the volatile impurities with it and leaving the tank from top (Fig. 3). The tank should be deep and smaller air bubbles should be passed through so that the removal is more efficient.

Air strippers work on the principle of Henry's law. The equation is shown as follows:

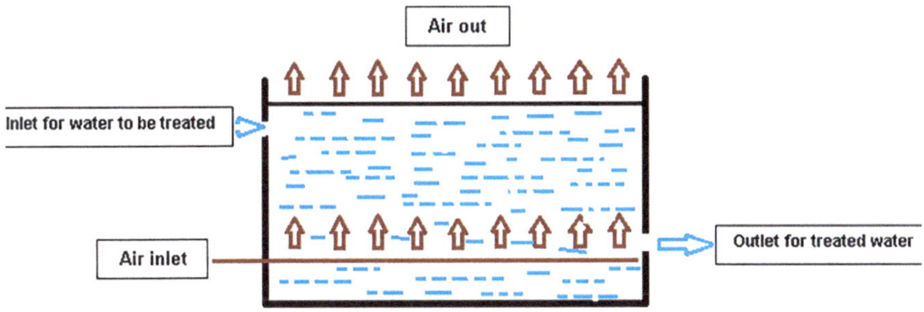

Fig. 3 Diffused-aeration air strippers

$$H = C_G / C_L \tag{2}$$

where H = Henry's constant; C_L = contaminant at equilibrium in liquid phase; C_G = contaminant at equilibrium in gaseous phase. Henry's law determines the tendency of transfer of a compound from liquid to gaseous phase. With the increase in Henry's constant the volatility also increases (Srinivasan, 2009; Fischer 2019).

Air stripping has been employed for the removal of various volatile impurities like benzene, toluene, xylenes, ethylbenzene, styrene and chlorinated hydrocarbons, etc. It has also shown good removal efficiency for DBPs (Fischer 2019). In a study (de Zamarron 2005) induced draft-stripping column has shown above 90% removal efficiency in removing few THMs at air-to-water ratios of around 60. It was also noticed that chloroform was most easily removed while bromoform was most difficultly removed. Another study (Wu and Wu 2009) employed packed column for THMs removal and an efficiency of above 90% has been reported at 24 °C temperature, hydraulic load 20 $m^3/(m^2 \cdot h)$ and air-to-water ratio above 60. Dichloromethanes showed higher removal as compared to tribromomethanes. Ultrasonic along with air stripping method were used to remove carbon tetrachloride and 1,1,1-trichloroethane using batch mode (Ayyildiz et al. 2005). A software tool was used to design the air stripping columns for calculating the removal of THMs (Canosa and Martí 2018). Removal of chloroform was done using this technique (Samadi et al. 2004).

3.3 Removal of DBPs Using Tank Mixing and Spray Aeration

Water storage tank mixing is a proven technique for treatment of DBPs and also maintaining the consistent disinfectant residual. Mixing helps in reducing water age by circulation and homogenization and in minimization of thermal and chemical stratification which promotes formation of DBPs. DBPs can be removed more effectively by spraying water through nozzles into the water tank. The principle is similar to that of air stripping, where the volatile impurities are transferred to air from water when they come in contact during mixing and spraying. The air can be blown through vapour space of the tank to enhance the process. This technique has been found to be less effective than the air stripping, however, it has been proven to provide enough removal of DBPs in order to meet the regulations (Dawe 2009; Kramer et al. 2014; IXOM Watercare 2016; Alicia 2019).

There are many tanks mixing system, the two most common are active tank mixing and passive tank mixing. Active mixing is the fluid disturbance through continuous stirring of water from top or bottom at any time and for any length of period using powered mixing apparatuses. Mixing at bottom promotes minimizing stratification in the tank, while mixing at the top of water surface introduces oxygen into water thus promoting the DBPs to volatilize and transferred to the air from water. The air containing volatile DBPs is vented out from tanks and then further treated. Passive mixing also involves fluid disturbance using external source of energy but it is dependent on external conditions and it cannot be applied at any time and for any period of time. Passive mixing inlets and outlets of tank are mechanically configured such as by installation of the duckbill mixing manifold at the bottom of the tank which forces inside water into jets promoting its mixing (Dawe 2009; IXOM Watercare 2016).

3.4 Removal of DBPs Using Advanced Oxidation Processes (AOPs)

Oxidation is the process of transferring electrons from reducing agent to the oxidizing agent, resulting in the chemically transforming both the species. Sometimes, during this transfer a highly unstable free radical species with unpaired electrons are formed. These are extremely reactive in order to get their electrons paired. This free radical then keeps on oxidizing other organic as well as inorganic species till a thermodynamically stable product is produced (Deng and Zhao 2015). Advanced oxidation process (AOP) destructs organic and inorganic contaminants present in water and wastewater into biodegradable and less toxic and even non-toxic substances with the help of ozone, hydrogen peroxide, UV radiation or combination of these. AOP was first proposed in 1980 for treating potable water and then was employed in treating various wastewaters. AOP relies on the formation of short-lived free radical intermediates like hydroxyl radicals ($OH^{\bullet}$) or sulphate radicals ($SO_4^{\bullet-}$) which help in removal of volatile and traceable organic contaminants and certain inorganic contaminants. This technique can also be used for pre-treatment prior to biological treatment (Deng and Zhao 2015).

Hydroxyl radicals are produced during application only in situ as they have a very short life span. They are produced through various methods involving combination of oxidants like H_2O_2 and O_3, UV light or ultrasound rays and catalysts like Fe^{2+}. Depending upon the method employed AOPs can be photochemical or non-photochemical. In photochemical method hydroxyl radicals are produced from UV radiations and O_3, H_2O_2 and/or Fe^{2+}, while, during non-photochemical method formation of hydroxyl radicals occurs in unclear light via either ozonation or Fenton process. Photochemical method employing UV and H_2O_2 is the most commonly adopted method for treating drinking water (Ghernaout 2019). Hydroxyl radicals are quite non-selective and react having rate constant of order 10^8 to 10^{10} M^{-1} s^{-1}. The four basic steps of reaction of hydroxyl radical with organic contaminants involve radical addition and combination, hydrogen abstraction, electron transfer. The neutral organic molecules, on addition of hydroxyl radicals, are converted to carbon centred free radical like $ROO^{\bullet}$, $R^{\bullet}$ or $R^{\bullet}$–OH, which further reacts to form H_2O_2 and super oxide $O_2^{\bullet-}$. These more reactive species chemically degrade organic compounds sometimes to the level of mineralization. Hydroxyl radical is generated in many ways (Gottschalk et al. 2009):

(i) With O_3, hydroxyl radicals are produced directly when ozone reacts preferably with ionic or dissociated form rather than neutral organic species.

$$3O_3 + H_2O \rightarrow 2OH^{''} + 4O_2 \tag{3}$$

The hydroxyl radical formation is improved by irradiation or other oxidizing agents like H_2O_2. In case of peroxone (O_3/H_2O_2) system, hydroxyl radical is produced as follows:

$$H_2O_2 \rightarrow HO_2^{-} + H^{+} \tag{4}$$

$$HO_2^{-} + O_3 \rightarrow OH^{''} + O_2^{-} + O_2 \tag{5}$$

When there is UV irradiation along with O_3, H_2O_2 is first generated by photolysis of O_3 and then hydroxyl radical is formed:

$$O_3 + H_2O + hv \rightarrow H_2O_2 + O_2 \tag{6}$$

$$H_2O_2 + hv \rightarrow 2OH^{''} \tag{7}$$

(ii) Generation of hydroxyl radicals also occurs by photons in presence of some oxidizing agent or catalyst. TiO_2 or a semiconductor of RO-type can act as catalyst in this reaction (Tang 2003).

$$TiO_2 + hv \rightarrow e^{-}_{cb} + hv^{+}_{vb} \tag{8}$$

In the valence band (vb), positive holes with oxidative tendency are formed and negative electrons having reductive tendency are generated in the conduction band (cb). Hydroxyl radicals are further generated from these holes or electrons when OH^-, H_2O and $O_2^{\bullet-}$ comes on TiO_2 surface as follows:

$$hv^{+}_{vb} + OH^{-}_{(surface)} \rightarrow OH^{''} \tag{9}$$

$$hv^{+}_{vb} + H_2O_{(absorbed)} \rightarrow OH^{''} + H^{+} \tag{10}$$

$$e^{-}_{cb} + O_{2(absorbed)} \rightarrow O_2^{''-} \tag{11}$$

More hydroxyl radicals are formed by H_2O_2 or O_3 along with UV irradiation, as shown below:

$$H_2O_2 + hv \rightarrow 2OH^{''} \tag{12}$$

Also, photolysis of H_2O at wavelength less than 242 nm results in the formation of hydroxy radical:

$$H_2O + hv \rightarrow OH^{''} + H^{''} \tag{13}$$

(iii) Hydroxyl radicals are generated through Fenton's process also where metal is required to activate H_2O_2 and iron is used most commonly (Pignatello et al. 2006).

$$Fe^{2+} + H_2O_2 \rightarrow Fe^{3+} + OH^{''} + OH^{-} \tag{14}$$

$$Fe^{3+} + H_2O_2 \rightarrow Fe^{2+} + HO^{''}_2 + H^{+} \tag{15}$$

$$OH^{''} + H_2O_2 \rightarrow HO^{''}_2 + H_2O \tag{16}$$

$$OH^{''} + Fe^{2+} \rightarrow Fe^{3+} + OH^{-} \tag{17}$$

$$Fe^{3+} + HO^{''}_2 \rightarrow Fe^{2+} + O_2H^{+} \tag{18}$$

$$Fe^{2+} + HO^{''}_2 + H^{+} \rightarrow Fe^{3+} + H_2O_2 \tag{19}$$

$$2HO^{''}_2 \rightarrow H_2O_2 + O_2 \tag{20}$$

Fe^{3+} formed generates sludge during water/wastewater treatment which needs separate disposal increasing the cost of operation. Also, during Fenton process, at acidic pH, effective formation of $OH^{\bullet}$ occurs, thus, restricting practical application of Fenton process in treatment of water/wastewater.

(iv) Other AOPs that have been employed in water treatment are irradiation using ultrasound and electron beam. During ultrasound irradiation under extreme conditions of around 4200–5000 K temperature and 200–500 atm pressure, the molecules of water in gaseous form are split inside microbubble generating hydroxyl radical (Pignatello et al. 2006):

$$H_2O \rightarrow OH^{''} + H^{''} \tag{21}$$

Similarly, irradiation using electron beam results in formation of free radicals through fragmentation of water:

$$H_2O + e^{-} \rightarrow 2.7\,OH^{''} + 2.7\,H_3O^{+} + 2.6e^{-} + 0.7\,H_2O_2 + 0.6\,H^{''} + 0.45\,H_2 \tag{22}$$

(v) Sulphate radical-based AOPs involves the formation of sulphate radicals from oxidant $S_2O_8^{2-}$ in the presence of heat, ultraviolet radiation, transition metal, or high pH as follows (Kolthoff and Miller 1951; House 1962):

$$S_2O_2^{-8} \rightarrow 2SO_4^{"-} \left(\text{in presence of heat / UV radiation}\right) \quad (23)$$

$$S_2O_2^{-8} + Mn^{+} \rightarrow SO_4^{"-} + SO_4^{2-} + Mn^{+1} \quad (24)$$

During degradation of various hydrocarbons, the $OH^{\bullet}$ breaks C=C or abstracts H from C-H bond, while $SO_4^{\bullet-}$ has the tendency to remove the electrons from organic compounds resulting in formation of carbon-based radical cations (Forsey 2004).

AOP has been employed for removing THMs (Hansen et al. 2013) and it is reported that the total amount of THM was reduced to low concentrations, i.e., from 21.91 ppb to 1.8 ppb with the removal efficiency of 91.78%. Other studies employed UV of intermediate pressure for treating DBPs present in swimming pool and UV and vacuum UV irradiation for THMs and chloramines removal (Zhang 2018). One more study (Buchanan et al. 2006) reported removal of THMFP and $CHCl_3$, $CHBrCl_2$ and $CHClBr_2$ using VUV and UVC at a dose >40 J cm^{-2} for UVC. Another study employed Ozone/UV along with biological activated carbon (BAC) filter for treating HAAs, THMs and total organic carbon (TOC) (Trang et al. 2014). For removal of THMs solar heterogeneous photocatalysis has also been employed (Pérez-Lucas et al. 2022).

3.5 *Removal of DBPs Using Heating/Boiling*

To ensure providing safe drinking water, household water treatment (HWT) becomes essential. For this purpose, heating/boiling has been found to be very successful method. Heating/boiling has also been found effective in removing and reducing various DBPs from drinking water at household level. Traditionally even before tea preparation water is boiled. This gradually has attracted attention towards treatment of DBPs (Liu et al. 2020).

The mechanisms during treatment of DBPs via heating/boiling are volatilization and thermal accelerated transformation of DBPs. At higher temperatures the DBPs show varying degree of volatility, which depends directly on the low or high boiling point of the DBPs itself. Hydrolysis and decarboxylation, on the other hand, implies that the DBPs have been thermally transformed (Wu et al. 2001; Ebrahim et al. 2016).

A report (Wang et al. 2023) shows that boiling water for a minute there is reduction in chloroform concentration of about 74.8% and after boiling for 3 min 84.8% and boiling for 5 min resulted in 100% removal. Many studies have been conducted which report that DBPs have been reduced and removed from drinking water via volatilization and thermal transformation (Wu et al. 2001; Ebrahim et al. 2016; Zhang et al. 2020; Sinha et al. 2021; Serajuddin et al. 2022; Li et al. 2021, 2023; Wang et al. 2023).

It is also, however, reported that thermal transformation also results in cleaving the larger halogenated DBPs to smaller halogenated molecules like THMs and HAAs and also dechlorination of organic compounds which must be avoided and taken care of. THMs are volatile and thus removed during boiling. HAAs are non-volatile but are broken down via chemical reaction at higher temperatures (Wu et al. 2001).

3.6 Removal of DBPs Using Adsorption

Adsorption is a promising method for separating and thus treating various contaminants from water and wastewater because it has simple design, requirement of less initial as well as operational investment (Crini 2005). It involves accumulation of liquid or gaseous adsorbate onto the surface of adsorbent (Kumar et al. 2014). Adsorption mechanism is classified as physical (physisorption) or chemical (chemisorption). Physisorption is non-specific, reversible, multilayer process and is governed by weak van der Waals forces. While chemisorption is specific, irreversible, monolayer process and involves covalent binding or electrostatic attraction of adsorbate onto the surface of adsorbent, adsorption is affected by process variables like adsorbate concentration, adsorbent dose, pH, temperature, etc. (Nik-Abdul-Ghani et al. 2021).

With the progression of process of adsorption, an equilibrium of solute (adsorbate) is established between solution and adsorbent. The uptake capacity of adsorbent at equilibrium (q_e, mg g^{-1}) is given by as follows:

$$q_e = (C_o - C_e) x V / M \quad (25)$$

where V is the volume of solution in L; M is the adsorbent mass in g; C_0 and C_e are the adsorbate concentrations initially and at equilibrium in mgL^{-1}, respectively.

For the purpose of adsorption activated carbon has always been a preferred choice. It is produced when highly carbonaceous material is activated via physical and chemical methods thus, increasing the pores and surface area. Conventionally, the use of activated carbon is done in powdered and granular forms. Its use in adsorption is associated with advantages like its stable chemical properties, large surface area, multiple pore sizes and structures along with efficient adsorption for a range of pollutants. It can be recycled and reused. Its use has been done in removal of tough pollutants which are hard to remove to a level of discharge standard by other treatment methods (Tchobanoglous et al. 2003). The cost efficiency of adsorption depends greatly on the accessibility and cost of the carbonaceous precursor material like wood, coal, etc. for the synthesis of activated carbon. Thus, to reduce overall process cost and keeping in mind the environmental obstacles, efforts are done to produce activated carbon using inexpensive carbonaceous material such as agricultural wastes (Isoda et al. 2014; Canales-Flores and Prieto-García 2020; Ali et al. 2018).

In removing DBPs and its precursors, granular activated carbon (GAC) is being used effectively as it is macroporous and has rough surface along with widely distributed fissures and ridges, which provide excellent removal sites (Moreno-Castilla 2004). GACs are well capable of removing compounds of lower and intermediate molecular weight and not higher molecular weight compounds because of sieving effect. GAC filters are used in removing contaminant organic compounds with intermediate molecular weights (Xing et al. 2008). A rapid sand filter has been used with GAC filter and studied for the removal of DOCs and DBPs. GAC filter was found to be more efficient than sand filters (Kim and Kang 2008). Modified polysaccharide adsorbents containing β-cylodextrin or chitosan were used for removing THMs (Guo et al. 2021). Activated carbon fibres have also been used for removing THMs. It showed better adsorption for THM with bromine than THM with chlorine (Nakamura et al. 2001). In another study GAC and modified GAC have been used to study the removal of $CHCl_3$, $CHBrCl_2$, $CHBr_2Cl$ and $CHBr_3$ from tap water (He et al. 2017). Removal of brominated HAAs was done using combined GAC filtration and biofiltration and average removal was reported to be 89.5% (Vines and Terry 2023). HAA adsorptive removal using GAC is also reported in Tung et al. (2006); Ghomshe et al. (2011)); Saengchut et al. (2020a, b). Performance of GAC postfilter-adsorber was tested in removing THMs, HAAs and DOCs (Babi et al. 2007; Tung and Xie 2008). Removal of DBPs trichloromethane (TCM) and trichloroacetic acid (TCA) has been studied in batch mode at different pH, initial concentration, adsorbent dose and temperature. Activated carbon was also tested for TCM and TCA removal from 8 tap water samples and it showed removal efficiency of about 98.1%–99.37% (Hasan and Makki 2021). GAC has also been deployed in the removal of intermediate aromatic halogenated DBPs from water in rapid small-scale columns (Jiang et al. 2017).

Colonization of GAC to a high degree by bacteria forms biological activated carbon (BAC) as activated carbon has rough surface with large surface area. That means GAC having bioactive surface and removes organic contaminants including DOCs and DBPs using process of biodegradation are BAC (Nishijima and Speitel 2004). Various studies have been conducted where BAC is employed in treating various DBPs (Zhou and Xie 2002; Xie and Zhou 2002; Tung et al. 2009; Tseng et al. 2012; Lou et al. 2014a, b, 2016; Tang and Xie 2016; Wang et al. 2019). Mechanisms involved in removing organic compounds present in water using BAC are adsorption and biodegradation. The fresh activated carbon utilizes adsorptive path for removal, however, with increasing volume of treated water, the adsorbent becomes exhausted and its adsorptive capacity declines. With this the bacteria starts growing and accumulating on activated carbon surface. Contaminants adsorbed on activated carbon macropores are retained long enough so that they are degraded by attached bacteria and biodegradation becomes the main mechanism for removal of organic compounds (Simpson 2008).

3.7 *Removal of DBPs Using Sand Filters*

Filtration involves passing the water contaminants through a porous material. Since ancient times filters like sand, gravel and charcoal are in use for the separation and removal of various water contaminants, the traditional sand filters are basically slow sand filters and rapid sand filters. Recently another type is in use, i.e., up-flow sand filters. For the efficient functioning of rapid and up-flow sand filters, chemicals are required for flocculation, while for slow sand filters chemicals are not required, instead it uses microbes as its top layer (Crini and Lichtfouse 2019).

3.7.1 Slow Sand Filters

Slow sand filtration, one of the oldest methods, has been in use for over 150 years for the treatment of water because of its simplicity and ease of maintenance. It is suitable to be used in small and remote water systems. It is a biological process as the top layer of filter comprises of microbes, which is employed in treating water. This microbial community layer or biofilm, referred to as 'Schumtzdecke', develops on top of sand filters over a sufficient period of time and effectively removes organic contaminants by feeding over them and digesting them. Prior to slow sand filtration sedimentation is required as pre-treatment. It is comprised of sand bed of thickness of 80 to 100 cm with effective size of 0.25 to 0.35 mm. Filter is supported with a gravel base of 30 to 75 mm depth. The installation is costly; however, operation and maintenance are cheaper. Filter requires cleaning in about 3 to 4 months which is a tedious job as it requires scrapping of Schumtzdecke as well as the sand layer (Crini and Lichtfouse 2019).

3.7.2 Rapid Sand Filters

Comparatively, the rapid sand filters are newer than slow sand filters. It is a physical process and is employed for treating the suspended water contaminants. It shows high rate of filtration between 4800 to 7200 l/h/sq.m and requires pre-treatment of coagulation, flocculation and sedimentation. Gravel base of 45 to 50 cm depth is used in four to five layers. The filter sand is 3 to 50 mm in effective size with the sand bed thickness of 60–75 cm. These filters are cheap to install but costlier to maintain and operate. Cleaning is simple and is required at every one to two days and is done by back washing with or without the use of compressed air agitation (Crini and Lichtfouse 2019).

In comparison with rapid sand filters, the slow sand filters are much in use for the treatment of water contaminants. It has been reported, however, that the filters alone are not very effective in removal of contaminants but have shown better removal efficiency when used in combination with other treatment methods (Mitchell and Ullman 2016). In removing various DBPs, the slow sand filter is used in combination with other methods specially filtration. A GAC filter is sandwiched between two layers of slow sand filters and used in removal of TOC and DBPs (Slow Sand Filtration n.d.). A study compared the use of sand filtration and ultrafiltration and sand filtration outperformed in reduction of formation of chlorites than ultrafiltration (Xu et al. 2019). A study reported biodegradation of HAA in traditional rapid sand filter practising pre-chlorination (Chuang et al. 2011). The control efficiency of various DBPs was evaluated employing different treatment methods like pre-ozonation, coagulation-sedimentation, O_3-BAC and sand filtration (Lin et al. 2020). A study investigated treatment of DBP precursors of type CX_3R from rainwater using conventional sand filtration and GAC filtration which outperformed (He et al. 2020). An investigation aimed at studying the removal of THM using GAC and sand as dual filter media having different depths and it showed 87% removal of THM (Mohamed et al. 2006). Slow sand filter was investigated for removing bacteria, *Giardia* and THMs (Pyper 1985). They are also utilized for the simultaneous formation and biodegradation of HAAs under different chlorine doses (Hsieh et al. 2014). A study also reports that while HAAs are reduced in less quantity using coagulation, flotation and rapid filtration units, but around 80% has been reduced during slow sand filtration (Saengchut et al. 2020a, b).

4 Conclusion

For removing DBPs from treated water various physico-chemical technologies are there like filtration and studies show that the pressure-driven filtration techniques, i.e., nanofiltration and reverse osmosis are more efficient in removing DBPs. Conventional sand filtration method is also employed for DBPs treatment and slow sand filtration has been found to remove DBPs more efficiently than the rapid. Advanced oxidation treatment of DBPs is also one of the approaches which destructs them to their less toxic or non-toxic forms. Advanced oxidation is also achieved via employing single oxidant or in combination with radiations like UV, O_3, H_2O_2 giving better results. Use of activated carbon specially GAC in batch as well as column mode is one of the most effective methods amongst all treatment methods in removing DBPs. GAC filtration has been used in combination with sand filtration or ultrafiltration for its excellent performance. The other methods available are air stripping, tank mixing and spray aeration and heating/boiling which are very easy to perform, do not require much skill and proving very good treatment option for controlling and removing DBPs from drinking water.

However, during study it was found that attention should be paid towards emerging DBPs as data related to their occurrence and hazards is lacking. Employing a single method for removal of DBPs is not likely the best way for their treatment and even in few cases it has not shown good removal efficiencies. Thus, emphasis should be given in using hybrid/coupled methodologies as they provide comparatively better removal efficiencies. In conclusion, the advanced drinking water treatments

(ADWTs) should be combining different methodologies and as every raw water has different characteristic the treatment scheme should be specifically adapted in order to provide improved and better removal efficiencies of DBPs.

References

Abeynayaka A, Visvanathan C, Monthakanti N, Hashimoto T, Katayama H (2012) Removal of DOM and THM formation potential of tropicalsurface water by ceramic microfiltration. Water Supply 12(6):869–877

Ali GAM, Habeeb OA, Algarni H, Chong KF (2018) CaO impregnated highly porous honeycomb activated carbon from agriculture waste: symmetrical supercapacitor study. J Mater Sci 54:683–692

Alicia QC (2019) 'The effects of ventilation on Trihalomethanes removal by spray aeration', UNLV Theses, Dissertations, Professional Papers, and Capstones 3745

American Membrane Technology Association (AMTA) (2018) 'Ceramic Membranes, 'America's Authority in Membrane Treatment

Ates N, Yilmaz L, Kitis M, Yetis U (2009) Removal of disinfection by-product precursors by UF and NF membranes in low-SUVA waters. J Membr Sci 328(1–2):104–112

Ayyildiz O, Anderson PR, Peters RW (2005) Laboratory batch experiments of the combined effects of ultrasound and air stripping in removing CCl_4 and 1 11-TCA from water. J Hazard Mater B120:149–156

Azmi AA, Jai J, Zamanhuri NA, Yahya A (2018) Precious metals recovery from electroplating wastewater: a review. IOP Conference Series: Materials Science and Engineering 358(1):012024

Babi KG, Koumenides KM, Nikolaou AD, Makri CA, Tzoumerkas FK, Lekkas TD (2007) Pilot study of the removal of THMs, HAAs and DOC from drinking water by GAC adsorption. Desalination 210(1–3):215–224

Bo'zym M, Wzorek M, Kłosok-Bazan I (2017) Stopie'n nara˙zenia korzystaja˛cych z pływalnina działanie wybranych toksycznych produkto'w dezynfekcji wody. Bromatol Chem Toksykol 1:55–64

Bodzek M (2013) 'Przegla˛d mo˙zliwo'sci wykorzystania technik membranowych w usuwaniu mikroorganizmo'w i zanieczyszcze'n organicznych ze ´srodowiska wodnego. In˙zynieria i Ochrona', ´Srodowiska, 16 (1):5–37

Bodzek M, Konieczny K (2005) Wykorzystanie proceso'w membranowych w uzdatnianiu wody. Oficyna Wydawnicza Projprzem-Eko, Bydgoszcz

Bond T, Goslan EH, Parsons SA, Jefferson B (2012) A critical review of trihalomethane and haloacetic acid formation from natural organic matter surrogates. Environmental Technology Reviews 1(1):93–113

Buchanan W, Roddick F, Porter N (2006) Formation of hazardous by-products resulting from the irradiation of natural organic matter: comparison between UV and VUV irradiation. Chemosphere 63(7):1130–1141

Canales-Flores R, Prieto-García F (2020) Taguchi optimization for production of activated carbon from phosphoric acid impregnated agricultural waste by microwave heating for the removal of methylene blue. Diam Relat Mater 109:108027

Canosa J, Martí V (2018) Design of Stripping Columns Applied to drinking water to minimize carcinogenic risk from Trihalomethanes (THMs). Toxics 6(1):18

Chalatip R, Chawalit R, Nopawan R (2009) Removal of haloacetic acids by nanofiltration. J Environ Sci 21(1):96–100

Chen B, Zhang C, Wang L, Yang J, Sun Y (2021) Removal of disinfection byproducts in drinking water by flexible reverse osmosis: efficiency comparison, fates, influencing factors, and mechanisms. J Hazard Mater 401:123408

Chuang YH, Wang GS, Tung HH (2011) Chlorine residuals and haloacetic acid reduction in rapid sand filtration. Chemosphere 85(7):1146–1153
Crini G (2005) Recent developments in polysaccharide-based materials used as adsorbents in wastewater treatment. Prog Polym Sci 30:38–70
Crini G, Lichtfouse E (2019) Advantages and disadvantages of techniques used for wastewater treatment. Environ Chem Lett 17:145–155
Dawe P (2009) Best management practices for the control of disinfection by-products in drinking water systems in newfoundland and Labrador. Government of Newfoundland & Labrador Department of Environment and Conservation Water Resources Management Division, St. John's, NL, A1B 4J6 Canada
De Munari A, Schäfer AI (2010) Impact of speciation on removal of manganese and organic matter by nanofiltration. J Water Supply Res Technol AQUA 59(2–3):152–163
Dembek ZF, Anderson EL (2018) 'Food, waterborne, and agricultural disease' medical aspects of biological warfare. US Army Medical Department Center and School, Borden Institute, pp. 71–92
Deng Y, Zhao R (2015) Advanced oxidation processes (AOPs) in wastewater treatment. Curr Pollution Rep 1:167–176
Durmish BH, Reka AA, Gjuladin-Hellon T, Ismaili M, Srbinovski M, Shabani A (2015) Disinfection of drinking water and trihalomethanes: a review. IJARCS 2:1145–1156
Dymaczewski Z, Jez-Walkowiak J, Michałkiewicz M, Soza'nski MM (2019) Znaczenie procesudezynfekcji w zapewnieniu bezpiecze'nstwa mikrobiologicznego wody przeznaczonejdo spo˙zycia przez ludzi Ochrona. Srodowiska 41(11):3–9
Ebrahim SJ, Bidarpoor F, Eslami A, Ebrahimzadeh L (2016) Removal of Trihalomethanes from drinking water via heating method. Biomed Pharmacol J 9(1):61
EPA (2012) 'A Citizen's guide to air stripping' U.S. Environmental Protection Agency, Office of Solid Waste and Emergency Response
Ersan MS, Ladner DA, Karanfil T (2016) The control of N-nitrosodimethylamine, Halonitromethane, and Trihalomethane precursors by Nanofiltration. Water Res 105: 274–281
Fischer D (2019) Air stripping can remove VOCs, THMs and CO2 to improve drinking water quality. ES&E Magazine 2019
Forsey SP (2004) In situ chemical oxidation of creosote/coal tar residuals: experimental and numerical investigation. University of Waterloo
Gao J, Sun SP, Zhu WP, Chung TS (2016) Green modification of outer selective P84 nanofiltration (NF) hollow fiber membranes for cadmium removal. J Membr Sci 499:361–369
Ghernaout D (2019) Electrocoagulation and Electrooxidation for disinfecting water: new breakthroughs and implied mechanisms. Appl Eng 3:125–133
Ghernaout D, Boucherit A (2015) Review of coagulation's rapid mixing for NOM removal. J Res Develop Chem 926518:32
Ghernaout D, Elboughdiri N (2020a) Disinfection by-products: presence and elimination in drinking water. Open Access Libr J 7(2)
Ghernaout D, Elboughdiri N (2020b) Magnetic field application: an underappreciated outstanding technology. Open Access Libr J 7:e6000
Ghomshe SMT, Mousavi SM, Soltanieh M, Kordi AKS (2011) Batch and column study of haloacetic acids adsorption onto granular activated carbon. Sci Res Essays 6(16):3553–3560
Gopal K, Tripathy SS, Bersillon JL, Dubey SP (2007) Chlorination byproducts, their toxicodynamics and removal from drinking water. J Hazard Mater 140(1):1–6
Gottschalk C, Libra JA, Saupe A (2009) Ozonation of water and waste water: a practical guide to understanding ozone and its applications. John Wiley & Sons
Guo R, Bharadwaj L, Wilson LD (2021) Adsorption studies of waterborne Trihalomethanes using modified polysaccharide adsorbents. Molecules 26(5):1431
Hansen KMS, Zortea R, Piketty A, Vega SR, Andersen HR (2013) Photolytic removal of DBPs by medium pressure UV in swimming pool water. Sci Total Environ 443:850–856

Hasan NI, Makki HF (2021) Disinfection by-product removal by activated carbon-using batch mode. IOP Conf. Series: Earth and Environmental Science 790:012035
He J, Shi M, Wang F, Duan Y, Zhao T, Shu S, Chu W (2020) Removal of CX_3R-type disinfection by-product precursors from rainwater with conventional drinking water treatment processes. Water Res 185:116099
He X, Elkouz M, Inyang M, Dickenson E, Wert EC (2017) Ozone regeneration of granular activated carbon for trihalomethane control. J Hazard Mater 326:101–109
House DA (1962) Kinetics and mechanism of oxidations by peroxydisulfate. Chem Rev 62(3):185–203
Hsieh ST, Tung HH, Wang GS (2014) Simultaneous formation and biodegradation of HAAs in slow sand filters: effects of residual chlorine in filter influent. Desalin Water Treat 52(28–30):5366–5373
Hua Z, Guo K, Kong X, Lin S, Wu Z, Wang L, Huang H, Fang J (2019) PPCP degradation and DBP formation in the solar/free chlorine system: effects of pH and dissolved oxygen. Water Res 150:77–85
Ishaq MS, Afsheen Z, Khan A, Khan A (2018) 'Disinfection methods', Photocatalysts, edited by Sher Bahadar Khan and Kalsoom Akhtar, Intech Open. https://www.intechopen.com/books/7478
Isoda N, Rodrigues R, Silva A, Gonçalves M, Mandelli D, Figueiredo F, Carvalho W (2014) Optimization of preparation conditions of activated carbon from agriculture waste utilizing factorial design. Powder Technol 256:175–181
Itoh M, Kunikane S, Magara Y (2001) Evaluation of nanofiltration for disinfection by-products control in drinking water treatment. Water Supply 1(5–6):233–243
IXOM Watercare (2016) Tank mixing & spray Aeration helps stop flushing for Dbp/Thm compliance. Ixom Watercare News & Views
Jiang J, Zhang X, Zhu X, Li Y (2017) Removal of intermediate aromatic halogenated DBPs by activated carbon adsorption: a new approach to controlling halogenated DBPs in chlorinated drinking water. Environ Sci Technol 51(6):3435–3444
Karthik NB, Bal Krishna KC, Khan SJ, Sathasivan A (2022) A new chloramine recovery method in nitrifying water without "chlorine burn". Process Saf Environ Prot 165:841–850
Kim J, Kang B (2008) DBPs removal in GAC filter-adsorber. Water Res 42(1–2):145–152
Kochkodan V, Hilal N (2015) A comprehensive review on surface modified polymer membranes for biofouling mitigation. Desalination 356:187–207
Kolthoff I, Miller I (1951) The chemistry of persulfate. II. The reaction of persulfate with mercaptans solubilized in solutions of saturated fatty acid soaps. J Am Chem Soc 73(11):5118–5122
Kramer A, Wagner MC, Gell R (2014) 'In-tank aeration to reduce disinfection byproducts', northern Kentucky, http://www.aquametrologysystems.com/wp-content/uploads/2014/11/Presentation-In-Tank-Aeration-to-Reduce-Disinfection-Byproducts.pdf
Kristiana I, Liew D, Henderson RK, Joll CA, Linge KL (2017) Formation and control of nitrogenous DBPs from Western Australian source waters: investigating the impacts of high nitrogen and bromide concentrations. J Environ Sci 58:102–115
Kumar S, Ahlawat W, Bhanjana G, Heydarifard S, Nazhad MM, Dilbaghi N (2014) Nanotechnology-based water treatment strategies. J Nanosci Nanotechnol 14:1838–1858
Li B, Chen B, Chen S, Yang M, Wu Q (2021) Removing chlorinated haloacetaldehydes from drinking water by household heating devices with and without chlorine: efficiency, influencing factors, and mechanisms. Chemosphere 284:131202
Li P, Zhuang Y, Shi B, Zhang K (2023) Effects of boiling on iron particles in drinking water. AQUA–Water Infrastructure, Ecosys Soc 72(1):83–95
Lin Q, Dong F, Miao Y, Li C, Fei W (2020) Removal of disinfection by-products and their precursors during drinking water treatment processes. Water Environ Res 92(5):698–705
Liu X, Chen L, Yang M, Tan C, Chu W (2020) The occurrence, characteristics, transformation and control of aromatic disinfection by-products: a review. Water Res 184:116076

Lou J, Chan HY, Han JY, Yang CY (2016) High removal of haloacetic acids from treated drinking water using bio-activated carbon method, vol 57. Environ Sci, p 25627

Lou J, Chan HY, Yang CY, Tseng WB, Han JY (2014b) Reducing and verifying haloacetic acids in treated drinking water using a biological filter system. J Environ Sci Health A Tox Hazard Subst Environ Eng 49:1693

Lou J, Chang CJ, Chen WH, Tseng WB, Han JY (2014a) Removal of Trihalomethanes and Haloacetic acids from treated drinking water by biological activated carbon filter Engineering Water. Air Soil Pollution 225

Mazhar MA, Khan NA, Ahmed S, Khan AH, Hussain A, Rahisuddin, Changani F, Yousefi M, Ahmadi S, Vambol V (2020) Chlorination disinfection by-products in municipal drinking water–a review. J Clean Prod 273:123159

Mazloomi S, Nabizadeh R, Nasseri S, Naddafi K, Nazmara S, Mahvi A (2009) Efficiency of domestic reverse osmosis in removal of trihalomethanes from drinking water. J Environ Health Sci Eng 6(4):301–306

Minnesota Department of Health (MDH) (2022) 'Disinfection and disinfection by products, drinking water disinfection: a public health success story', Drinking Water Protection Program, 651–201-4700. https://www.health.state.mn.us/communities/environment/water/factsheet/ddbp.html. Accessed 29 April 2023

Mitchell SM, Ullman JL (2016) Removal of phosphorus, BOD, and pharmaceuticals by rapid rate sand filtration and ultrafiltration systems. J Environ Engg 142:11

Miyashita Y (2007) Removal of N-nitrosamine by Nanofiltration and reverse osmosis membranes.', A Thesis,. Georgia Institute of. Technology

Miyashita Y, Park SH, Hyung H, Huang CH, Kim JH (2009) Removal of N-nitrosamines and their precursors by nanofiltration and reverse osmosis membranes. J Environ Eng 135(9):788–795

Mohamed MA, Hassan AH, El Messiry MA, Hazzaa RA (2006) Removal of Trihalomethanes by dual filtering media (GAC-sand) at El-Manshia water purification plant. J Egypt Public Health Assoc 81(3–4):241–258

Mohshim DF, Mukhtar HB, Man Z, Nasir R (2013) Latest development on membrane fabrication for natural gas purification: a review. J Eng 2013:1–7

Moreno-Castilla C (2004) Adsorption of organic molecules from aqueous solutions on carbon materials. Carbon 42:83–94

Moslemi M, Davies SH, Masten SJ (2012) Rejection of bromide and bromate ions by a ceramic membrane. Environ Eng Sci 29:1092–1096

Nakamura T, Kawasaki N, Araki M, Yoshimura K, Tanada S (2001) Trihalomethane removal by activated carbon fiber. J Environ Sci Health A Tox Hazard Subst Environ Eng 36(7):1303–1310

Namiesnik J, Chrzanowski W, Szpinek P (2003) 'Nowe horyzonty i wyzwania w analityce i monitoringu srodowi-skowym' Centrum Doskonałosci Analityki i Monitoringu Srodowiskowego (CEERM), Gdansk

Nik-Abdul-Ghani NR, Jami MS, Alam MZ (2021) The role of nanoadsorbents and nanocomposite adsorbents in the removal of heavy metals from wastewater: a review and prospect. Pollution 7(1):153–179

Nishijima W, Speitel GE (2004) Fate of biodegradable dissolved organic carbon produced by ozonation on biological activated carbon. Chemosphere 56:113–119

Olszak N (2020) The history of reverse osmosis. Complete Water Solutions

Orecki A, Tomaszewska MU, Morawski AW (2006) Treatment of natural waters by nanofiltration. Przemysl Chemiczny 85(8):1067–1070

Pérez-Lucas G, Martínez-Menchón M, Vela N, Navarro S (2022) Removal assessment of disinfection by-products (DBPs) from drinking water supplies by solar heterogeneous photocatalysis: a case study of trihalomethanes (THMs). J Environ Manag 321:115936

Pignatello JJ, Oliveros E, MacKay A (2006) Advanced oxidation processes for organic contaminant destruction based on the Fenton reaction and related chemistry. Crit Rev Environ Sci Technol 36(1):1–84

Plumlee MH, López-Mesas M, Heidlberger A, Ishida KP, Reinhard MN (2008) Nitrosodimethylamine (NDMA) removal by reverse osmosis and UV treatment and analysis via LC–MS/MS. Water Res 42(1):347–355
Pyper GR (1985) 'Slow sand filter and package treatment plant evaluation operating costs and removal of bacteria, giardia and Trihalomethanes', project, Water Engineering Research Laboratory Office Of Research And Development U.S. Environmental Protection Agency
Rajamohan R, Venugopalan V, Debasis M, Usha N (2014) Efficiency of reverse osmosis in removal of total organic carbon and trihalomethane from drinking water. Res J Chem Environ 18: 1–6
Ram'ırez-Castillo FY, Loera-Muro A, Jacques M, Garneau P, Avelar-Gonz'alez FJ, Harel J et al (2015) Waterborne pathogens: detection methods and challenges. Pathogens 4:307–334
Regulations.gov (2006). https://www.regulations.gov/document/EPA-HQ-OW-2002-0043-0688. Accessed 24 May 2023
Richardson SD, Fasano F, Ellington JJ, Crumley FG, Buettner KM, Evans JJ et al (2008) Occurrence and mammalian cell toxicity of iodinated disinfection byproducts in drinking water. Environ Sci Technol 42(22):8330–8338
Richardson SD, Plewa MJ, Wagner ED, Schoeny R, DeMarini DM (2007) Occurrence, genotoxicity, and carcinogenicity of regulated and emerging disinfection by-products in drinking water: a review and roadmap for research. Mutation Res/Rev Mutation Res 636(1–3):178–242
Robeson LM, Liu Q, Freeman BD, Paul DR (2015) Comparison of transport properties of rubbery and glassy polymers and the relevance to the upper bound relationship. J Membr Sci 476:421–431
Saengchut P, Pentamwa P, Karuchit S (2020a) Removal of Haloacetic acids (Haas) in water supply by coagulation–flocculation and activated carbon adsorption processes. Eng J Res Develop 31(4)
Saengchut P, Pentamwa P, Karuchit S (2020b) Removal of haloacetic acids (HAAs) in water supply by coagulation–flocculation and activated carbon adsorption processes. Engg J Res Dev 31(04):173–184
Samadi MT, Nasseri S, Mesdaghinia AR, Alizadefard MR (2004) Removal of chloroform ($CHCl_3$) from Tehran drinking water by GAC and air stripping columns. Iranian J Env Health Sci Eng 1(1):5–12
Samadi MT, Nasseri S, Mesdaghinia A, Alizadefard MR (2006) A comparative study on THMS removal efficiencies from drinking water through nanofiltration and air stripping packed-column. Iranian J Water Wastewater 57:14–21
Serajuddin M, Hasan MM, Haque ME (2022) Trihalomethanes removal from drinking water by heating: a case study. Int J Res–GRANTHAALAYAH 10(10):22–32
Simpson DR (2008) Biofilm processes in biologically active carbon water purification. Water Res 42(12):2839–2848
Sinha R, Gupta AK, Ghosal PS (2021) A review on Trihalomethanes and Haloacetic acids in drinking water. Global status, health impact, insights of control and removal technologies. J Environ Chem Engg 9(6):106511
Skotnicka-Pitak J (2006) Problem dezynfekcji wody w polskich publikacjach ostatnich kilku lat. Gospodarka Wodna 11:439–440
Slow Sand Filtration (n.d.) MECC, https://water.mecc.edu/exam_prep/slowsand.htm. Accessed 29 April 2023
Solecka M (2009) Dezynfekcja 'scieko'w miejskich za pomoca̧ kwasu nadoctowego. Gaz. Wodai Technika Sanitarna 5:8–11
Srinivasan A, Chowdhury P, Viraraghavan T (2009) 'Air stripping in industrial wastewater treatment', Water and wastewater treatment technologies, Vol. 1, Edited by Saravanamuthu Vigneswaran, ©Encyclopedia of Life Support Systems (EOLSS)
Tang H, Xie YF (2016) Biologically active carbon filtration for haloacetic acid removal from swimming pool water. Eng Sci Total Environ 541:58
Tang WZ (2003) Physicochemical treatment of hazardous wastes. CRC Press

Tchobanoglous G, Burton FL, Stensel HD (2003) Wastewater engineering: treatment and reuse.4th edn. McGraw-Hill, Boston

Thabo B, Okoli BJ, Modise SJ, Nelana S (2021) Rejection capacity of Nanofiltration membranes for nickel, copper, silver and palladium at various oxidation states. Membranes 11(9):653

Trang VN, Dan NP, Phuong LD, Thanh BX (2014) Pilot study on the removal of TOC, THMs, and HAAs in drinking water using ozone/UV–BAC. Desalin Water Treat 52(4–6):990–998

Tseng W, Lou J, Wu MC, Fang H (2012) Reducing humic acid and disinfection by-products in raw water using a bio-activated carbon filter. Engineering

Tung HH, Unz RF, Xie YF (2006) HAA removal by GAC adsorption. Journal (American Water Works Association) 98(6):107–112

Tung HH, Unz RF, Xie YF (2009) Evidences of HAA biodegradation in GAC filtration. Engineering, Biology

Tung HH, Xie YF (2008) 'Disinfection by-product removal', Water Conditioning and Purification International Magazine

Uyak V, Koyuncu I, Oktem I, Cakmakci M, Toroz I (2008) Removal of trihalomethanes from drinking water by nanofiltration membranes. J Hazard Mater 152:789–794

Valero F, Arbós R (2010) Desalination of brackish river water using Electrodialysis reversal (EDR): control of the THMs formation in the Barcelona (NE Spain) area. Desalination 253(1):170–174

Visser T, Wessling M (2007) When do sorption-induced relaxations in glassy polymers set in? Macromolecules 40:4992–5000

Vines M, Terry LG (2023) Removal of Haloacetic acids via adsorption and biodegradation in a bench-scale filtration system. Water 15:1445

Wang W, Xie YF, Tang H (2019) The haloacetic acid leap in effluent of a biologically active carbon filter experiencing a disinfectant switch. Engineering Chemosphere 244:125435

Wang Y, Peng F, Zhao R, Dong X, Yang Z, Li H (2023) Removal and transformation of disinfection by-products in water during boiling treatment. Chemosphere 326:138426

Waniek A, Bodzek M, Konieczny K (2002) Trihalomethane removal from water using membrane processes. Pol J Environ Stud 11(2):171–178

Watson K, Farré M, Knight N (2012) Strategies for the removal of halides from drinking water sources, and their applicability in disinfection by-product minimisation: a critical review. J Environ Manag 110:276–298

Weinberg HS, Krasner SW, Richardson SD, Thruston Jr. AD (2002) 'The occurrence of disinfection by-products (DBPs) of health concern in drinking water: results of a Nationwide DBP occurrence study', EPA/600/R-02/068

Weiner ER (2012) Applications of environmental aquatic chemistry: a practical guide. CRC Press, Florida, USA

Wells W, Benjamin MM, Korshin GV (2001) Effective of thermal treatmentan halogenated disinfection byproduct in drinking water. Water Res 35:3545–3550

WesTech (2020) 'Strategies for Reducing Disinfection Byproducts', WesTech Engineering, https://www.westech-inc.com/blog/strategies-for-reducing-disinfection-byproducts. Accessed 28 April 2023

WHO (2000) 'Disinfectants and disinfectant by-products', environmental health criteria 216, International Programme On Chemical Safety, United Nations Environment Programme, International Labour Organisation, World Health Organization, Geneva. https://www.inchem.org/documents/ehc/ehc/ehc216.htm#SectionNumber:1.3. Accessed 26 April 2023

WHO (2006) 'A compendium of drinking-water quality standards in the Eastern Mediterranean Region', World Health Organization, Regional Office for the Eastern Mediterranean Regional centre for Environmental Health Activities (CEHA), pp. 1–44

World Health Organization (WHO) (2015) Progress on sanitation and drinking-water: 2010, Update Report. World Health Organization (WHO), Geneva, Switzerland

Wu F, Wu S (2009) Removal of Trihalomethanes from drinking water by air stripping. In: International conference on energy and environment technology, Guilin, China, pp 695–698

Wu WW, Benjamin MM, Korshin GV (2001) Effects of thermal treatment on halogenated disinfection by-products in drinking water. Wat Res 35(15):3545–3550
Xie YF, Zhou H (2002) Use of BAC for HAA removal—part 2, Column Study. Engineering 94:126
Xing W, Ngo HH, Kim SH, Guo WS, Hagare P (2008) Adsorption and bioadsorption of granular activated carbon (GAC) for dissolved organic carbon (DOC) removal in wastewater. Bioresour Technol 99:8674–8678
Xu D, Bai L, Tang X, Niu D, Luo X, Zhu X, Li G, Liang H (2019) A comparison study of sand filtration and ultrafiltration in drinking water treatment: removal of organic foulants and disinfection by-product formation. Sci Total Environ 691:322–331
Yang L, She Q, Wan MP, Wang R, Chang VW-C, Tang CY (2017a) Removal of haloacetic acids from swimming pool water by reverse osmosis and nanofiltration. Water Res 116:116–125
Yang L, Zhou J, She Q, Wan MP, Wang R, Chang VW-C et al (2017b) Role of calcium ions on the removal of haloacetic acids from swimming pool water by nanofiltration: mechanisms and implications. Water Res 110:332–341
Yang Y, Komaki Y, Kimura SY, Hu H-Y, Wagner ED, MariBnas BJ et al (2014) Toxic impact of bromide and iodide on drinking water disinfected with chlorine or chloramines. Environ Sci Technol 48(2):12362–12369
Yang Z, Sun Y, Shi N, Hu H (2015) Formation and variation of brominated disinfection by-products in a combined ultrafiltration and reverse osmosis process for seawater desalination. Huan Jing Ke Xue 36(10):3706–3714
Zainudin FM, Hasan MA, Abdullah SRSA (2018) An overview of the technology used to remove trihalomethane (THM), trihalomethane precursors, and trihalomethane formation potential (THMFP) from water and wastewater. J Ind Eng Chem 57:1–14
de Zamarron los AS (2005) 'Trihalomethane reduction through air stripping', ETD Collection for University of Texas, El Paso, AAI1425897
Zazouli MA, Kalankesh LR (2017) Removal of precursors and disinfection byproducts (DBPs) by membrane filtration from water; a review. J Environ Health Sci Eng 8:15–25
Zhang D, Wang F, Duan Y, Chen S, Zhang A, Chu W (2020) Removal of trihalomethanes and haloacetamides from drinking water during tea brewing: removal mechanism and kinetic analysis. Water Res 184:116148
Zhang K (2018) 'Degradation of trihalomethanes and chloramines by UV and VUV irradiation', A Thesis, Lyles School of Civil Engineering West Lafayette, Faculty of Purdue University, Indiana
Zhou H, Xie YF (2002) Using BAC for HAA removal part 1: batch study. Engineering 94:194

Removal of Disinfection By-Products from Drinking Water: Influencing Factors and Physico-Chemical Treatment Processes

Surya Singh and Sarvesh Rai

Abstract Disinfection is one of the most essential procedures in drinking water treatment. Among the various methods of disinfection, chlorination is among the most common methods being economical, reliable, easily-operable, easily-measurable, and efficient enough to maintain residual disinfecting effects for sufficient time period. However, it has been reported that chlorine does react chemically with the naturally available carbon-based compounds in water and produces disinfection by-products (DBPs) such as trihalomethanes. Chloroform ($CHCl_3$) is one of the most common trihalomethanes. It has been reported that both short-term and long-term exposure to trihalomethanes result in adverse health impacts. As per WHO standards, the permissible limits for various trihalomethanes span from 60 μg/L to 300 μg/L. Formation and concentration of DBPs depend upon type and characteristics of source water, type of disinfectants used, operational conditions, temperature, pH, and contact time of chlorinated water with DBPs precursors in water treatment plants and water distribution network. Various treatment techniques have been evolved for the removal of trihalomethanes, such as enhanced coagulation, membrane separation, reverse osmosis, adsorption, filtration, advanced oxidation, etc.; however, each process has its own advantages as well as shortcomings. In this chapter, summary of various physico-chemical processes for the removal of DBPs shall be discussed, with major focus on the two best available techniques, viz. enhanced coagulation and activated carbon adsorption. Moreover, the applicability of alternative disinfectants shall also be explored.

Keywords Chlorination · Disinfection by-products · Physico-chemical processes · Water contamination

S. Singh
Division of Environmental Monitoring and Exposure Assessment (Water & Soil), ICMR–National Institute for Research in Environmental Health, Bhopal, India
e-mail: suryasingh.nireh@icmr.gov.in

S. Rai (✉)
Central Pollution Control Board, Regional Directorate, Lucknow, India
e-mail: srai.cpcb@nic.in

S. Madhav et al. (eds.), *Drinking Water Disinfection By-products*,
https://doi.org/10.1007/978-3-031-49047-7_11

1 Introduction

Availability of clean water is one of the primary requirements for survival and development of humans. Moreover, United Nations' Sustainable Development Goals (SDGs) also emphasize over the clean water and sanitation (Goal no. 6) in order to achieve overall development of nations (SDGs 2015). Understanding the importance of this issue, multiple techniques have been developed in the past in order to treat and/or remove various physico-chemical and microbiological contaminants from drinking water. A conventional drinking water treatment plant fundamentally involves 3-stage process, viz. primary, secondary, and tertiary treatment (Pakharuddin et al. 2021). Primary treatment mainly helps to remove the coarser size particles, grit, sand, etc. Secondary treatment involves the removal of organic impurities from water. Tertiary or the final treatment helps to disinfect the water to make it free from any microbiological impurity. There are various methods to employ for the tertiary treatment, such as ozonation, bromination, chlorination, UV treatment, filtration, etc. (Tsitsifli and Kanakoudis 2018).

In order to minimize microbial contamination in water, disinfection is one of the most preferred techniques incorporated in drinking water treatment plants. This process holds importance as it helps to remove various disease-causing microbes such as *Salmonella typhi*, *Vibrio cholera*, *Shigella dysenteriae*, etc. in the potable water supply (Postigo et al. 2018). As the water treated in drinking-water treatment plants takes considerable time in reaching up to the end-user, those disinfecting chemicals are preferred which can maintain their disinfecting effect for long duration, such as halogenating agents (Ndlangamandla et al. 2018). For the water disinfection purpose, these halogenating agents generally refer to the chlorinating and brominating agents. Moreover, easy availability and low cost further make chlorinating agents a preferred choice as disinfectants. These chlorinating agents include various oxidants, e.g., chlorine, chloramines, chlorine dioxide, etc.

Here it is notable that although residual effect of these oxidants helps to control the re-emergence of disease-causing organisms in water, it can result in formation of various disinfection by-products (DBPs) too (Chaukura et al. 2020). The discovery of disinfection by-products goes long back when trihalomethanes (THMs) were identified as the first class of halogenated DBPs in the treated water (Bellar et al. 1974). Later, U.S. Environmental Protection Agency (EPA) carried out various monitoring activities and reported the presence of chloroform in all the drinking water that has undergone chlorine treatment (Symons et al. 1975). This study confirmed the association of chloroform with the water disinfection process using chlorination. Subsequently, bromide and hypobromous acid (formed after the oxidation of bromide by water) were also found in treated drinking water, thereby confirming the formation and presence of brominated THM species.

These DBPs are potentially toxic to human physiological system and hence their formation, toxicity level, and removal mechanisms need to be understood. Generally, as the most common route, humans get exposed to DBPs by ingesting chlorinated and/or brominated drinking water. Inhalation is another route of exposure which takes place by breathing in the air having DBPs. Absorption of DBPs through skin while bathing and swimming activities is the dermal route of exposure. Human health effects owing to low amount of exposure to DBPs are largely unknown; however, excess exposure may result into hepatic and nervous system abnormalities (CDC 2009). A number of epidemiological studies have shown significant association between the exposures to chlorinated water and undesirable health outcomes, such as urinary bladder cancers (Costet et al. 2011), pre-mature births (Wright et al. 2017; Grellier et al. 2010), and miscarriages (Waller et al. 1998). One of the studies has shown 47% higher prevalence of urinary bladder cancer among European males who were exposed to THM concentration higher than 50 μg/L, compared to males exposed to THM concentration less than 5 μg/L (American Cancer Society 2017; Li and Mitch 2018). In context of India, cancer risk due to DBPs was estimated to be 8.99×10^{-4} and 8.92×10^{-4} for males and females, respectively, in some selected Indian cities (Mishra et al. 2014).

There are certain regulations as well, given by USEPA, for monitoring and control of DBPs, such as total trihalomethane rule, information collection rule, Stage 1 disinfectant and DBPs rule, Stage 2 disinfectant and DBPs rule, and 'upcoming' Stage 3 disinfectant and DBPs rule, etc. (Fig. 1). Apart from these, certain other regulations were also brought into action specifically to maintain the balance between anionic/cationic and bacterial/viral risks in water. These are shown chronologically in Fig. 1. Among all these rules, Stage 1 disinfectant and DBPs rule covers monitoring, reporting, and best available technologies for DBP control (US EPA 2010). The stage 2 rules offer stricter compliance for the monitoring, and target public water supply systems having higher risk (US EPA 2022). This chapter will discuss briefly about various physico-chemical processes available for the elimination of DBPs from water, with detail of the two best available techniques.

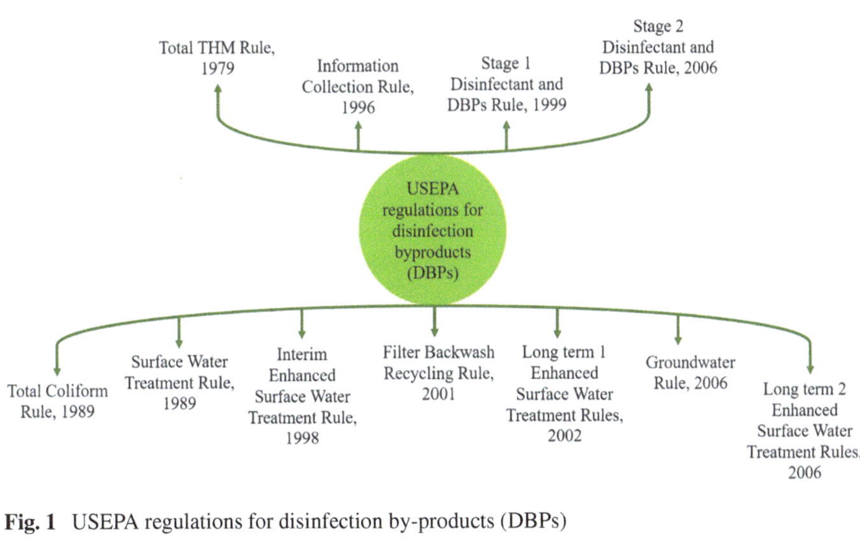

Fig. 1 USEPA regulations for disinfection by-products (DBPs)

2 Factors Influencing the Formation of Disinfection by-Products

Studies have shown that development and persistence of DBPs in water are affected by a number of parameters (Fig. 2); e.g., temperature, material of the pipe, pH, other contaminants present in water, residual-free chloride and bromide levels, etc.

2.1 *Temperature*

High temperature results in higher rate of reactions owing to increased kinetic energy of the molecules. Therefore, high concentration of various DBPs has been found in the months of summer compared to winter (Wei et al. 2010; Zhang et al. 2015). Moreover, summer season (i.e., high temperature) also results in increased organic load in the water bodies, which further contribute to DBPs. Leenheer et al. have reported that fluctuating temperature in combination with changing organic foliage and microbiological activities results in varying amounts of natural organic matter in the rivers (Leenheer et al. 2007). This water would therefore result in varying amount of DBPs in different seasons upon being used for chlorination. Thus, it can be further expected that low amount of DBPs shall be produced in the months of winter when the temperature is low (Srivastav and Kaur 2020).

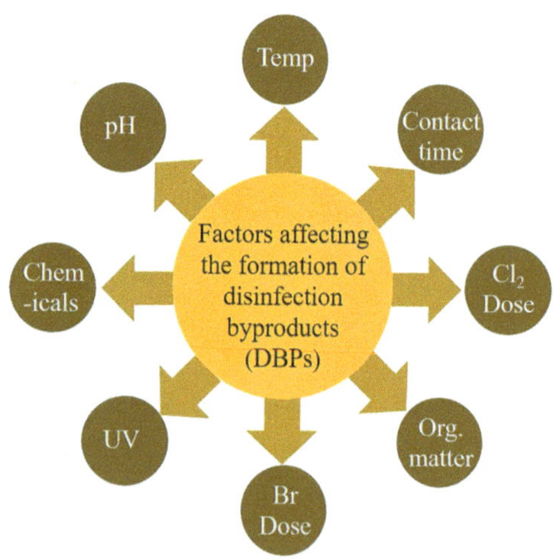

Fig. 2 Factors influencing the formation of disinfection by-products

2.2 pH

pH has an important function in various processes of water treatment, especially the aerobic processes where treatment is carried out with the help of various microbes. In order to ensure the optimum functionality of microbes, a certain pH range needs to be maintained. Though, disinfection process does not involve microbes directly; these may affect the amount of natural organic matter present in water, and thus indirectly affect the formation of DBPs. It has been seen that elevated pH (> 8) results in decrease in the amount of haloketones and haloacetonitriles by 5 times and 1.5 times, respectively, because these DBPs hydrolyze in the basic conditions (Singer 1994). Studies have shown that DBP formation is higher during the chlorination process, as compared to chloramination (Sedlak and Von Gunten 2011). However, some studies are contradictory to this fact and report that there is no effect of pH on the formation of trihalomethanes, haloacetic acids, haloketones, and haloacetonitriles (Zhou et al. 2019).

2.3 Organic Matter in Water

Natural organic matter is the precursor of DBPs, therefore, it is evident that excess of it would lead to more DBP formation and vice versa. Singer has reported that high amount of organic matter in water reacts with the halogenated (e.g., chlorinated) compounds resulting in a variety of DBPs, such as trihalomethanes, haloacetic acids, halonitromethanes, haloacetonitriles, etc. (Singer 1994; Souaya et al. 2015; Yang et al. 2018). Fabris et al. proved the relation between organic matter and trihalomethane formation (Fabris et al. 2008). Rise in the amount of trihalomethanes and haloacetic acids upon increasing the organic matter has also been described by Lu et al. (Lu et al. 2009). Kim further supported this fact by finding the increased amount of trihalomethanes and haloacetic acids upon increasing the organic content in water (Kim 2009). Similarly, Uyak and Toroz and Scheili et al. also reported that formation of trihalomethanes and haloacetic acids gets enhanced upon rising organic carbon content in water (Uyak and Toroz 2007; Scheili et al. 2015).

2.4 Effect of Free Residual Chlorine and Chlorine Dosage

Free residual chlorine is generally added in water for preventing microbial contamination during the supply phase. As this is the precursor of DBPs in water, it is evident to get more amount of DBPs at high concentrations of free residual chlorine (Singer 1994; Richardson and Postigo 2015; Srivastav and Kaur 2020). As per World Health Organization's guidelines, the prescribed amount of free residual chlorine in water lies in the range of 0.2–0.5 mg/L (WHO 2011). It is reported that least concentration of haloacetic acids and haloketones is obtained when free residual chlorine is in the WHO's prescribed range (Srivastav and Kaur 2020). Moreover, DBPs continue to form till all the free residual chlorine is exhausted; therefore, amount of free residual chlorine affects DBPs formation up to significant extent.

2.5 Contact Time

Souaya et al. reported proportional relationship between the contact time of water and disinfectant. Increase in the amount of DBPs, such as trihalomethanes and haloacetic acids, was found upon increasing the contact time of chlorine and water (Souaya et al. 2015). Lu et al. have also shown the similar findings (Lu et al. 2009).

There are certain other factors as well which affect the development of chlorinated, brominated, and/or nitrogenous DBPs in water, e.g., UV irradiation (Guo et al. 2016), presence of bromide (Heller-Grossman et al. 1999), chemicals present in cosmetic products, etc. (Bottoni et al. 2014).

3 Removal Processes for Disinfection By-Products

There may be several strategies for controlling the DBPs in drinking water. On a broader scale, the obviation techniques for DBPs may be clubbed into three, viz. source control of DBP precursor (viz. organic matter), use of chlorine substitutes (alternative disinfectants), and elimination of DBPs. Currently one of the most used disinfecting agents is chlorine and therefore majority of the removal processes are targeted towards the chlorinated DBPs.

One of the preliminary attempts to minimize the formation of DBPs is to control the occurrence of natural organic matter in water, so that growth of algae and other organic matter can be minimized (Oliver and Shindler 1980; Hoehn et al. 1980; Wachter and Andelman 1984; Karimi and Singer 1991). There could be several management strategies for this, such as detention basins for trapping nutrient-rich organic matter. Moreover, watersheds which are used for accumulating drinking water supply needs to be controlled with proper land-use. Nutrient control can also be done through nutrient cycling with the use of aeration systems mounted at the base of the water body (Hao et al. 2021) and/or harvesting program for limiting the aquatic growth (Zhu et al. 2022). Further, in order to minimize disinfecting products originated from bromide, prevention of mixing of brine into the drinking water source is suggested (Sun et al. 2013). This process is specifically useful where extensive amount of brine is generated such as during oil and natural gas production. Here, it is important to understand that DBPs generally have different precursors, which lead to different set of reactions. Hence, it is imperative to first identify the precursor of DBP in the water undergoing treatment and accordingly selection of the removal process needs to be made. Researchers have also suggested that aquifer storage and recovery can be one of the efficient methods for controlling DBPs in treated water (Singer 1994). Storage of treated/disinfected water in aquifers for a sufficient period before supplying it to the consumers may significantly reduce the concentration of halogenated DBPs (Maliva and Missimer 2010). However, this process is highly dependent upon the geological characteristics of the aquifer used for storage and the type of compound (DBP) (Pavelic et al. 2005). For example, chloroform was found to be highly persistent DBP, while bromoform the least during the aquifer storage.

Considering the second approach, disinfectants which are substitutes of chlorine, such as chloramines, have been used in various water treatment facilities with the intention of curtailing the formation of haloacetic acids, trihalomethanes, and other DBPs having carbonaceous origin. However, chloramines have been found to produce nitrogenous DBPs such as cyanogen chloride, N-nitrosodimethylamine (Yang et al. 2018; Chuang et al. 2019), which nullify the advantage of curtailing the development of other DBPs. The processes having no use of halogens such as treatment with ultraviolet radiation and/or ozone are other alternative methods for eliminating DBPs, since these are free from halogen compounds to act with natural organic matter resulting in DBPs (Yang et al. 2018). However, these methods have the disadvantage of not providing any residual disinfecting effect in the water (Chen et al. 2018). For treating the water having algal contamination, UV radiation has been proven to be a satisfactory option (Chen et al. 2018). Another benefit of utilizing UV method is its suitability for easy installation in the existing water treatment setup.

Apart from preventing DBPs formation, the removal of DBPs may also be adopted after their formation, as one of the control strategies (Chaukura et al. 2020). There are various techniques which may be employed for the removal of disinfection by-products, such as adsorption processes, ion exchange, coagulation, pre-oxidation, membrane filtration, biological treatment, etc. Each of the identified processes has its own advantages and shortcomings in the removal of DBPs. Moreover, each of the process performs well in a certain specified reaction condition.

Air stripping is a fascinating technique which can remove the volatile DBPs, such as trihalomethanes. This process is convenient to use in small-scale water treatment plants (Wang et al. 2015; Valdivia-Garcia et al. 2019). Valdivia-Garcia et al. have found that this process is affected negatively due to high concentration of brominated DBPs (Valdivia-Garcia et al. 2019). However, efficacy of the process is increased in the summer season, when trihalomethane formation is high, owing to high distribution constant. Another advantage of applying air stripping is that it doesn't produce any harmful by-products (Wang et al. 2015).

Membrane technologies have also gained sound momentum in the area of DBPs removal. Membranes are a kind of barriers which separate undesired molecules employing the size exclusion and diffusion mechanisms. Based on the size of pores, membranes may be of several types, such as microfiltration, nanofiltration, ultrafiltration, reverse osmosis, etc. As membranes may be of various pore sizes and removal potential, the properties (viz., size, electrostatic potential, aromatic character) of contaminants influence the selection of membranes (Chaukura et al. 2020).

Reverse osmosis is a membrane separation process which works under high pressure; however, it cannot remove trihalomethanes under normal conditions except at low concentrations. Nanofiltration may be considered as the low pressure reverse osmosis process, which is able to remove the natural organic matter, organic molecules, and other DBP precursors (Marais et al. 2018). Further, it is a cost-effective and easily-operable process. The only drawback is the requirement of pre-treatment of water, energy, and safe disposal of reject (Yang et al. 2019). Similarly, ultrafiltration membranes are good at reducing solids which are in suspension and trihalomethane precursors. But these membranes are not efficient to remove humic substances having more potential to form haloacetic acids and trihalomethanes (Xu et al. 2019). Another type of membranes are microfiltration membranes, which are generally used to remove particulate matter. Therefore, these membranes are often utilized for pre-treating the water before introducing it for nano- or ultrafiltration (Chellam and Sari 2016). However, the major disadvantage of employing membranes in the DBPs removal is the fouling of membranes with time, which affects the efficiency, selectivity, flow rate, and pressure drop across the membrane. Therefore, it is often suggested to pre-treat the water before passing it through the membranes. Some studies have also reported that impregnation of membranes with nanoparticles have the potential to reduce fouling (Daels et al. 2015; Guo et al. 2019).

Advanced oxidation process is another impressive technique for water purification. These processes utilize hydroxyl radicals as oxidant to react with the organic contaminants in water. Thus, various precursors may be used to undergo this process, such as ozone, hydrogen peroxide (H_2O_2) with ozone, H_2O_2 with UV, TiO_2, Fenton process, ultrasound, etc. (Oller et al. 2011; Miklos et al. 2018; Aziz et al. 2019). Advanced oxidation processes are efficient in the sense that these completely mineralize the organic matter existing in water at ambient conditions; but it requires considerable input of energy and chemicals (Sharma et al. 2018; Matilainen and Sillanpaa 2010).

Application of zero-valent iron (ZVI) is also one of the important techniques to address the issue of DBPs and its precursors (Chiu 2013). Though, ZVI has largely been used in the permeable reactive barriers to treat the contaminated underground water (Plessl et al. 2022), its application for DBPs removal is also reported. ZVI being the reducing agent can consume free chlorine molecules (Bedner et al. 2004). Moreover, ZVI has been shown to degrade not only chlorinated DBPs rather ozonation DBPs as well, such as haloacetic acids (Hozalski et al. 2001; Zhang et al. 2004), chloropicrin (Pearson et al. 2005), bromate (Xie and Shang 2005), and N-nitrosodimethylamine (Odziemkowski et al. 2000). Besides removing the DBPs, ZVI can also be employed to remove natural organic matter such as humic acids through the formation of iron oxides or hydroxides (Gu et al. 1994; Giasuddin et al. 2007).

4 Best Available Techniques for Removal of Disinfection By-Products

4.1 Enhanced Coagulation

Enhanced coagulation is one of the best available techniques which can be utilized for successful elimination of DBPs by utilizing the conventional water treatment setup (US EPA 1998). This process involves the addition of excess coagulant for the elimination of DBP precursors, as proposed by the USEPA in the Disinfectants/Disinfection By-products (D-DBP) Rule. In order to evaluate the process of enhanced coagulation, TOC removal is generally considered as it can be easily monitored and measured.

As a pre-requisite of enhanced coagulation, specific UV adsorption (SUVA) is measured to know the treatability of the water through this process. SUVA is defined as the UV adsorption (at 254 nm) per mg/L of dissolved organic carbon (DOC), as shown in Eq. 1. SUVA value less than 2 $L{\cdot}mg\text{-}m^{-1}$ denotes that water is difficult-to-treat, while higher SUVA represents easy-to-treat water.

$$SUVA = UV / DOC \quad (1)$$

where SUVA = Specific UV adsorption expressed in $L{\cdot}mg\text{-}m^{-1}$.

Al and Fe salts are the most frequently used coagulants in drinking water treatment plants for enhanced coagulation. For the elimination of natural organic matter, the major mechanisms involve charge neutralization, adsorption, complexation, and sweep flocculation. The enhanced coagulation process differs from the conventional coagulation in terms of objective, mechanism, and operative conditions. Whereas conventional coagulation is targeted to remove turbidity of the water, enhanced coagulation serves the purpose of removing natural organic matter (i.e., DBP precursors). The characteristics of raw water also influence the process efficiency of enhanced coagulation, such as alkalinity, pH, etc. US EPA DBP Rules have proposed the requirements for enhanced coagulation as given in Tables 1 and 2. The first requirement states that the treatment system should meet the TOC removal criteria as mentioned in Table 1, for complying with the requirements of enhanced coagulation. Since higher alkalinity creates hindrance in achieving better TOC removal, low TOC removal is required for the highly alkaline water (Table 1). However, for the water treatment systems which do not meet the criteria listed in

Table 1 Enhanced coagulation requirement of the Disinfectants/Disinfection By-products Rule for percent total organic carbon removal (US EPA 1998)

	Alkalinity, mg/L as $CaCO_3$ in source water		
Total organic carbon, mg/L in source water	<60	60. – 120.	>120
2–4	35%	25%	15%
4–8	45%	35%	25%
>8	50%	40%	30%

Table 2 Equivalent dose of coagulants under the step 2 requirement (mg/L) (US EPA 1999)

Regular grade alum, $Al_2(SO_4)_3.14H_2O$ (mg/L)	Reagent grade alum, $Al_2(SO_4)_3.18H_2O$ (mg/L)	Ferric chloride, $FeCl_3.6H_2O$ (mg/L)	Ferric chloride, $FeCl_3$ (mg/L)	Ferric sulfate, $Fe_2(SO_4)_3.9H_2O$ (mg/L)	Ferrous sulfate, $FeSO_4.7H_2O$ (mg/L)
10	11.2	9.1	5.5	9.5	9.4
20	22	18	11	19	19
30	34	27	16	28	28
40	45	36	22	38	37
50	56	46	27	47	47
60	67	55	33	57	56
70	78	64	38	66	66
80	90	73	44	76	75
90	101	82	49	85	84
100	112	91	55	95	94

Table 3 Target pH values under the step 2 requirement for enhanced coagulation (US EPA 1999)

Alkalinity (mg/L)	0–60	>60–120	>120–240	>240
Target pH	5.5	6.3	7.0	7.5

Table 1, an alternative method has been proposed for determining the TOC removal percentage. In this case, jar tests are conducted for one year by adding alum, $Al_2(SO_4)_3.14H_2O$, or its equivalent doses of Fe^{+3} salts as mentioned in Table 2. This procedure is carried out at an interval of 10 mg/L until the pH is lowered down to the targeted value. The targeted values of pH are shown in Table 3 (US EPA 1999). This step 2 procedure is set at the point of diminishing returns (PODR). PODR is the point on the graph of TOC removal vs. coagulant addition, at which slope changes from >0.3 mg/L of TOC removal for every 10 mg/L of alum (or its equivalent) added to <0.3 mg/L of TOC removal for every 10 mg/L of alum (or its equivalent) added, and remains so until the target pH is achieved (US EPA 1999). In case, any water does not meet the PODR condition, then such type of water is not amenable to undergo the process of enhanced coagulation.

4.1.1 Factors Affecting the Enhanced Coagulation

There are various water-based parameters, e.g., alkalinity, pH, temperature, TOC concentration in source water, origin and type of NOM, etc. which affect the process of TOC removal and thereby enhanced coagulation. Moreover, design and operating conditions, viz. pH, coagulant type and dosage, mixing time, sedimentation, sludge handling, etc. also affect the process (Fig. 3). Among these, there are two operational conditions that affect the process of TOC removal to a significant extent.

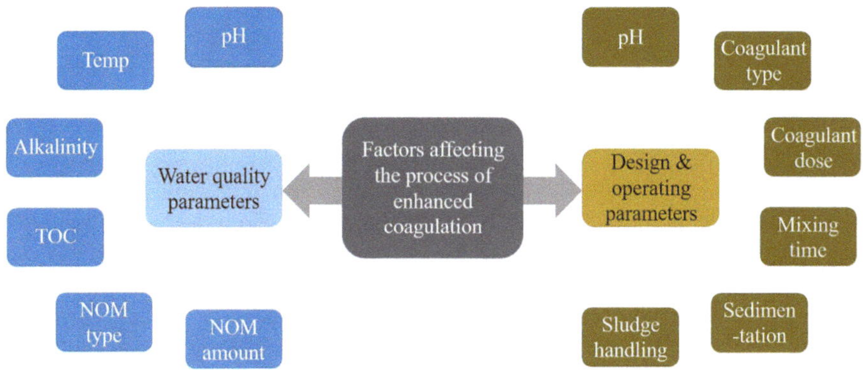

Fig. 3 Factors affecting the enhanced coagulation

4.1.1.1 pH Preadjustment

Preadjustment of pH is a fundamental requisite to carry out enhanced coagulation. Considering this, US EPA has defined the target pH values for enhanced coagulation process in the step 2 requirements, as shown in Tables 2 and 3. Generally, alum or Fe^{+3} salts are used to achieve the defined pH values; however, in some cases acids are also added to maintain the pH. Among acids, sulfuric acid (H_2SO_4) is generally used for this purpose.

4.1.1.2 Coagulants

In the water treatment procedures, most of the enhanced coagulation processes are carried out using alum or ferric salts. Besides enhanced coagulation, these coagulants also perform well for the removal of color, turbidity, and various other water contaminants. Sometimes, organic and/or inorganic polymers are also used in coagulation processes of water, though these are not recommended by the USEPA. As per the experiments conducted by Xie, in order to achieve TOC removal of 35% under the step 1 requirements, only 40 mg/L (equivalent dose) of polymerized aluminium chloride is needed, as compared to the 60 mg/L of alum (Xie 2004). Thus, low dose of coagulants will in turn help in lowering down the cost of sludge handling and disposal. However, polymerized aluminium chloride doesn't help in maintaining the pH of the water unlike that of alum or ferric chloride and therefore acid addition is required. Moreover, cost is another issue as polymerized salts are significantly costly. A trade-off, therefore, needs to be set between the high cost of polymerized salts and low sludge handling and disposal cost.

4.2 Adsorption Through Activated Carbon

Adsorption through carbon has the advantage that it helps to reduce both DBPs and DBP precursors. For the water treatment procedures, both granular and powdered activated carbon have been used (Xie 2004). Since activated carbon is efficient for the removal of odor, color, and few other chemicals too, along with improving the taste; powdered activated carbon is generally used in drinking water treatment plants. Efficiency of powdered activated carbon has been seen in Ohio River water and 50% reduction in the trihalomethane formation potential was observed (Najm et al. 1991). Trihalomethane removal has also been studied using granular activated carbon (GAC), while studies on the elimination of other DBPs are rather limited.

In case of use of GAC for the carbon adsorption, it has been seen that adsorption process is efficient in the beginning of the process; however, it reduces with time owing to depletion of GAC. It happens due to the saturation of active sites on the surface of GAC for adsorption, which leads to reduction in adsorption. Efficiency was found to be reduced from 100% to 15% after the operation of 50 days (Cummings and Summers 1994). According to D-DBP rule (USEPA 2010), GAC is one of the best available techniques for the removal of DBPs and its precursors with an empty bed contact time (EBCT) of 10 min and reactivation frequency of no more than 6 months (Xie 2004). EBCT is calculated as the ratio of GAC volume and flow rate of water, as shown in Eq. 2.

$$\mathrm{EBCT} = V_{\mathrm{GAC}} / Q_{\mathrm{water}} \tag{2}$$

where

EBCT = empty bed contact time (s).

V_{GAC} = GAC volume (m^3).

Q_{water} = flow rate of water (m^3/s).

The performance of GAC is generally indicated through a breakthrough curve, as depicted in Fig. 4. The Y-axis indicates the organic carbon ratio of effluent-to-influent (C_e/C_0), while X-axis denotes the bed volume feed. Since some amount of total organic carbon is non-adsorbable, the curve does not start from zero. After initial increase in the removal, the profile becomes constant with the increase in bed volumes. Once the curve becomes flat, it is assumed that adsorption is complete. Thus, the adsorptive capacity of GAC is determined through the amount of water required to reach the state of a flat curve.

4.2.1 Factors Affecting the Adsorption Through Activated Carbon

The ability of GAC for eliminating DBP precursors relies upon various parameters such as physico-chemical properties of carbon, properties of natural organic matter, characteristics of water, temperature, presence of bromide, EBCT, etc. It has been reported that GAC having basic characteristics are more suitable for the adsorption

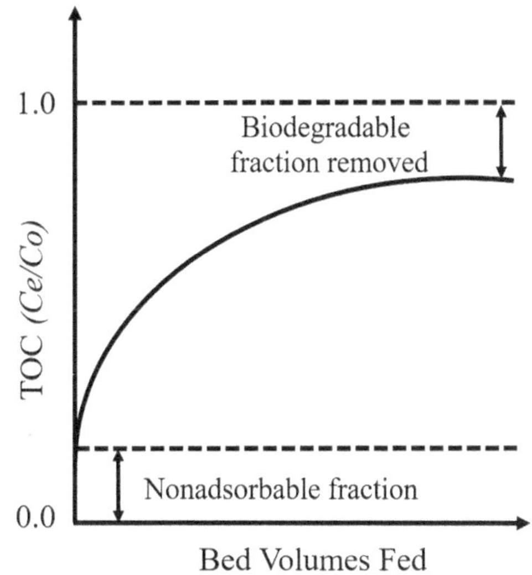

Fig. 4 A characteristic breakthrough curve showing total organic carbon (TOC) elimination using GAC

of halogenated DBP precursors, as compared to GAC with acidic characteristics (Erdem et al. 2020). Moreover, porosity of the carbon and bromide concentration also affects the performance of GAC. Increased adsorption of trihalomethane precursors has been noticed in both primary and secondary micropores of activated carbon filters upon pre-chlorination in the water. Erdem et al. demonstrated that upon increasing the bromide concentration from 200 μg/L to 800 μg/L, speciation of trihalomethane shifted from trichloromethane to tribromomethane, which ultimately affects the removal efficiencies (Erdem et al. 2020). Increase in temperature and EBCT is also reported to enhance the dissolved organic carbon removal (Liu et al. 2022). Moona et al. have shown that longer EBCT (> 30 min) helped in the removal of dissolved organic carbon and microbial organic matter (Moona et al. 2021). Further, adsorption process also gets affected by the biological activities of microbes in due course of time. It has been shown that during early phase, adsorption is the main mechanism for removal; while upon aging, biodegradation becomes prominent (Kim and Kang 2008). As temperature affects the microbial activities, effect of increasing temperature on haloacetic acid removal was shown by Kim and Kang. Upon enhancing the temperature from 2 °C to 25 °C, significant rise in the haloacetic acid removal was noticed; however, similar effect could not be seen for trihalomethane removal (Kim and Kang 2008). Thus, performance of the GAC for DBPs removal can be modulated by optimizing various operational parameters.

Wang et al. reported that GAC can be efficiently used in the point-of-use devices for removal of haloacetic acids and trihalomethanes (Wang et al. 2015). The occurrence of bioactivity on the surface of GAC helps significantly in the removal of biodegradable DBPs, viz. haloacetic acids, aldehydes, and ketones (Chowdhury 2013; Wang et al. 2015). Other reaction media conditions, such as higher water temperature, long contact time, mixing of native and activated GAC further enhance the removal efficiency. Thus, GAC offers a good opportunity for the removal of

DBPs. However, need of frequent change of GAC makes the process little unacceptable. Zhang et al. tested the performance of GAC for removing the DBPs in a water treatment facility using batch as well as column test. Coconut shell-based GAC was found to have the highest adsorption capacity of 2.83 mg/g, compared to bituminous coal-based carbons (Zhang et al. 2022). Adsorption experiment was run up to 96 h and reduction in chlorine residual was found to be decreased from 10 mg/L to 0.4 mg/L at the end of experiment. Small-scale column tests also concluded that life of GAC increases with increase in the empty bed contact time (Liu et al. 2022). However, adsorption of trihalomethanes is rather poor onto the activated carbon surface. Freundlich *K* values of 93 μg/g and 241 μg/g for the chloroform and bromodichloromethane, respectively, indicate that these compounds are not fairly adsorbed onto the carbon (Speth and Miltner 1998). Therefore, frequent reactivation of GAC becomes necessary for effective removal of trihalomethanes.

Regeneration of GAC needs to be carried out on a frequent basis, *esp.* in the media where EBCT is short. Several researches have been carried out to look for the elimination of haloacetic acids and trihalomethanes using GAC. However, contradictory results were obtained as few researchers reported that adsorption of haloacetic acids is more compared to trihalomethanes (Xie 2004), while others reported vice versa (Xie and Zhou 2002). Xie and Zhou reported that at least 40% removal of five haloacetic acids (viz. monochloroacetic acid, dichloroacetic acid, trichloroacetic acid, monobromoacetic acid, and dibromoacetic acid) could be seen in the initial 60 days, while approximately complete removal could be seen after 60 days (Xie and Zhou 2002).

It is further suggested that GAC adsorption process along with the biological degradation of DBPs can be effectively employed for controlling the DBPs. Moreover, it is also recommended that replacement of GAC should be carried out in the month of summer, as it helps to control trihalomethanes along with promoting the biological activity in warmer temperatures (Xie 2004).

5 Conclusion

Water disinfection is the fundamental part of any treatment process if the water is intended to be used for drinking, bathing, or for any such purposes; in order to minimize the harmful impacts of pathogenic organisms. Chlorination is generally used for water disinfection, being the simple and cost-effective process for large-scale water supplies. However, post-treatment formation of disinfection (chlorination) by-products is another challenge to deal with, as these products pose significant risk to the human health. Therefore, removal of DBPs and/or its precursors is of paramount importance. Though there are various chemical and biological processes, physico-chemical processes of DBPs removal might play a substantial role in this respect. Enhanced coagulation and adsorption through activated carbon have been identified as some of the best available physico-chemical techniques. Moreover, both of these techniques are employable in the existing water treatment facilities which make their adaptation easier for the public health engineers. Nevertheless, both these processes can be made more efficient by optimizing various process parameters, design, and operating conditions in order to enhance the desired output. Research and development in this direction, along with the effective policy formulation are further needed to accomplish the target of sustainable safe drinking water supplies to everyone.

References

American Chemical Society, Key statistics for bladder cancer. Available online at https://www.cancer.org/cancer/bladder-cancer/about/key-statistics.html. Accessed Sep 5 2017

Aziz KHH, Miessner H, Mahyar A, Mueller S, Kalass D, Moeller D, Omer KM (2019) Removal of dichloroacetic acid from aqueous solution using non-thermal plasma generated by dielectric barrier discharge and nano-pulse corona discharge. Sep Purif Technol 216:51–57. https://doi.org/10.1016/j.seppur.2019.01.074

Bedner M, MacCrehan WA, Helz GR (2004) Making chlorine greener: investigation of alternatives to sulfite for dechlorination. Water Res 38:2505–2514. https://doi.org/10.1016/j.watres.2004.03.010

Bellar TA, Lichtenberg JJ, Kroner RC (1974) The occurrence of organohalides in chlorinated drinking water. J Am Water Works Assoc 66:703. https://doi.org/10.1002/j.1551-8833.1974.tb02129.x

Bottoni P, Bonadonna L, Chirico M, Caroli S, Zaray G (2014) Emerging issues on degradation by-products deriving from personal care products and pharmaceuticals during disinfection processes of water used in swimming pools. Microchem J 112:13–16. https://doi.org/10.1016/j.microc.2013.09.001

CDC (2009) Disinfection by-products (DBPs). Available online at https://www.cdc.gov/biomonitoring/pdf/THM-DBP_FactSheet.pdf. Accessed Aug 5 2023

Chaukura N, Marais SS, Moyo W, Mbali N, Thakalekoala LC, Ingwani T, Mamba BB, Jarvis P, Nkambule TTI (2020) Contemporary issues on the occurrence and removal of disinfection byproducts in drinking water–a review. J Environ Chem Eng 8:103659. https://doi.org/10.1016/j.jece.2020.103659

Chellam S, Sari MA (2016) Aluminium electrocoagulation as pretreatment during microfiltration of surface water containing NOM: a review of fouling, NOM, DBP, and virus control'. J Hazardous Material 304:490–501. https://doi.org/10.1016/j.jhazmat.2015.10.054

Chen S, Deng J, Li L, Gao N (2018) Evaluation of disinfection by-product formation during chlor(am)ination from algal organic matter after UV irradiation. Environ Sci Pollut Res 25:5994–6002. https://doi.org/10.1007/s11356-017-0918-x

Chiu PC (2013) 'Applications of zero-valent iron (ZVI) and nanoscale ZVI to municipal and decentralized drinking water systems–a review', novel solutions to water pollution, chapter 14. ACS Symp Ser 1123:237–249. https://doi.org/10.1021/bk-2013-1123.ch014

Chowdhury S (2013) Trihalomethanes in drinking water: effect of natural organic matter distribution. Water SA 39:1–17. https://doi.org/10.4314/wsa.v39i1.1

Chuang YH, Szczuka A, Shabani F, Munoz J, Aflaki R, Hammond SD, Mitch WA (2019) Pilot-scale comparison of microfiltration / reverse osmosis and ozone / biological activated carbon with UV/hydrogen peroxide or UV/ free chlorine AOP treatment for controlling disinfection byproducts during wastewater reuse. Water Res 152:215–225. https://doi.org/10.1016/j.watres.2018.12.062

Costet N, Villanueva CM, Jaakkola JJK, Kogevinas M, Cantor KP, King WD, Lynch CF, Nieuwenhuijsen MJ, Cordier S (2011) Water disinfection by-products and bladder cancer: is there a European specificity? A pooled and meta-analysis of European case-control studies'. Occup Environ Med 68:379–385. https://doi.org/10.1136/oem.2010.062703

Cummings L, Summers RS (1994) Using RSSCTs to predict field-scale GAC control of DBP formation. J Am Water Works Assoc 86. Available online at https://www.osti.gov/biblio/7081987. Accessed Feb 6 2023

Daels MRN, Radetic M, De Clerck K, Van Hulle SWH (2015) Electrospun nanofiber membranes functionalized with TiO_2 nanoparticles: evaluation of humic acid and bacterial removal from polluted water. Sep Purif Technol 149:488–494. https://doi.org/10.1016/j.seppur.2015.06.016

Erdem CU, Ateia M, Liu C, Karanfil T (2020) Activated carbon and organic matter characteristics impact the adsorption of DBP precursors when chlorine is added prior to GAC contactors. Water Res 184:116146. https://doi.org/10.1016/j.watres.2020.116146

Fabris R, Chow CWK, Drikas M, Eikebrokk B (2008) Comparison of NOM character in selected Australian and Norwegian drinking waters. Water Res 42:4188–4196. https://doi.org/10.1016/j.watres.2008.06.023

Giasuddin ABM, Kanel SR, Choi H (2007) Adsorption of humic acid onto nanoscale zerovalent iron and its effect on arsenic removal. Environ Sci Technol 41:2022–2027. https://doi.org/10.1021/es0616534

Grellier J, Bennett J, Patelarou E, Smith RB, Toledano MB, Rushton L, Briggs DJ, Nieuwenhuijsen MJ (2010) Exposure to disinfection by-products, fetal growth, and prematurity: a systematic review and meta-analysis. Epidemiology 21:300–313. https://doi.org/10.1097/EDE.0b013e3181d61ffd

Gu B, Schmitt J, Chen Z, Liang L, McCarth JF (1994) Adsorption and desorption of natural organic matter on iron oxide: mechanisms and model. Environ Sci Technol 28:38–46. https://doi.org/10.1021/es00050a007

Guo J, Khan S, Cho SH, Kim J (2019) ZnS nanoparticles as new additive for polyethersulfone membrane in humic acid filtration. J Ind Eng Chem 79:71–78. https://doi.org/10.1016/j.jiec.2019.05.015

Guo ZB, Lin YL, Bin X, Hu CY, Huang H, Zhang TY, Hai CW, Yun GN (2016) Factors affecting THM, HAN, and HNM formation during UV-chlor(am)ination of drinking water. Chem Eng J 306:1180–1188. https://doi.org/10.1016/j.cej.2016.08.051

Hao A, Kobayashi S, Xia D, Mi Q, Yan N, Su M, Lin A, Zhao M, Iseri Y (2021) Controlling eutrophication via surface aerators in irregular-shaped urban ponds. Water 13:3360. https://doi.org/10.3390/w13233360

Heller-Grossman L, Idin A, Limoni-Relis B, Rebhun M (1999) Formation of cyanogen bromide and other volatile DBPs in the disinfection of bromide-rich lake water. Environ Sci Technol 33:932–937. https://doi.org/10.1021/es980147e

Hoehn R, Barnes DB, Thompson B, Randall CW, Grizzard T, Shaffer P (1980) Algae as sources of trihalomethane precursors. J Am Water Works Assoc 72:344. https://doi.org/10.1002/J.1551-8833.1980.TB04528.X

Hozalski RM, Zhang L, Arnold WA (2001) Reduction of haloacetic acids by Fe^0: implications for treatment and fate. Environ Sci Technol 35:2258–2263. https://doi.org/10.1021/es001785b

Karimi AA, Singer PC (1991) Trihalomethane formation in open reservoirs. J Am Water Works Assoc 83:84–88. https://doi.org/10.1002/j.1551-8833.1991.tb07120.x

Kim J (2009) Fate of THMs and HAAs in low TOC surface water. Environ Res 109:158–165. https://doi.org/10.1016/j.envres.2008.11.003

Kim J, Kang B (2008) DBPs removal in GAC filter-adsorber. Water Res 42:145–152. https://doi.org/10.1016/j.watres.2007.07.040

Leenheer JA, Aiken GR, Woodside G, O'Connor-Patel K (2007) DOM in the recharge waters of the Santa Ana River basin. J Am Water Works Assoc 99:118–131. https://doi.org/10.1002/j.1551-8833.2007.tb07960.x

Li XF, Mitch WA (2018) Drinking water disinfection byproducts (DBPs) and human health effects: multidisciplinary challenges and opportunities. Environ Sci Technol 52:1681–1689. https://doi.org/10.1021/acs.est.7b05440

Liu Z, Mills EC, Mohseni M, Barbeau B, Berube PR (2022) Biological ion exchange as an alternative to biological activated carbon removal: impact of temperature and empty bed contact time (EBCT). Chemosphere 288:132466. https://doi.org/10.1016/j.chemosphere.2021.132466

Lu J, Zhang T, Ma J, Chen Z (2009) Evaluation of disinfection by-products formation during chlorination and chloramination of dissolved natural organic matter fractions isolated from a filtered river water. J Hazard Mater 162:140–145. https://doi.org/10.1016/j.jhazmat.2008.05.058

Maliva RG, Missimer TM (2010) Aquifer storage and recovery: developing sustainable water supplies. IDA J Desalination Water Reuse 2:74–80. https://doi.org/10.1179/ida.2010.2.2.74

Marais SS, Ncube EJ, Msagati TAM, Mamba BB, Nkambule TTI (2018) Comparison of natural organic matter removal by ultrafiltration, granular activated carbon filtration, and full scale conventional water treatment. J Environ Chem Eng 6:6282–6289. https://doi.org/10.1016/j.jece.2018.10.002

Matilainen A, Sillanpaa M (2010) Removal of natural organic matter from drinking water by advanced oxidation processes. Chemosphere 80:351–365. https://doi.org/10.1016/j.chemosphere.2010.04.067

Miklos DB, Remy C, Jekel M, Linden KG, Drewes JE, Hubner U (2018) Evaluation of advanced oxidation processes for water and wastewater treatment: a critical review. Water Res 139:118–131. https://doi.org/10.1016/j.watres.2018.03.042

Mishra BK, Gupta SK, Sinha A (2014) Human health risk analysis from disinfection by-products (DBPs) in drinking and bathing water of some Indian cities. J Environ Health Sci Eng 12:73–82. https://doi.org/10.1186/2052-336X-12-73

Moona N, Holmes A, Wunsch UJ, Pettersson TJR, Murphy KR (2021) Full-scale manipulation of the empty bed contact time to optimize dissolved organic matter removal by drinking water biofilters. ACS EST Water 1:1117–1126. https://doi.org/10.1021/acsestwater.0c00105

Najm IN, Snoeyink VL, Lykins BW, Adams JQ (1991) Using powdered activated carbon: a critical review. J Am Water Works Assoc 83:65–76. https://doi.org/10.1002/j.1551-8833.1991.tb07087.x

Ndlangamandla NG, Kuvarega AT, Msagati TAM, Mamba BB, Nkambule TTI (2018) A novel photodegradation approach for the efficient removal of natural organic matter (NOM) from water. Physics and Chemistry of the Earth, Parts A/B/C 106:97–106. https://doi.org/10.1016/j.pce.2018.05.011

Odziemkowski MS, Gui L, Gillham RW (2000) Reduction of N-nitrosodimethylamine with granular iron and nickel-enhanced iron. 2. Mechanistic studies. Environ Sci Technol 34:3495–3500. https://doi.org/10.1021/es9909780

Oliver BG, Shindler DB (1980) Trihalomethanes from the chlorination of aquatic algae. Environ Sci Technol 14:1502–1505. https://doi.org/10.1021/es60172a004

Oller I, Malato S, Sanchez-perez JA (2011) Combination of advanced oxidation processes and biological treatments for wastewater decontamination: a review. Sci Total Environ 409:4141–4166. https://doi.org/10.1016/j.scitotenv.2010.08.061

Pakharuddin NH, Fazly MN, Sukari SHA, Tho K, Zamri WFH (2021) Water treatment process using conventional and advanced methods: a comparative study of Malaysia and selected countries. IOP Conference Series: Earth Environ Sci 880:012017. https://doi.org/10.1088/1755-1315/880/1/012017

Pavelic P, Nicholson BC, Dillon PJ, Barry KE (2005) Fate of disinfection byproducts in groundwater during aquifer storage and recovery with reclaimed water. J Contam Hydrol 77:351–373. https://doi.org/10.1016/j.jconhyd.2005.04.001

Pearson CR, Hozalski RM, Arnold WA (2005) Degradation of chloropicrin in the presence of zero-valent iron. Environ Toxicol Chem 24:3037–3042. https://doi.org/10.1897/04-614Ra.1

Plessl K, Russ A, Vollprecht D (2022) Application and development of zero-valent iron (ZVI) for groundwater and wastewater treatment. Int J Environ Sci Technol 20:6913. https://doi.org/10.1007/s13762-022-04536-7

Postigo C, Emiliano P, Barcelo D, Valero F (2018) Chemical characterization and relative toxicity assessment of disinfection byproduct mixtures in a large drinking water supply network. J Hazard Mater 359:166–173. https://doi.org/10.1016/j.jhazmat.2018.07.022

Richardson SD, Postigo C (2015) 'Formation of DBPs: state of the science'. Chapter 11. In: Karanfil T, Mitch B, Westerhoff P, Xie Y (eds) Recent advances in disinfection byproducts. American Chemical Society, Washington DC, pp 189–214. https://doi.org/10.1021/bk-2015-1190.ch011

Scheili A, Rodriguez MJ, Sadiq R (2015) Seasonal and spatial variations of source and drinking water quality in small municipal systems of two Canadian regions. Sci Total Environ 508:514–524. https://doi.org/10.1016/j.scitotenv.2014.11.069

SDGs (2015) United Nations Sustainable Development Goals. Available online at https://sdgs.un.org/goals. Accessed Aug 5 2023

Sedlak DL, Von Gunten U (2011) The chlorine dilemma. Science 331:42–43. https://doi.org/10.1126/science.1196397

Sharma A, Ahmad J, Flora SJS (2018) Application of advanced oxidation processes and toxicity assessment of transformation products. Environ Res 167:223–233. https://doi.org/10.1016/j.envres.2018.07.010

Singer PC (1994) Control of disinfection by-products in drinking water. J Environ Eng 120:727–744. https://doi.org/10.1061/(ASCE)0733-9372(1994)120:4(727)

Souaya EM, Abdullah AM, Mossad M (2015) Factors affecting on formation of DBPs in greater Cairo drinking water. Org Chem Curr Res S5:002. https://doi.org/10.4172/2161-0401.S5-002

Speth TF, Miltner RJ (1998) Technical note: adsorption capacity of GAC for synthetic organics. J Am Water Works Assoc 90:72. Available online at https://www.osti.gov/biblio/599601. Accessed Feb 5 2023

Srivastav AL, Kaur T (2020) Factors affecting the formation of disinfection by-products in drinking water: human health risk. In: Disinfection by-products in drinking water. Elsevier, pp 433–450. https://doi.org/10.1016/B978-0-08-102977-0.00019-6

Sun M, Lowry GV, Gregory KB (2013) Selective oxidation of bromide in wastewater brines from hydraulic fracturing. Water Res 47:3723–3731. https://doi.org/10.1016/j.watres.2013.04.041

Symons JM, Bellar TA, Carswell JK, DeMarco J, Kropp KL, Robeck GG, Seeger DR, Slocum CJ, Smith BL, Stevens AA (1975) National organics reconnaissance survey for halogenated organics in drinking water. J Am Water Works Assoc 67:634

Tsitsifli S, Kanakoudis V (2018) Disinfection impacts to drinking water safety–a review. PRO 2:603. https://doi.org/10.3390/proceedings2110603

US EPA (1998) Stage 1 disinfectant and disinfection byproduct rule. Washington, DC. Available online at https://www.epa.gov/dwreginfo/stage-1-and-stage-2-disinfectants-and-disinfection-byproducts-rules. Accessed Feb 2 2023

US EPA (1999) Enhanced coagulation and enhanced precipitative softening guidance manual. Available online at https://nepis.epa.gov/Exe/ZyPDF.cgi/200021WV.PDF?Dockey=200021WV.PDF. Accessed Feb 8 2023

US EPA (2010) Comprehensive disinfectants and disinfection byproducts rules (Stage 1 and Stage 2): quick reference guide. Available online at https://nepis.epa.gov/Exe/ZyPDF.cgi?Dockey=P100C8XW.txt. Accessed Feb 9 2023

US EPA (2022) Surface water treatment rules. Available online at https://www.epa.gov/dwreginfo/surface-water-treatment-rules. Accessed Feb 9 2023

Uyak V, Toroz I (2007) Investigation of bromide ion effects on disinfection by-products formation and speciation in an Istanbul water supply. J Hazard Mater 149:445–451. https://doi.org/10.1016/j.jhazmat.2007.04.017

Valdivia-Garcia M, Weir P, Graham DW, Werner D (2019) Predicted impact of climate change on trihalomethanes formation in drinking water treatment. Sci Rep 9:9967. https://doi.org/10.1038/s41598-019-46238-0

Wachter JK, Andelman JB (1984) Organohalide formation on chlorination of algal extracellular products. Environ Sci Technol 18:811–817. https://doi.org/10.1021/es00129a002

Waller K, Swan SH, DeLorenze G, Hopkins B (1998) Trihalomethanes in drinking water and spontaneous abortion. Epidemiology 9:134–140

Wang X, Mao Y, Tang S, Yang H, Xie YF (2015) Disinfection byproducts in drinking water and regulatory compliance: a critical review. Front Environ Sci Eng 9:3–15. https://doi.org/10.1007/s11783-014-0734-1

Wei J, Ye B, Wang W, Yang L, Tao J, Hang Z (2010) Spatial and temporal evaluations of disinfection by-products in drinking water distribution systems in Beijing, China'. Sci Total Environ 408:4600–4606. https://doi.org/10.1016/j.scitotenv.2010.06.053

WHO (2011) Guidelines for drinking water, 4th edition. World Health Organization, Geneva. Available online at https://apps.who.int/iris/bitstream/handle/10665/44584/9789241548151_eng.pdf. Accessed Jan 27 2011

Wright JM, Evans A, Kaufmann JA, Rivera-Nunez Z, Narotsky MG (2017) Disinfection by-product exposures and the risk of specific cardiac birth defects. Environ Health Perspect 125:269–277. https://doi.org/10.1289/EHP103

Xie L, Shang C (2005) Role of humic acid and quinone model compounds in bromate reduction by zero-valent iron. Environ Sci Technol 39:1092–1100. https://doi.org/10.1021/es049027z

Xie YF (2004) Disinfection byproducts in drinking water: formation, analysis, and control'. Lewis Publishers, CRC Press, Boca Raton, Florida

Xie YF, Zhou H (2002) Use of BAC for HAA removal: part II, column study. Journal of American Water Works Association 94:126–134. Available online at https://www.jstor.org/stable/41298639. Accessed Feb 7 2023

Xu D, Bai L, Tang X, Niu D, Luo X, Zhu X, Li G, Liang H (2019) A comparison study of sand filtration and ultrafiltration in drinking water treatment: removal of organic foulants and disinfection by-product formation. Sci Total Environ 691:322–331. https://doi.org/10.1016/j.scitotenv.2019.07.071

Yang L, Chen X, She Q, Cao G, Liu Y, Chang VWC, Tang CY (2018) Regulation, formation, exposure, and treatment of disinfection byproducts (DBPs) in swimming pool waters: a critical review. Environ Int 121:1039–1057. https://doi.org/10.1016/j.envint.2018.10.024

Yang Z, Zhou Y, Feng Z, Rui X, Zhang T, Zhang Z (2019) A review on reverse osmosis and nanofiltration membranes for water purification. Polymers 11:1252. https://doi.org/10.3390/polym11081252

Zhang L, Arnold WA, Hozalski RM (2004) Kinetics of haloacetic acid reactions with Fe(0). Environ Sci Technol 38:6881–6889. https://doi.org/10.1021/es049267e

Zhang XL, Yang H, Wang X, Karanfil T, Xie YF (2015) Trihalomethane hydrolysis in drinking water at elevated temperatures. Water Res 78:18–27. https://doi.org/10.1016/j.watres.2015.03.027

Zhang Y, Silverman I, Shayan SI, Zhang Q, Mulford L, Iranipour GT (2022) Removal of disinfection byproduct precursors by granular activated carbon: Hillsborough County water treatment facility case study. Case Studies Chem Environ Eng 6:100254. https://doi.org/10.1016/j.cscee.2022.100254

Zhou X, Zheng L, Chen S, Du H, Raphae BMG, Song Q, Wu F, Chen J, Lin H, Hong H (2019) Factors influencing DBPs occurrence in tap water of Jinhua region in Zhejiang Province, China. Ecotoxicol Environ Safety 171:813–822. https://doi.org/10.1016/j.ecoenv.2018.12.106

Zhu S, Wu X, Zhou M, Ge X, Yang X, Wang N, Lin X, Li Z (2022) Effects of harvesting intensity on the growth of *Hydrilla verticillata* and water quality. Sustainability 14:15390. https://doi.org/10.3390/su142215390

Availability and Minimization of Nitrogenous Disinfectant By-Products in Drinking Water

Richa Saxena, Himanshu Gupta, and Soniya Dhiman

Abstract In view of high genotoxicity and cytotoxicity, the existence of nitrogenous disinfection by-products (N-DBPs) like haloacetonitriles, haloacetamides, nitrosamines, halonitromethanes, and nitrosamines is of higher concern as compared to regulated disinfectant by-products (DBPs). Due to exposure of DBPs the risk of bladder cancer is reported to be increased. The application of chlorine and other disinfectants, around 600 DBPs in drinking water was reported. Type of disinfectant is a main factor in formation of nitrogenous disinfectant by-products, nitrogen can be derived from chloramination, from the disinfectants or organic precursors such as organic nitrogen in dissolved state. The existing analytical methodologies detect three major groups of N-DPB Haloacetonitriles, Haloacetamides, and Halonitromethanes. Among these, Haloacetonitriles are available in higher concentration. Particularly, the hydrophobic acid formation is reported as the main source of trihalomethane precursors and may be removed by coagulation. Coagulation is useful for removal of colloids, but it cannot be applied to neutral fractions and similar reactive hydrophilic base. Filtration is also found to be ineffective for water treatment. Therefore, single method cannot be used to control all nitrogenous disinfectants by-products. This chapter describes the strategies to control disinfectant by-product formation based on water-specific assessment of precursor sources. The application of membranes and biodegradation to reduce DBPs is also discussed.

Keywords Nitrogenous disinfection by-products · Drinking water · Haloacetonitriles · Precursor sources · Water treatment

---

R. Saxena
Department of Physics, School of Sciences, IFTM University,
Moradabad, Uttar Pradesh, India

H. Gupta (✉)
Department of Chemistry, School of Sciences, IFTM University,
Moradabad, Uttar Pradesh, India

S. Dhiman
Department of Biochemical Engineering and Biotechnology, Indian Institute of Technology
Delhi, New Delhi, India

S. Madhav et al. (eds.), *Drinking Water Disinfection By-products*,
https://doi.org/10.1007/978-3-031-49047-7_12

1 Introduction

Disinfection is an important process for treatment of drinking water to ensure water safety in urban water supply system. Pathogenic microorganism can be made inactive effectively by reducing the possibility of water borne diseases by which people get infected. Unintentionally disinfectant by-products (DBPs) are generated by the reaction of disinfectants like chloramine with anthropogenic contaminants, chlorine, natural organic matter, and halides in water (Richardson et al. 2007; Mazhar et al. 2020). DBPs have exerted negative impact on human health so they have been attractive concerns. In the recent years nitrogenous disinfection by-products have been one of the most concerned disinfection. The toxicity of nitrogenous by-products has been widely detected in drinking water that is generally higher than that of regulated disinfectant by-products. The application of chlorine and other disinfectants, particularly ozone, chlorine dioxide, and chloramines results in the detection of more than 600 disinfectants by products in simulated laboratory disinfection tests or in drinking water (Chu et al. 2013). One of the most critical treatment processes for drinking water is chlorination due to its effectiveness against various kinds of pathogens as well as usefulness for public health issues related to drinking water. Chlorination is mostly utilized for the treatment of sewage. Unregulated nitrogenous DBPs (N-DBPs) like haloacetonitriles and haloacetamides are formed because of the use of combined chlorine. N-DBPs are reported to be higher geno and cytotoxic in comparison with general disinfectant by-products.

The continuous efforts seek to control regulated disinfectant by-products such as haloacetic acid (HAAs) and trihalomethanes (THMs) mostly by the use of other disinfectants. Most of the facilities for drinking water utilize secondary disinfectant, i.e. chloramine (Krasner et al. 2009; Cowman and Singer 1995). Various substances in different forms have been utilized for remediation of water from different contaminants (Gupta and Gupta 2016; Singh et al. 2018; Kumar and Gupta 2020; Singh et al. 2020; Gupta 2018; Gupta and Gupta 2018; Singh et al. 2022; Kumar et al. 2023). But chlorinated water has expanded the generation of N-DBPs, which are potentially toxic compounds. Drinking water disinfection has prevented major water borne diseases like typhoid and cholera. There is a lack of treatment of water in most developing countries. Fragile group such as children, adult, and people with weak immune system, under medication for different diseases or surgery are more prone to the toxicity. In comparison with regulated disinfectant by-products, N-DBPs that are reported in the drinking water are more frequently genotoxic, cytotoxic, and developmentally toxic (Richardson et al. 2003; Zhao et al. 2010; Zhai and Zhang 2011). As nitrogen source is a prerequisite for the formation of nitrogen-disinfectant by-products in drinking water, influence of sewage and algae blooms on water sources may be responsible for the increase in the occurrence of nitrogenous disinfectant by products. As early as in the 1980s, dihaloacetonitriles (DHAN) was quantified as the first N-DBPs group in raw chlorinated water (Oliver 1983; Trehy and Bieber 1981). Based on the reaction condition and compound, type of disinfectant is an essential factor in the formation of N-DBPs (Fig. 1). The nitrogen may be

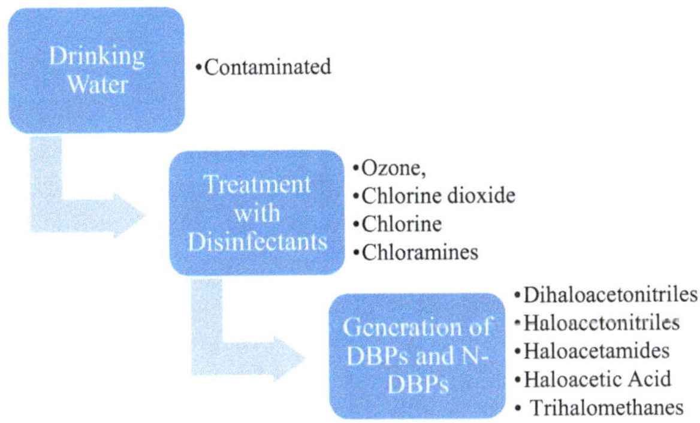

Fig. 1 Schematic representation of generation of DBPs and N-DBPs

derived either through the dissolved organic nitrogen (DON), organic precursors, or chloramination from disinfectant.

Human activities such as release of wastewater effluent and algal activity have great impact on drinking water (Mitch et al. 2009). As these two are DON, which enriched their presence may lead to increase in the concentration of various N-DBPs. Based on the available reports, amino acids are referred to as precursors of Cyanogen halides, Haloacetamides (HAcAms), and Haloacetonitriles (HANs) (Hirose et al. 1988; Ram 1985; Reckhow et al. 2001; Trehy et al. 1986). Therefore, understanding the availability and minimizing the occurrence of N-DBPs in the water industry is currently a challenging task. To study correlation between cancer mortality and drinking water chlorination, numerous reports are available in literature. Researchers have reported the clear connection between occurrence of urinary bladder cancer and chlorinated drinking water exposure. Adverse reproductive outcomes are also caused by the exposure to by-products of chlorination in water. Drinking water is used for various purposes including bathing, showering, laundering, washing and cleaning, etc. The exposure and uptake potential are high due to contact with skin, inhalation with air, and by drinking water. Water needs appropriate treatment methodology for removing all disease-causing germs because of contamination from various natural sources and manmade interventions in drinking water.

Although N-DBPs plays very important role in carcinogenic potential to humans, however none of species of N-DBPs are generally regulated for drinking water through the large governmental bodies anywhere in the world. Even though researchers worldwide are playing an important role in investigating the availability and N-DBPs toxicity, organic matter precursor, mechanism of formation, important technology to reduce their formation and health risk for human are not clear.

2 Types of Nitrogenous By-Products

In order to kill microorganisms like viruses, protozoa and bacteria drinking water is disinfected by use of chlorine, which may cause serious illness and sometimes results in death. Nowadays, chemical disinfectants are most commonly employed for treatment of drinking water. It is discovered by scientists that disinfectant by-products are generated by the reaction of organic matter with chlorine in freshwater during the water treatment process (Bellar et al. 1974). Due to the reaction of chemical disinfectants with disinfectant by-products precursors disinfectant by products are formed. Bromine serves as an inorganic precursor while natural organic matter serves as the organic precursor measured by total organic carbon. Various chlorinated DBPs like bromoform, haloacetic acids, 3-chloro-4-(dichloromethyl)-5-hydroxy-2(5H)furanone (MX), bromodichloro methane (BDCM), trihalomethanes (THMs), chloral hydrate, chloroform, dibromochloro methane (DBCM) and haloacetonitriles are generally formed (Singer et al. 1995; Twort and Law 1994; Xie and Reckhow 1993; Drinking water chlorination 2003; Janda and Svecova 2000). THMs are found to be major by-product of chlorination and potentially hazardous to human health.

In recent years, nitrogenous disinfection by-products like haloacetamides (HAcAms), nitroamines, haloacetonitriles (HANs), halonitromethanes (HNMs), cyanogen halides are reported in a number of reports (Shah and Mitch 2011; Bond et al. 2011; Mitch et al. 2009). These groups are found in lower concentration in drinking water than Trihalomethanes and haloacetic acid. Emergence of haloacetic acid and Trihalomethanes is reduced by using chloramines instead of chlorine as final disinfectant, but it may lead to the increased formation of other N-DBPs like cyanogen chloride (CNCl) and N-nitrosodimethylamine (NDMA). Dissolved organic nitrogen (DON) in surface water acts as a significant N-DBPs precursor and has an adverse effect on security of drinking water (Lee et al. 2007). The precursors of DON which are studied most extensively are amino acids. Conventional drinking water treatment methods are not sufficient to remove amino acids that are widespread in surface water. In 1980 from chlorinated water, the dihaloacetonitriles (DHAN) have been reported. Proteins, peptides, and algae organic compounds consist of aspartic acid (Asp), an important component and is naturally occurring amino acid in DON (Liu et al. 2015; Chu et al. 2012a, b). From the twenty free amino acids tested, aspartic acid had one of the highest yield of N-DBPs (Yang et al. 2012). Four HANs, dichloroacetonitrile (DCAN), bromochloroacetonitrile (BCAN) and trichloroacetonitrile (TCAN), chloropicrin and cyanogen chloride included in the survey of thirty-five treatment facilities of water in later decade (Krasner et al. 1989).

These nitrogenous by-products were also observed in 1997–1998 in drinking water treatment plants in survey of USA (McGuire et al. 2002). In drinking water, HAcAms were first reported in USA survey of 2000–2002. From the hydrolysis of HANs, DBPs group named, HAcAms are produced and they themselves degrade to corresponding HAAs (Glezer et al. 1999; Reckhow et al. 2001). In Chinese raw treated water nine HAcAm have also been investigated (Chu et al. 2013; Chu et al.

2014). Chloropicrin has also been observed in drinking water early research of DBPs (Merlet et al. 1985). In the USA nationwide 2000–2002 DBPs survey that sampled 12 drinking water treatment plants (DWTPs) receiving high precursor load a total of eight HNMs were reported (Krasner et al. 2006; Weinberg et al. 2002). Five HNMs, six HANs and CNCl were included in 11 USA DWTPs survey receiving water that were impacted by algal and waste water (Mitch et al. 2009). It has been noticed that upon reaction of monochloramine with chloroacetaldehyde, the formation of N, 2-dichloroacetamide is observed. In different parts of Europe four HANs have been reported in drinking water (Goslan et al. 2009; Goslan et al. 2014). As compared to chlorinated analogues, brominated DBPs are typically more genotoxic and cytotoxic. Other N-DBPs that exists as ozone disinfection by products in drinking water are cyanoformaldehyde, heptanentrile, and benzeneacetonitrile. In Israeli water other N-DBP was bromide with a concentration of 2 mgL^{-1} and chlorine dioxide disinfectants such as 3-bromopropanenitrile and 2,3,5-tribromopyrrole (Richardson et al. 2003). In aquatic environment, the nitrogenous organic matter which is most important and abundant was amino acids including peptides, proteins, and free amino acids (FAAs). These are originated from aquatic microorganisms, domestic wastewater and industrial effluents. It is reported by researchers that the nitrosamine precursor contribution from Haloacetic acid (HAAs) and microorganisms or their nitrogen containing components is negligible (Mitch and Sedlak 2004). The earlier studies have been focused mainly on amino acids, microorganism, proteins, and plasma in pristine states. The important substrate for microorganism is nitrogenous matter. These are observed in complex biochemical reactions like deamination, decarboxylation, and hydrolysis in natural state and can be transformed into other chemicals or amines.

3 Availability of Nitrogenous By-Products in Drinking Water

The existing analytical methodologies detect five major groups of N-DPBs namely Haloacetonitriles, Haloacetamides and Halonitromethanes, cyanogen halides and nitrosamines (Fig. 2). Among these, Haloacetonitriles are available in higher concentration.

3.1 Haloacetonitriles (HANs)

In a US survey, HANs, HNMs, HAcAms were reported in WTPs and HANs was found to be at higher concentration having median 1 $\mu g\ L^{-1}$ and 10 $\mu g\ L^{-1}$ (Shen 2021). During a survey in USA and Canada reported data for the assessment of formation, availability and control of the DBPs emerging do not have any regulation in the USA during the 1990s and 2000s. The different parts of the world including in France, Australia, and Spain were related to problems of medical odor. In Greece,

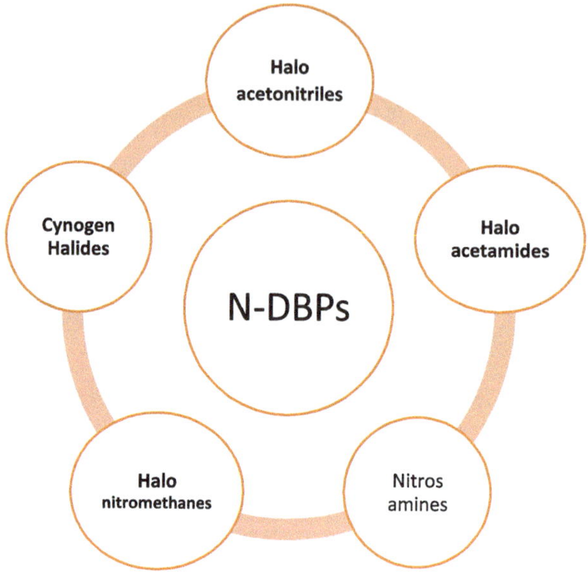

Fig. 2 Five major groups of nitrogenous disinfection by-products

various emerging DBPs have been studied. During chlorination and chloramination iodinated THMs occurs. Iodinated THM formation is a function of disinfection scenario and precursor. For chloraminated system in plants that don't use chlorine or ozone as preoxidant, the yield of iodinated THM was highest. Iodinated acids were detected. A study reported in the USA and Canada that in chlorinated and chloraminated drinking water, availability of two iodinated THMs and five iodinated acids was reported. Out of 759 analysis of samples, the most frequently reported NDBPs were BCAN, DCAN, and DBAN. In all disinfected samples, the median concentration of DCAN was found to be 0.9 μg L^{-1}. At level below World Health Organization guidelines, the measured HANs and other reported N-DBPs of 20 mg L^{-1} for DCAN and 70 mg L^{-1} for DBAN. For HANs species, DCAN and BCAN were the most prevalent species with concentrations 1 μg L^{-1} and 0.6–12 μg L^{-1} and 11 μg L^{-1}, respectively. For HAcAms species, dibromochloroacetamide (DBAcAm) and dichloroacetonitrile (DCAcAm) were prevalent species with median concentrations 0.6 μg L^{-1} and 1.3 μg L^{-1}, whereas maximum concentrations of 5.6 μg L^{-1} and 2.8 μg L^{-1}, respectively (Bond et al. 2011). Bromodichloronitromethane, dibromochloronitromethane, chloropicrin, and bromopicrin were prevalent species with maximum concentrations in the range 2 to 5 μg L-1 for HNMs species. NDMA was the critical species with maximum concentration in the range 45–1000 ng L^{-1} and other species were also found in water in low amounts (Bond et al. 2011).

3.2 Haloacetamides (HAcAms)

DCAcAm was found to be most prevalent species in US survey in 2000–2002 having 1.3 μg L^{-1} as median concentration. The maximum and median concentrations of sum of HAcAms were 7.4 mg L^{-1} and 1.4 mg L^{-1}, respectively. The median concentration for the two dihalogenated HAcAms-DCAcAm and DBAcAm were 0.6 and 0.7 mgL^{-1} having equivalent maxima being 4.5 and 5.1 mg L^{-1}. In China the concentration of Dichloroacetonitrile (DCAcAm) was compared with pre-chloramination DWTP reached a maximum of 1.8 mg L^{-1} (Chu et al. 2011). As compared to trichloroacetonitrile (TCAN) the detection of 2,2,2-Trichloroacetamide (TCAcAm) was common up to 1.7 mg L^{-1}. Upon application of chlorine dioxide before chloramine/chlorine at three sites, HAcAms were frequently identified.

3.3 Cyanogen Halides (CNX)

CNX formation takes place by post-chloramination of formaldehyde, a direct halogenation of amides results in the formation of HAcAms. HAcAms can be formed through the hydrolysis of HANs at alkaline pH (Ueno et al. 1996; Glezer et al. 1999) and through the hydrolysis and subsequent chlorination of cyanoacetic acid (Reckhow et al. 2001). CNX are generally formed at plants, which cause CNCl formation and chlorination. The CNCl median concentrations in treatment with free chlorine and chloroamination were 0.4 mg L^{-1} and 2.2 mg L^{-1}, respectively. In 2006–2007 survey of N-DBP, CNX formation was generally observed only in plants with chloramination having maximum and median formation as 7.8 mg L^{-1} and 2.6 mg L^{-1}, respectively. The formation of CNX is found low in plant in which ozone was employed before the biofiltration and chloramination/chlorination, indicating CNX precursors, formaldehyde and other biological treatment can effectively remove by-products that results from ozonation (Krasner et al. 2007).

3.4 Halonitromethanes (HNMs)

In 2000–2002 and 2006–2007, in 12 water treatment works (WTWs), the HNMs sum on median basis was 1 μg L^{-1} and the most frequently reported HNM is chloropicrin and was quantified at 0.2–0.5 μg L^{-1}, respectively. For chloropicrin wastewater does not serve as significant precursor source, as the levels observed in wastewater treatment works (WWTW) of US were rather low ranging from 0–0.6 and 0–0.7 μg L^{-1}, respectively, in chloraminated and chlorinated effluents. Chloropicrin yield from Microcystis aeruginosa and chlorella vulgaris were 0.1 to 0.8 μg mg C^{-1} (Bond et al. 2012; Yang et al. 2011), whereas in another study the formation of chloropicrin from Microcystis aeruginosa was found to be 1.1 μg L^{-1} (Fang et al.

2010). Except for one WWTW that generate chloropicrin at 0.9–1.5 μg L^{-1}, the concentration of 09 HNMs in effluent of 06 WWTWs of US was found lesser than minimum reporting level or non-detectable. HNM level was formed to be insignificant by same effluent of UV and ozone disinfection. Chloropicrin formed through ozonation followed by chlorination of a eutrophic lake source has been found to be considerably greater compared to mesotrophic source on treatment in equivalent manner.

3.5 Other N-DBPs

In chloraminated drinking water, hydrazine is carcinogenic and is detected at 0.5–2.6 ng L^{-1}. The ozone disinfection by-products have been detected as Benzeneacetonitrile, heptanentrile and cyanoformaldehyde (Richardson et al. 1999). In Israeli water, bromide and chlorine dioxide, 2, 3, 5-tribromopyrrole and 3-bromopropanenitrile nitrogenous by-products were recorded. Many pollutants are generated as by-products, intermediates or metabolites during degradation of are contaminants (Gupta and Dhiman 2023; Kaur et al. 2023; Gupta and Gupta 2015; Gupta 2016). There may be many other DBP, some of which may contain nitrogen that is also present in drinking water which can be identified on basis of chemical and toxicological models in future research. Table 1 shows the concentration of various nitrogenous disinfectant by-products in distribution samples, water samples and disinfected samples used in different reported studies. The median concentration of HAN4 was slightly enhanced in relation to final concentrations in water as 2.6 to 2.8 mg L^{-1}, respectively, whereas highest concentration of HAN was observed in the supply system distribution sample disinfected by chlorination-ozonation. However, BAN and CAN have been only observed above the MDLs, their maxima in distribution were 3.8 and 1.9 mg L^{-1}. The availability of the 02 monohalogenated HANs has not been reported to be linked with higher concentrations of other HANs.

4 Mechanism of Formation N-DBPs in Drinking Water

In order to formulate N-DBPs, nitrogen source is a prerequisite. The essential source is naturally dissolved organic nitrogen (DON). DON acts as complex organic macromolecule that consists of larger chemical groups. The occurrence of dissolved organic nitrogen was reported from the metabolism of plants, animals, and algae. DON involves pyrimidines, pyrroles, purines, nitriles, amino sugars, amides, proteins, amino acids widespread in water (Westerhoff and Mash 2002). The concentration of wastewater enhances, that consists of extensive dissolved organic nitrogen than that of natural raw water. A critical role is played by precursor for CNX and HANs. Kynurenic acid, aspartic acid, and tryptophan demonstrated the production

Table 1 Summary of various nitrogenous disinfectant by-products in different samples used in their studies (Content taken from original article by Bond et al. 2015 with permission through RightsLink)

Compound	Distribution samples			Water samples			Disinfected Samples			References
	Unit: μg L^{-1}									
	Median	**75%ile**	**Max**	**Median**	**75%ile**	**Max**	**Median**	**75%ile**	**Max**	
Chloronitromethane	0.7	1.4	3.2	1.3	1.6	3.5	0.8	1.5	3.5	Mitch et al. (2009)
Bromochloronitromethane	0.2	0.2	0.3	0.3	0.2	0.3	0.2	0.3	0.3	
Bromonitromethane	0.1	0.2	3.8	0.2	0.2	0.5	0.2	0.2	0.3	
Dichloronitromethane	0.2	0.2	0.3	0.1	0.2	0.2	0.1	0.2	0.3	
Chloropicrin	0.1	0.2	0.5	0.5	0.2	0.4	0.1	0.2	0.5	Krasner et al. (2006)
Dibromonitromethane	0.1	0.2	0.3	0.3	0.1	0.2	1.3	0.2	0.3	Mitch et al. (2009)
Chloroacetonitrile	1.0	1.2	1.9	0.6	0.6	1.0	1.0	1.2	1.9	Goslan et al. (2014); Krasner et al. (2006)
Bromoacetonitrile	0.2	0.4	3.8	0.7	1.3	3.2	0.2	0.3	3.8	
Dichloroacetonitrile	0.9	1.7	4.4	1.0	1.3	2.9	0.9	1.6	4.4	
Dibromochloroacetonitrile	1.0	1.9	8.1	0.2	0.3	0.7	0.2	0.2	0.7	
2,2 dichloroacetamide	0.7	1.2	4.5	2.1	0.7	1.2	0.6	1.1	4.5	

of DCAN. For the formulation of HANs, HAcAms, CNX and nitrosamines, a route is provided by aldehyde and amides. The reaction of monochloramine and acetaldehyde led to formation of DCAN. HAcAms are formed by direct halogenation of amides. The production of CNX takes place by post chlorination of formaldehyde. At alkaline pH, the hydrolysis of HANs can produce HAcAms (Ueno et al. 1996; Glezer et al. 1999). Increased algal organic matter (AOM) and organic matter (EfOM) in drinking water supplies usually increase formation of N-DBPs. For HNMs like chloropicrin in chlorination water, precursor is thought to be nitromethane moiety natural organic molecules (NOM), whereas extensive hydrophilic NOM substances such as nucleic acids, primary amines, amino sugars, and amino acids may produce a small amount in chloropicrin waters (Shan et al. 2012). Dimethylamine (DMA) may produce asymmetric 1,1-dimethylhydrazine and is precursor for nitroamines. For N-nitrosodimethylamine (NDMA), wastewater is found to be precursor as wastewater in chlorination increased the yield of NDMA. It is shown in previous studies that nitriles may be produced by aldehyde and monochloramine reaction and chlorination and ozonation of drinking water may produce common DBPs (Pedersen et al. 1999; Cloirec and Martin 1985).

4.1 Effect of pH

The sensitivity to pH in water of five N-DBPs is different, but in acidic pH most of them are stable. Disinfection at lower pH minimizes THMs, whereas disfavors other DBP groups at higher pH (Stevens et al. 1989). At pH 5, DCAN was stable as reported during a pilot investigation. With time DCAN decreases at pH 7 and it hardly formed at pH 9.4. After chlorination of aspartic acid at pH 5, 7, and 9 for 3 h, DCAN yield was found to be highest at pH 5. At the pH of 5, the formulation of DCAcAm was 0.2%, and 0.49% at pH 9 (Chu et al. 2010), showing that hydrolysis of DCAN is not only the formulation pathway. In US survey 2000–2002, DCAcAm was observed in highest pH range, i.e. 6.8–7.1. With an increase of pH in chlorinated water, an enhancement in CNX decomposition was also observed. The pH value of 7–8 serves as optimum condition for the formation of NDMA from UDMH (unsymmetrical 1,1-dimethylhydrazine).

4.2 Effect of Temperature

With an increase in temperature DCAN and DCAcAm concentration was reported to be decreased. With increase in temperature, TCMN concentration increases. The formation rate and decomposition rate of N-DBPs increase with an increase in temperature. Increasing the temperature may increase N-DBPs formation and decomposition rates (Zhang et al. 2013). With the increase in temperature, the yields of

DCAcAm and DCAN decrease, therefore at higher temperature, rate of decomposition were relatively higher compared to formation rates (Chen et al. 2017).

4.3 *Effect of Bromide*

Equivalent chlorine species are not effective as the bromine species. Brominated N-DBPs presents high frequency of toxicity and genotoxicity than chlorinated N-DBPs. After spiking with bromide molar DHAN formation increases to 73% for water source which is heavily allochthonous and 101% for water source which is largely autochthonous. The distribution of DHAN shifts the DCAN to BCAN and afterwards into DBAN. In DBP formation, bromamines plays an important role. In chloraminated drinking water, 2–2 Bromoiodoacetamide was detected (Plewa et al. 2008; Richardson et al. 2007).

4.4 *Effect of Chloramination*

The detection of NDMA is reported when chloramination exists in solution. The DCAN yield from chloramines was higher than by chlorine, which is found consistent with previous reports. In free chlorine solutions DCAN was not stable as compared in monochloramine solutions but during chloramination it may undergo hydrolysis to form DCAcAm, the rate of hydrolysis of DCAcAm is much lower in monochloramine solution than that in chlorine solutions (Chu et al. 2012a, b) with increase of contact time of free chlorine DCAcAm decreased. In monochloramine solutions, TCNM was not detected this shows that trichloronitromethane (TCNM) was not generated because of the reaction between Asp and monochloramine.

5 Detection of Nitrogenous By-Products

Chromatography related methods are used for pre-treatment of sample and identification or determination of disinfectant by-products. Pre-treatment of sample involves preservation, enrichment of sample as well as DBP extraction. It is useful for stabilization of disinfection by-products prior to the analysis. It is also useful for increasing analytical method sensitivity, potential interference reduction or elimination as well as to obtain more suitable form of analytes for better characterization. In public health protection and drinking water management, water disinfection serves as fundamental process as infectious water borne disease proliferation is prevented by it. After consumption or prolonged exposure, some of the DBP may cause adverse impact on human health. It is due to the applied disinfection techniques like ozonation, chloramination, and chlorination (Rook 1976; Evlampidou et al. 2020;

Villanueva et al. 2015). Upon application of methodologies which are non-targeted more than 700 disinfectants by products have been identified till now (Richardson and Plewa 2020).

In batch scale reactors, reservoir water samples were chlorinated and characterization of profile of dissolved organic matter was performed, which results in detection of more than 650 m/z signals. The potabilization trains of drinking water typically includes dissolved organic matter removal as well as filtration process in order to mitigate the final DBP concentration and remove precursors of potential DBPs (Williams et al. 2019). The techniques which are most widely used are liquid chromatography (LC) and gas chromatography (GC) for detecting DBPs present in water samples. The detectors which are most frequently applied are mass selective detectors (MS) coupled with GC and electron capture detectors (ECD). In comparison with MS detectors, ECD are more sensitive but these were not commonly employed for identification of unknown DBPs due to poor sensing. LC/MS method are applied in order to detect thermally unstable DBPs as well as DBPs with high-molecular-weight and polar nature. Total organic halides (TOX) methods are employed for quantification of unknown and known DBPs available in the drinking water. Accurate method for detection should be used for analysis of DBPs that are important for removal of DBPs from water. Various techniques are used for analysis of DBPs. GC analysis is the most commonly used method for detection of unknown DBPs in drinking water. For halogen compounds, ECD is mostly used as it is cheaper and has good sensitivity. MS is preferred for higher sensitivity, selectivity in detecting DBPs. LC technique is used for high molecular weight and low thermal stability compounds. The pretreatment DBP containing samples was performed by SPE. During pretreatment, prior to analyte extraction mixed mode solid phase extraction (SPE) was employed as preconcentration method with recovery of 80%.

For control of DBPs common strategies include removal of DBP precursor as well as application of alternative disinfectants. Various treatment processes like activated carbon adsorption, enhanced coagulation, nanofiltration, advanced oxidation process, and anion exchange can be used for removal of DBP precursors. To decrease turbidity, removal of the suspended particles, color, and pathogens in drinking water usually coagulation is applied. Activated carbon is very useful material for removal of contaminant from wastewater because of ability to hold and capture the liquid and gas phase particles due to their porous structure. Activated carbon has some special properties including easy maintenance, easy integration to current method, operation cost medium and regeneration, so the activated carbon can be used again for the same purpose. Mercury vapor lamps with low and high pressure are used for water treatment techniques that are based on ultraviolet light.

Many researchers have reported that hydrogen peroxide and UV light combination improves the removal efficiency of DBPs. An indirect reduction of NOM in water results in DBP production. A strong oxidant, ozone is usually employed for disinfection of wastewater, water and DBPs in water with smaller molecular weight. Ozone combination with hydrogen per oxide or ultraviolet provides control over oxidative process and high reactivity. A recently emerged technology is used for water treatment. Because of stability and lower energy band gap TiO_2 is the most widely used photo catalyst. This advanced oxidation technique has the ability to inactivate the microorganisms that are present in water.

During filtration, water passes through a membrane filtration process or different bed materials based porous structure. Some microorganisms available in water may be retained through the filter depending on pore size, suspended particles. Filters should be cleaned regularly to keep it more efficient so that contaminant cannot be added to water. Reverse osmosis, Microfiltration NF, and Ultrafiltration are used for achievement of highly efficient DBP removal. Various techniques and instrumental methods are utilized for detection and estimation of disinfection by-products in drinking water.

5.1 Analytical Treatment for N-DBPs

During disinfection in order to decrease the production of disinfection by-products like haloacetic acids and trihalomethanes in drinking water treatment plants, a shift to chloramination from chlorination was taken place, which is also useful to get water quality index. Basic analysis methods are essential for detection and quantification of DBPs and their precursors in water samples of laboratory and real water. These are used for studying the formation, occurrence, and treatment of nitrogenous disinfection by-products. Three steps involved for quantification and analytical determination of DBPs are pretreatment, separation of compound by chromatography and identification with detectors.

In authentic water, the concentration of N-DBPs is low and is below the detecting limits of analytical instruments (Bond et al. 2015; Bei et al. 2016). It is necessary to concentrate and isolate analytes from complex matrices for pre-treatment. In two different immiscible liquids usually an organic solvent and water, based on relative solubilities, separation of compound was performed using liquid–liquid extraction. Drinking water treatment plants (DWTPs) that consist of high organic nitrogen content with natural organic matter exploit source waters. This may lead to the enhancement in amount of some N-DBPs for insufficient addition of inorganic nitrogen content and for removal of organic nitrogen content. It is reported in previous studies that for the identification and detection of HAcAms and THMs, ethyl acetate, n-hexane and methyl tert-butyl ether, organic solvents may be utilized as extraction solvents (Chen et al. 2015; Park et al. 2016). The estimation of the production and decay of HAcAms and HANs in a distributed system both chloraminated and chlorinated system SDS test can be effectively used. This test is useful not only for the prediction of concentration of HANs and HAcAms and it is used for their speciation. The drawback of liquid–liquid extraction (LLE) is that it is time consuming, gives low recovery and for monitoring regular DBPs it gives high detection limit and low recovery, although it is a simple and very inexpensive method (Kinani et al. 2016). Solid phase extraction method replaces the LLE method for the determination of the target analytes at trace level. To separate analytes liquid/gas chromatography method is used. The isolation, concentration, and purification of samples SPE may be applied for analysis.

In many studies for identification and detection of other DBPs including HAcAms and NAs in water samples, SPE method was used (Chu et al. 2012a, b; Chu et al. 2015; Pérez-Ruiz et al. 2005). High-performance liquid chromatography (HPLC) is reported as beneficial instrument for detection of N-nitroso compounds in trace quantities as well as for N-NO bond cleavage based specific post-column approaches. Using hydrolysis with mineral acids and heating (Lee and Field 1987) or by denitrosation with hydrobromic acid-acetic anhydride (Wang et al. 1992; Fu et al. 1993) or UV irradiation (Stefan and Bolton 2012; Mitch et al. 2003) the cleavage can be carried out. To complete the particular type of separation, in high-performance liquid chromatography (HPLC) at high pressure, liquid is used to force a sample through stationary phase containing column that usually consists of spherically or irregularly shaped particles whereas in gas chromatography (GC) carrier gas usually an inert gas like helium or an unreactive gas such as nitrogen is used to force the sample through stationary phase containing column generally consists of microscopic layer of polymer or liquid on an inert solid support. For separation of low molecular weight, volatile, semi-volatile and thermally stable N-DBPs, gas chromatography is used. Because of nature of HPLC, it is more compatible and better in comparison with GC for aqueous matrices in order to separate polar, thermally unstable DBPs, high molecular weight DBPs and highly polar N-DBPs. The most frequently used detectors are mass spectrometry (MS) coupled with LC, diode array and ultraviolet detectors. Ketones, alcohol, aldehydes, carboxylic acids, amides, amines, amino acids and phenols may be converted to fluorescent derivatives. CL has become more attractive because of its high sensitivity, simple instrumentation, and high linear range.

For determining variety of compounds having amine functionality CL method has become adaptable. GC is coupled with electron capture detectors (ECD) and mass selective detectors (MS). Because of less reproducibility, some significant limitations like existence of interfering species, the complex optimization make ELC less attractive than conventional CL, although it has large theoretical potential. The application of chemical derivation coupled with GCMS analysis is more frequent in research for identifying new products. Depending on the availability of functional group, a number of derivation reagents can be used. In order to reduce the mass spectra complexity and to increase the sensitivity, mass spectrometry is usually preceded by online chromatographic separation. The most commonly used separation technology used for this purpose is liquid and gas chromatography. In order to analyze high molecular weight and polar organohalogen compound the ideal tool is LC-MS. For the separation of wide range of compound chromatography results in poor OXBP separation, the resolution of HPLC is not sufficient as compared to GC. One of the approaches for minimizing the limitation involves the use of LC column packed with sub-2 μm stationary phase particles. Usually reversed phase columns are employed to accomplish chromatographic separations.

5.2 Analytical Method for HNMs

During GC/MS or GC/ECD analysis, HNM's consisting of three halogens on the alpha carbon and nitro group, unstable thermally, and decompose at injection port temperature, largely caused by low bond dissociation energy (Chen et al. 2002). In the study of samples of 30 mL of water involving with 3 mL of MTBE with the presence of 11 g of sodium sulfate and 1 g of copper sulfate for GC/MS analysis and quantitation examined the trihalonitromethanes. THMs found as major decomposition products by thermally generated trihalomethyl radicals, from the abstraction of a hydrogen atom from solvent. Because of the addition of sodium sulfate to water the aqueous phase ionic strength was increased. Studies reported that HNMs are extremely sensitive for decay and unstable due to various usual preservatives. For determination of chloropicrin, quenching agent is ammonium chloride and to minimize the sample degradation, samples need to be immediately analyzed (Kristiana et al. 2014). First method was developed by Montesinos et al. for determination of 9 HNMs using GC-MS technique (Montesinos and Cardador 2011). Huang et al. (2012) developed liquid–liquid extraction method, which is sensitive and fast for analysis of HNMs with adequate precision, high sensitivity. The ionization efficiency of the functional groups in electrospray ionization like ketones and aldehydes is usually low as they are neutral compounds. Detection sensitivity to the target analytes can be increased by introduction of a permanently charged or proton affinitive moiety. The combining chemical derivatization strategy with GC-MS or LC-MS provides various advantages for the structure elucidation and identification of unknown compounds.

5.3 Analytical Method of HANs and HAcAms

The nitrile group bonded alpha carbon is bonded to three halogens in HANs, whereas chlorinated and chloraminated waters produce the most abundant HANs, dihaloacetonitriles (Huang et al. 2012). In basic and neutral conditions, due to base catalysis HAAs and HAcAms are formed by hydrolysis of inherently unstable HANs. The stability of HANs decreases with the increase in halogen atom number as well as with enhancement in chlorine atom substitution by atoms of bromine. It is due to the fact that electron withdrawing halogens have an activating effect on reactivity of nitrile, resulting in decrease in the stability of HAN with increase in degree of halogenation. It is due to the opposite stabilizing effect on HANs caused by shift from chlorine atom to bromine atom (Nikolaou et al. 2002). In order to determine HANs in drinking water, four analytical methods have been compared, that were based on P& T- GC/MS, headspace—GC/MS (HS-GC/MS), and LLE--GC/S (Yang et al. 2019; Liberatore et al. 2017). In US survey of 2000–2002, for the first time HAcAms were reported. These consist of amide group bonded alpha carbon which is bonded to three halogens, whereas chlorinated and chloraminated waters produce the most abundant HANs, dihaloacetonitriles. It is reported in earlier reports that in aqueous solutions, HAcAms are inherently unstable, hydrolyzing to their corresponding HAAs via base catalysis in neutral and basic conditions. Four quenching agents, ascorbic acid, ammonium chloride, sodium sulfite, and sodium thiosulphate were investigated by Chu et al. on the stabilities of HAcAms (Chu et al. 2010, 2013). Chu et al. (2013) investigated a newer methodology for rapid screening of five HAcAms in pool water and drinking water, employing salt-assisted LLE coupled with injection port. In 2015, the study of HAN hydrolysis to form HAcAms using SPE-GC/MS method was demonstrated (Chinn et al. 2007).

Using modified EPA methods four THMs, seven HANs, six HAMs and nine HAAs were measured. LLE followed by GC/ECD may be used to conduct separation and analysis of samples. Portable spectrophotometer based on HACH method can be used to detect residues and chlorine.

5.4 Analytical Method for Other Emerging N-DBPs

Various non-target identification methods are used with advanced instrumentation facilities which revealed a significant progress in order to identify various known and unknown disinfection by-products. Zhang and associates applied precursor ion scan method to examine chlorinated saline sewage effluents and identified three new halonitrophenols as well as 2,6-dichloro-4-nitrophenol. The UPLC/MS/MS was applied in ESI mode for saline sewage effluents during chlorination to identify highly toxic disinfection by-products such as tribromopyrrole, tribromoiodopyrrole, tribromochloropyrrole, and tetrabromopyrrole. A total of sixty five disinfection by-products were identified including two chlorinated, eight di-chlorinated, and twenty mono-chlorinated N-DBPs (Kodamatani et al. 2008). Simulated distribution systems (SDS) was applied for prediction of information of HAAs and THMs using an inexpensive and simple laboratory based method. The American water utilities were also suggested through United States Environmental Protection Agency for examination of their distributive system for availability of DBP levels and collection of DBPs level information. There is very less information available for HANs and HAcAms behavior in the distributive system. However, it is well established that HAcAms may be formed by hydrolysis of HANs, and the formed HAcAms may undergo hydrolysis to produce dihaloacetic acids (DHAAs). The aim of SDS test is the stimulation of water quality, residues of disinfectants as well as water ages for real distribution system in order to allow comfortable sample collection during the regular intervals of time to analyze DBP formation. To consider the effect of potential variation in distribution practices, these tests are found to be useful tool. In UK for both real distribution systems and SDS tests to examine HAcAms and HANs formation, the four water treatment plants were evaluated using a sampling survey. The extraction of the samples was performed by LLE techniques, whereas electron capture detector equipped gas chromatography technique was used to analyze the samples. For detection of HAAs the most common detection methods employ post-column derivatization by methylation, applying boron trifluoride in methanol, acidic methanol or diazomethane as reagents. An alternative approach is provided by ion chromatography and hydrophilic interaction liquid chromatography, which enables polar stationary phases to efficiently separate smaller and polar compounds. The electrospray ionization is well suited in order to analyze substances with high polarity such as HAAs. The suitable technique for online, continuous, and direct measurement of volatile molecules, in particular, is membrane introduction mass spectrometry. This technique is based on the application of semi-permeable membrane that works like an interface between mass spectrometer and aqueous samples. Table 2 represents the sample extraction and detection methods of various DBPs (Hariganesh et al. 2020).

6 Methods of Minimization of N-DBPs in Drinking Water

The control on disinfectant by-products may be done by removal of precursors of DBPs and by modification of disinfection practices and a better quality of drinking water can be obtained. The life of human beings is affected due to nitrogenous disinfectant by-products. There are various evidences that show that variety of nitrogenous disinfectant by-products significantly affect human health. In order to keep the drinking water safe various regulations are required to control the number of nitrogenous by-products, which remove the potential cancer risks. The amount of N-DBPs is not high as the DBPs, but N-DBPs are of higher concern to authorities of water management in different nations across the globe, as N-DBPs are found to be more highly toxic in comparison with DBPs. Natural organic matter can be removed by using coagulation, membranes, ozone bio filtration, and granular activated carbon. Instead of disinfectants, technologies like activated charcoal may be applied for removal of organic matter dissolved in drinking water. HAN formation is seen to be enhanced due to preoxidation with chlorine dioxide and post disinfection at acidic level. Chloropicrin formation increases because of ozonation or UV irradiation to drinking water. Therefore, alternative effective methods can be used to control HNMs. The formation of CNX may be minimized by operation of disinfection without secondary chloramination (Bond et al. 2011). Chloramination, use of certain polymers exchange or resins and presence of EfOM are the factors that are responsible for NDMA formation in drinking water. Free chlorine can control NDMA formation. Filtration and coagulation, common water treatment processes are not effective in removing DON. It is important to identify various N-DBPs than their precursors and their pattern of formation is required to be studied. System engineering approach must be effectively taken for reduction of concentration of these groups in drinking water.

Table 2 The sample extraction and detection methods of various DBPs

DBPs	Reaction Type	Instrument	Detection Method	Reference
N-DBPs and C-DBPs Haloacetonitriles Haloacetamides Haloacetic acid	Chloramination;	Gas Chromatography High resolution quadrupole time of flight (GC-QTOF-MS) (Agilent 5975 C GC- MSD and Agilent 7200 accurate mass GC-QTOF)	EPA method 551; EPAMethod 552.2	Nihemaiti et al. (2017)
Iodinated DBPs	Chlorine-ammonia treated; Chloramination; chlorination	Liquid chromatography (UPLC) ultra-performance liquid chromatography coupled with ESI mass spectrometer having triple quadrupole (ESI- tqMS)	A gradient elution using CH_3CN/H_2O was applied. Product ion scans and selective ion recording gave iodinated by-products	Ding and Zhang (2009)
Haloacetaldehydes, Haloketones, and other DBPs	Organic matter; bromide, iodide	(GC)-TOF (Agilent 7890B coupled with pagasus BT TOF MS), GC-single quadrupole (SQ) MS (Aligent 7890B coupled with 5977 A SQ-MS)	Calibration curve method	Kimura et al. (2019)
Brominated DBPs	Chlorination; bromination	UPLC/ESI-tqMS	Multi-channel product ion scan	Pan et al. (2014)
Phenylacetonitrile	Chlorination	GC-MS(focus GC-DSQII MS, Thermo scientific)	EPA method 551.1	Ma et al. (2016)
Aromatic halogenated DBPs	Chlorination; Bromination	UPLC/ESI –tqMS	Product ion scan	Pan and Zhang 2013
Carbonaceous DBPs and nitrogenous DBPs	Chlorination	GC-MSMS (Thermo TSQ quantum XLS triple quadrupole MS)	EPA method 551.1; EPA method 552.2	Zhang et al. (2016); Zhou et al. (2015)
N-Nitroso dimethylamine	Chloramination	HPLC-UV	A gradient elution using CH_3OH/H_2O was applied and photodiode array detector was used	Huang et al. (2018)

7 Conclusions

The lives of people are affected by nitrogenous disinfectant by-products. There is range of evidences that show that DBPs and N-DBPs significantly affect human health. The increase in formation of N-DBPs is due to the shift from chlorination to chloramination. Better quality of drinking water can be obtained by controlling disinfectants by-products through modifications of disinfection practice and by removal of DBP precursor. The study of different reports concludes that various methods can be used for minimizing N-DBPs treatment of drinking water. In drinking water, the concentration of N-DBPs is below the detection limit of the analytical equipment. Due to identification of N-DBPs groups having disparate precursor and formation mechanism control parameters which are group specific and can show opposite impact on other N-DBPs, ammonium chloride and ascorbic acid are the best quenching agents for determination of HNMs, HANs, and HAcms and immediately samples need to be analyzed for minimization of degradation and hydrolysis. Two analytical methods that are frequently used are LLE-GC/MS and LLE-GC/ECD. In comparison with EPA, LLE procedures, SPE, SPME pretreatment procedures are better. In order to determine volatile N-DBPs, HS and P&T pretreatments are used. NAs can be analyzed by SPE followed by LC or GC with the help of MS detectors. Although research is going on to develop new method for treatment for the reduction of levels of by-products that are formed during disinfection. Moreover, this is established that generally the risk to health by use of by-products at levels in which these are available in drinking water is small as compared to risk formed by inadequate disinfection. Therefore single method cannot be used for all N-DBPs. Dissolved organic matter can be removed by using activated charcoal. Currently, none of the countries have regulations for N-DBPs. In order to remove N-DBPs, it is necessary to identify different N-DBPs in drinking water. Thereafter, the formation pattern of N-DBPs as well as their precursors is required to be focused. Finally, to reduce the formation of N-DBPs in drinking water system engineering approach must be taken. The real sample analysis must be performed in order to quantify the role played by different disinfection by-products, particularly N-DBPs during the pathways of reaction.

References

Bei E, Shu Y, Li S, Liao X, Wang J, Zhang X, Chen C, Krasner S (2016) Occurrence of nitrosamines and their precursors in drinking water systems around mainland China. Water Res 98:168–175

Bellar TA, Lichtenberg JJ, Kroner RC (1974) The occurrence of organohalides in chlorinated drinking waters. J Am Water Works Assoc 66(12):703–706

Bond T, Huang J, Templeton MR, Graham N (2011) Occurrence and control of nitrogenous disinfection by-products in drinking water e a review. Water Res 45(15):4341–4354

Bond T, Templeton MR, Graham N (2012) Precursors of nitrogenous disinfection by-products in drinking water—a critical review and analysis. J Hazard Mater 235-236:1–16

Bond T, Templeton MR, Mokhtar Kamal NH, Graham N, Kanda R (2015) Nitrogenous disinfection byproducts in English drinking water supply systems: occurrence, bromine substitution and correlation analysis. Water Res 85:85–94
Chen PH, Richardson SD, Krasner SW, Majetich G, Glish GL (2002) Hydrogen abstraction and decomposition of bromopicrin and other trihalogenated disinfection byproducts by GC/MS. Environ Sci Technol 36:3362–3371
Chen TL, Tzing SH, Ding WH (2015) Rapid screening of haloacetamides in water using salt-assisted liquid-liquid extraction coupled injection-port silylation gas chromatography-mass spectrometry. J Chromatogr A 1422:340–344
Chen W, Liu Z, Tao H, Xu H, Yanmei G, Chen Z, Yub J (2017) Factors affecting the formation of nitrogenous disinfection by-products during chlorination of aspartic acid in drinking water. Sci Total Environ 575:519–524
Chinn R, Lee T, Krasner S, Dale M, Richardson SD, Pressman JG, Speth TF, Miltner RJ, Simmons JE, Solid-phase extraction of 35 DBPs with analysis by GC/ECD and GC/MS, In Proceeding, 2007 Water Quality Technology Conference, Charlotte, NC, American Work Association, 2007; pp. 1–20
Chu WH, Gao NY, Deng Y (2010) Formation of haloacetamides during chlorination of dissolved organic nitrogen aspartic acid. J Hazard Mater 173:82–86
Chu W, Gao N, Yin D, Krasner SW, Templeton MR (2012a) Trace determination of 13 haloacetamides in drinking water using liquid chromatography triple quadrupole mass spectrometry with atmospheric pressure chemical ionization. J Chromatogr A 1235:178–181
Chu W, Gao N, Yin D, Krasner SW (2013) Formation and speciation of nine haloacetamides, an emerging class of nitrogenous DBPs during chlorination or chloramination. J Hazard Mater 260:806–812
Chu W, Gao N, Yin D, Krasner SW, Mitch WA (2014) Impact of UV/H_2O_2 preoxidation on the formation of haloacetamides and other nitrogenous disinfection byproducts during chlorination. Environ Sci Technol 48(20):12190–12198
Chu W, Gao N, Krasner SW, Templeton MR, Yin D (2012b) Formation of halogenated C-, N-DBPs from chlorination and UV irradiation of tyrosine in drinking water. Environ Pollut 161:8–14
Chu WH, Gao NY, Deng Y, Templeton MR, Yin DQ (2011) Formation of nitrogenous disinfection by-products from pre-chloramination. Chemosphere 85(7):1187–1191
Chu W, Li D, Gao N, Templeton MR, Tan C, Gao Y (2015) The control of emerging haloacetamide DBP precursors with UV/persulfate treatment. Water Res 72:340
Cloirec CL, Martin G (1985) Water chlorination chemistry, environmental impact and health effects. In: Jolley RL (ed) Evolution of Amino acids in water treatment plants and the effect of chlorination on amino acids, vol 5. Lewis Publishers, Inc, Chelsea, MI, pp 821–834
Cowman GA, Singer PC (1995) Effect of bromide ion on haloacetic acid speciation resulting from chlorination and chloramination of aquatic humic substances. Environ Sci Technol 30(1):16–24
Ding G, Zhang X (2009) A picture of polar iodinated disinfection byproducts in drinking water by (UPLC/)ESI-tqMS. Environ Sci Technol 43:9287–9293
Drinking water chlorination white paper: a review of disinfection practices and issues. Chlorine Chemistry Council, 2003
Evlampidou I, Font-Ribera L, Rojas-Rueda D, Gracia-Lavedan E, Costet N, Pearce N, Vineis P, Jaakkola JJK, Delloye F, Makris KC (2020) Trihalomethanes in drinking water and bladder cancer burden in the European Union. Environ Health Perspect 128:107001–107014
Fang J, Ma J, Yang X, Shang C (2010) Formation of carbonaceous and nitrogenous disinfection by-products from the chlorination of Microcystis aeruginosa. Water Res 44:1934–1940
Fu C, Xu H, Wang Z (1993) Sensitivity assay system for nitrosamines utilizing high-performance liquid chromatography with peroxyowalate chemiluminescence detection. J Chromatogr A 634:221–227
Glezer V, Harris B, Tal N, Iosefzon B, Lev O (1999) Hydrolysis of haloacetonitriles: linear free energy relationship, kinetics and products. Water Res 33(8):1938–1948

Goslan EH, Krasner SW, Bower M, Rocks SA, Holmes P, Levy LS, Parsons SA (2009) A comparison of disinfection by-products found in chlorinated and chloraminated drinking waters in Scotland. Water Res 43(18):4698–4706

Goslan EH, Krasner SW, Villanueva CM, Carrasco Turigas G, Toledano MB, Kogevinas M, Stephanou EG, Cordier S, Grazuleviciene R, Parsons SA, Nieuwenhuijsen MJ (2014) Disinfection by-product occurrence in selected European waters. J Water Supply Res Technol AQUA 63(5):379–390

Gupta H (2016) Photocatalytic degradation of phenanthrene in the presence of akaganeite nanorods and the identification of degradation products. RSC Adv 6(114):112721–112727

Gupta B, Gupta H (2015) Iron oxide mediated degradation of mutagen pyrene and determination of degradation products. Int J Environ Sci Dev 6(12):908

Gupta H (2018) Anthracene removal from water onto activated carbon derived from vehicular Tyre. Sep Sci Technol 53(4):613–625

Gupta H, Dhiman S (2023) Synthetic nanoparticle-based remediation of soils contaminated with polycyclic aromatic hydrocarbons. In: Handbook of green and sustainable nanotechnology. https://doi.org/10.1007/978-3-030-69023-6_102-1

Gupta H, Gupta B (2016) Adsorption of polycyclic aromatic hydrocarbons on banana peel activated carbon. Desalin Water Treat 57(20):9498–9509

Gupta H, Gupta B (2018) Vehicular tire as potential adsorbent for the removal of polycyclic aromatic hydrocarbons. Polycycl Aromat Compd 38(4):354–368

Hariganesh S, Vadivel S, Maruthamani D, Rangabhashiyam S (2020) Disinfection by-products in drinking water: detection and treatment methods. In: Disinfection by-products in drinking water, pp 279–304. https://doi.org/10.1016/B978-0-08-102977-0.00013-5

Hirose Y, Maeda N, Ohya T, Nojima K, Kanno S (1988) Formation of cyanogen chloride by the reaction of amino acids with hypochlorous acid in the presence of ammonium ion. Chemosphere 17:865–873

Huang H, Wu QY, Hu HY, Mitch WA (2012) Dichloroacetonitrile and dichloroacetamide can form independently during chlorination and chloramination of drinking waters, model organic matters and wastewater effluents. Environ Sci Technol 46:10624–10631

Huang EH, Huang S, McCurry DL (2018) Re-examining the role of dichloramine in high-yield N-Nitrosodimethylamine formation from N,N-Dimethyl-α-arylamines. Environ Sci Technol 5:154–159

Janda V, Svecova M (2000) By-products in drinking water disinfection. Chemické Listy 94(10):905–908

Kaur A, Gupta H, Dhiman S (2023) Biogenic metallic nanoparticles: synthesis and applications using medicinal plants. Handbook of green and sustainable. Nanotechnology. https://doi.org/10.1007/978-3-030-69023-6_101-1

Kimura SY, Cuthbertson AA, Byer JD, Richardson SD (2019) The DBP exposome: development of a new method to simultaneously quantify priority disinfection by-products and comprehensively identify unknowns. Water Res 148:324–333

Kinani A, Kinani S, Bouchonnet S (2016) Formation and determination of organohalogen by-products in water. Part III, characterization and quantitative approaches. Trends Anal Chem 85:295–305

Kodamatani H, Yamazaki S, Saito K, Amponsaa-Karikari A, Kishikawa N, Kuroda N, Tomiyasu T, Komatsu Y (2008) Highly sensitive method for determination of N-nitrosamines using high performance liquid chromatography with online UV irradiation and luminol chemiluminescence detection. J Chromatogr A 1216:92–98

Krasner SW, McGuire MJ, Jacangelo JG, Patania NL, Reagan KM, Marco Aieta E (1989) Occurrence of disinfection by-products in US drinking water. J Am Water Works Assoc 81(8):41–53

Krasner SW, Sclimenti MJ, Mitch WA, Westerhoff P, Dotson AA (2007) Wastewater and algal derived N-DBPs. AWWA, Denver, CO, USA. AWWA Annual Conference

Krasner SW, Westerhoff P, Chen B, Rittmann BE, Amy G (2009) Occurrence of disinfection byproducts in United States wastewater treatment plant effluents. Environ Sci Technol 43(21):8320–8325
Krasner SW, Weinberg HS, Richardson SD, Pastor SJ, Chinn R, Sclimenti MJ, Onstad GD, Thruston AD (2006) Occurrence of a new generation of disinfection byproducts. Environ Sci Technol 40(23):7175–7185
Kristiana I, Lethorn A, Joll C, Heitz A (2014) To add or not to add: the use of quenching agents for the analysis of disinfection by-products in water samples. Water Res 59:90–98
Kumar A, Gupta H (2020) Activated carbon from sawdust for naphthalene removal from contaminated water. Environ Technol Innov 20:101080
Kumar N, Kumar B, Gupta H, Gupta A (2023) Development and evaluation of cellulose/graphene-oxide based composite for removing phenol from aqueous solutions. Polymers 15(3):572
Lee SH, Field LR (1987) Fluorescence detection of some nitrosoamines in high performance liquid chromatography after post–column reaction. J Chromatogr A 386:137–148
Lee W, Westerhoff P, Croué J (2007) Dissolved organic nitrogen as a precursor for chloroform, dichloroacetonitrile, N-Nitrosodimethylamine, and trichloronitromethane. Environ Sci Technol 41(15):5485–5490
Liberatore HK, Plewa MJ, Wagner ED, VanBriesen JM, Burnett DB, Cizmas LH (2017) Identification and comparative mammalian cell cytotoxicity of new iodophenolic disinfection byproducts in chloraminated oil and gas wastewaters. Environ Sci Technol Lett 4:475–480
Liu C, Wang J, Chen W, Dong C, Li C (2015) The removal of DON derived from algae cells by cu-doped TiO_2 under sunlight irradiation. Chem Eng J 280:588–596
Ma X, Deng J, Feng J, Shanaiah N, Smiley E, Dietrich AM (2016) Identification and characterization of phenylacetonitrile as a nitrogenous disinfection byproduct derived from chlorination of phenylalanine in drinking water. Water Res 102:202–210
Mazhar MA, Khan NA, Ahmed S, Khan AH, Hussain A, Rahisuddin, Changani F, Yousefi M, Ahmadi S, Vambol V (2020) Chlorination disinfection by-products in municipal drinking water–a review. J Clean Prod 273:123159
McGuire MJ, McLain JL, Obolensky A (2002) Information collection rule data analysis. Awwa RF and AWWA, Denver, CO, USA
Merlet N, Thibaud H, Dore M (1985) Chloropicrin formation during oxidative treatments in the preparation of drinking water. Sci Total Environ 47:223–228
Mitch WA, Krasner SW, Westerhoff P, Dotson A (2009) Occurrence and formation of nitrogenous disinfection by products. Water Research Foundation, D, CO, USA
Mitch WA, Sedlak DL (2004) Characterization and fate of N-nitrosodimethylamine precursors in municipal wastewater treatment plants. Environ Sci Technol 38(5):1445–1454
Mitch WA, Sharp JO, Trussell R, Valentine RL, Cohen LA, Sedlak DL (2003) N-Nitrosidimethylamine (NDMA) as a drinking water contaminant: a review. Environ Eng Sci 20:389–404
Montesinos I, Cardador MJ (2011) Gallego M, determination of halonitromethanes. J Chromatogr A 1218:2497–2504
Nihemaiti M, Le Roux J, Hoppe-Jones C, Reckhow DA, Croue J-P (2017) Formation of haloacetonitriles, haloacetamides, and nitrogenous heterocyclic byproducts by chloramination of phenolic compounds. Environ Sci Technol 51:655–663
Nikolaou AD, Lekkas TD, Golfinopoulos SK, Kostopoulou MN (2002) Application of different analytical methods for determination of volatile chlorination by-products in drinking water. Talanta 56(4):717–726
Oliver BG (1983) Dihaloacetonitriles in drinking water: algae and fulvic acid as precursors. Environ Sci Technol 17(2):80–83
Pan Y, Zhang X (2013) Four groups of new aromatic halogenated disinfection byproducts: effect of bromide concentration on their formation and speciation in chlorinated drinking water. Environ Sci Technol 47:1265–1273

Pan Y, Zhang X, Wagner ED, Osiol J, Plewa MJ (2014) Boiling of simulated tap water: effect on polar brominated disinfection byproducts, halogen speciation, and cytotoxicity. Environ Sci Technol 48:149–156

Park KY, Choi SY, Lee SH, Kweon JH, Song JH (2016) Comparison of formation of disinfection by-products by chlorination and ozonation of wastewater effluents and their toxicity to Daphnia magna. Environ Pollut 215:314–321

Pedersen EJ, Urbansky ET, Marinas BJ (1999) Formation of cyanogen chloride from the reaction of monochloramine with formaldehyde. Environ Sci Technol 33:4239–4249

Pérez-Ruiz T, Martínez-Lozano Tomás CV, Martín J (2005) Automated solid phase extraction and high performance liquid chromatographic determination of nitroamines using post-column photolysis and tris(2,2 bipyridyl) ruthenium(III) chemiluminescene. J Chromatogr A 1077:49

Plewa MJ, Muellner MG, Richardson SD, Fasano F, Buettner KM, Woo YT, McKague AB, Wagner ED (2008) Occurrence, synthesis, and mammalian cell cytotoxicity and genotoxicity of haloacetamides: an emerging class of nitrogenous drinking water disinfection byproducts. Environ Sci Technol 42:955–961

Ram NM (1985) A review of the significance and formation of chlorinated N-organic compounds in water supplies including preliminary studies on the chlorination of alanine, tryptophan, tyrosine, cytosine, and syringic acid. Environ Int 11:441–451

Reckhow DA, MacNeill AL, Platt TL, MacNeill AL, McClellan JN (2001) Formation and degradation of dichloroacetonitrile in drinking waters. J Water Supply Res Technol AQUA 50(1):1–13

Richardson SD, Plewa MJ (2020) To regulate or not to regulate? What to do with more toxic disinfection by-products? J Environ Chem Eng 8(4):103939

Richardson SD, Plewa MJ, Wagner ED, Schoeny R, DeMarini DM (2007) Occurrence, genotoxicity, and carcinogenicity of regulated and emerging disinfection byproducts in drinking water: a review and roadmap for research. Mutat Res Rev Mutat Res 636:178–242

Richardson SD, Thruston AD, Caughran TV, Chen PH, Collette TW, Floyd TL, Schenck KM, Lykins BW, Sun GR, Majetich G (1999) Identification of new ozone disinfection byproducts in drinking water. Environ Sci Technol 33:3368–3377

Richardson SD, Thruston AD, Rav-Acha C, Groisman L, Popilevsky I, Juraev O, Glezer V, Mckague AB, Plewa MJ, Wagner DE (2003) Tribromopyrrole, brominated acids and other disinfection byproducts produced by disinfection of drinking water rich in bromide. Environ Sci Technol 37(17):3782–3793

Rook JJ (1976) Haloforms in drinking water. J-Am Water Work. Assoc 68(3):168–172

Shah AD, Mitch WA (2011) Halonitroalkanes, halonitriles, haloamides, and nitrosamines: a critical review of nitrogenous disinfection byproduct formation pathways. Environ Sci Technol 46(1):119–131

Shan J, Hu J, Kaplan-Bekaroglu SS, Song H, Karanfil T (2012) The effects of pH, bromide and nitrite on halonitromethane and trihalomethane formation from amino acids and amino sugars. Chemosphere 86(4):323–328

Shen Y (2021) Formation of nitrogenous disinfection by products in drinking water: emerging concerns and current issue. IOP Conference Series: Earth Environ Sci 801:012015

Singer PC, Obolensky A, Greiner A (1995) DBPs in chlorinated North-Carolina drinking waters. J-Am Water Work Assoc 87(10):83–92

Singh S, Gupta H, Dhiman S, Sahu NK (2022) Decontamination of cationic dye brilliant green from the aqueous media. Appl Water Sci 12(4):61

Singh S, Kumar A, Gupta H (2020) Activated banana peel carbon: a potential adsorbent for rhodamine B decontamination from aqueous system. Appl Water Sci 10(8):1–8

Singh S, Parveen N, Gupta H (2018) Adsorptive decontamination of rhodamine-B from water using banana peel powder: a biosorbent. Environ Technol Innov 12:189–195

Stefan MI, Bolton JR (2012) UV direct photolysis of N-Nitrosodimethylamine (NDMA): kinetic and product study. Helv Chim Acta 85:1416

Stevens AA, Moore LA, Miltner RJ (1989) Formation and control of non-trihalomethane disinfection by-products. J-Am Water Work Assoc 81:54–60

Trehy ML, Bieber TI (1981) Detection, identification and quantitative analysis of dihaloacetonitriles in chlorinated natural waters. In: Advances in the identification & analysis of organic pollutants in water, vol 2, pp 941–975

Trehy ML, Yost RA, Miles CJ (1986) Chlorination byproducts of amino acids in natural waters. Environ Sci Technol 20:1117–1122

Twort AC, Law FM (1994) Water supply.4th edn. Edward Arnold, London

Ueno H, Moto T, Sayato Y, Nakamuro K (1996) Disinfection by-products in the chlorination of organic nitrogen compounds: by-products from kynurenine. Chemosphere 33(8):1425–1433

Villanueva CM, Cordier S, Font-Ribera L, Salas LA, Levallois P (2015) Overview of disinfection by-products and associated health effects. Curr Environ Health Rep 2(1):107–115

Wang Z, Xu H, Fu C (1992) Sensitive fluorescene detection of some nitrosamines by precolumn derivatization with dansyl chloride and high-performance liquid chromatography. J Chromatogr A 589:349–352

Weinberg HS, Krasner SW, Richardson SD, Thruston AD (2002) The occurrence of disinfection by-products (DBPs) of health concern in drinking water, results of a nationwide DBP occurrence study. U.S. Environmental Protection Agency, National Exposure Research Laboratory, Athens, GA, USA

Westerhoff P, Mash H (2002) Dissolved organic nitrogen in drinking water supplies: a review. J Water Supply Res Technol AQUA 51:415–448

Williams CJ, Conrad D, Kothawala DN, Baulch HM (2019) Selective removal of dissolved organic matter affects the production and speciation of disinfection byproducts. Sci Total Environ 652:75–84

Xie YF, Reckhow DA (1993) Identification of trihaloacetaldehydes in ozonated and chlorinated fulvic-acid solutions. Analyst 118(1):71–72

Yang M, Liberatore HK, Zhang X (2019) Current methods for analyzing drinking water disinfection byproducts. Curr Opin Environ Sci Health 7:98–107

Yang X, Guo W, Shen Q (2011) Formation of disinfection byproducts from chlor(am)ination of algal organic matter. J Hazard Mater 197:378–388

Yang X, Shen Q, Guo W, Peng J, Liang Y (2012) Precursors and nitrogen origins of trichloronitromethane and dichloroacetonitrile during chlorination/chloramination. Chemosphere 88(1):25–32

Zhai H, Zhang X (2011) Formation and decomposition of new and unknown polar brominated disinfection byproducts during chlorination. Environ Sci Technol 45(6):2194–2201

Zhang X, Yang H, Wang X, Fu J, Xie YF (2013) Formation of disinfection by-products: effect of temperature and kinetic modeling. Chemosphere 90(2):634–639

Zhang Y, Shao Y, Gao N, Chu W, Sun Z (2016) Removal of microcystin-LR by free chlorine: identify of transformation products and disinfection by-products formation. Chem Eng J 287:189–195

Zhao YL, Qin F, Boyd JM, Anichina J, Li XF (2010) Characterization and determination of chloro- and bromo-benzoquinones as new chlorination disinfection byproducts in drinking water. Anal Chem 82(11):4599–4605

Zhou S, Zhu S, Shao Y, Gao N (2015) Characteristics of C-, N-DBPs formation from algal organic matter: role of molecular weight fractions and impacts of pre-ozonation. Water Res 72:381–390

GIS-Based Trends Analysis of THMs Compounds in Indian Drinking Water Supplies

Jaydev Kumar Mahato, Shivam Saw, B. D. Yadav, and Sunil Kumar Gupta

Abstract The study explored the GIS-based trends analysis for various trihalomethane (THMs) compounds, namely chloroform (CF), dibromochloromethane (DBCM), bromodichloromethane (BDCM), and bromoform (BF) in Indian drinking supplies. Five water treatment plants (WTPs) were selected to establish the concentration and spatial distribution. THMs compounds were estimated using the liquid–liquid extraction method by vigorous pentane: water mixture agitation, with gas chromatography equipped with electron-capture detection (ECD). Higher concentrations of CF (319.85 to 380.72 μg/l) were observed when source waters were subject to elevated organic matter levels. GIS-based spatial distribution of various THMs compounds indicated CF was most prevalent in chlorinated drinking water. The formation of THMs is then rooted in operational and geographic factors. The observed trends analysis will help the researchers and scientists, including water professionals, to confiscate the problems of THMs and improve the water quality in the WTPs from future perspectives.

Keywords Drinking water · THMs compounds · Trends analysis · GIS mapping

1 Introduction

The universal scenario of potable water presents a dismal picture, where most developing nations are fighting with scarcity. According to a report by WHO (2004), approx. 700.85 million people globally lack water for their basic needs. Health problems brought by contaminated drinking water affect over half of the population in underdeveloped countries (Zainudin et al. 2018). India is the seventh largest country, with 4% of the world's water resources, and it supports around 17% of the

J. K. Mahato · S. Saw · B. D. Yadav · S. K. Gupta (✉)
Department of Environmental Science and Engineering, Indian Institute of Technology (ISM), Dhanbad, India
e-mail: sunil@iitism.ac.in

S. Madhav et al. (eds.), *Drinking Water Disinfection By-products*,
https://doi.org/10.1007/978-3-031-49047-7_13

earth's population (Chakraborty 2017). India also dropped to the second last position in the safe drinking water index out of 123 countries worldwide (Thokchom et al. 2020). Surface resources are India's primary source of drinking water for human consumption. However, in raw water supplies, the occurrence of natural organic matter (NOM) is a grave concern (Krzeminski et al. 2019). It contains a versatile range of NOM (2.1–44.1 mg/L), which raises a significant challenge of forming potentially hazardous disinfection by-products (DBPs), especially Trihalomethane (THMs), during the chlorination process (Karimi Pasandideh et al. 2016; Park et al. 2019). These THMs compounds include chloroform ($CHCl_3$) (CF), dibromochloromethane ($CHClBr_2$) (DBCM), bromodichloromethane ($CHCl_2Br$) (BDCM), and bromoform ($CHBr_3$) (BF), which increased the risk of carcinogenicity and mutagenicity in human beings and identified as Class B carcinogens (Rook 1974; Clark et al. 1986). Many researchers reported an elevated range of THMs in the drinking water of Kolkata (466 μg/l), Dhanbad (503 μg/l) (Minashree Kumari 2014), Delhi (311–377 μg/l) (Hasan et al. 2010), Varanasi (380.9 μg/l), Raipur (324.3 μg/l), and Bhubaneswar (319.7 μg/l) (Mahato and Gupta 2020), Bokaro (594 μg/l) (Mishra et al. 2014), which exceeded the guideline value of USEPA (80 μg/), WHO (300 μg/l), and BIS (200 μg/l). The widespread occurrence and carcinogenic nature of THMs warrant in-depth attention to the Indian regulatory agencies and management authorities for controlling their levels in water to minimize potential health risks. In this context, Geographic Information Systems (GIS) could play a crucial role in analyzing and interpreting the THMs' spatial distribution in drinking water. Globally, it is a widely accepted technology to assess the geospatial distribution of not only contaminated water but also for soil and air for mitigation and management.

In light of the above, the present research was undertaken for GIS-based trends analysis of various THMs compounds in the drinking water supplies of five states with the help of advanced mapping techniques. The study also described the Hierarchical Cluster Analysis (HCA) of THMs compounds. This study is crucial for drinking water treatment and supplies and can be used to support decision-making bodies for the control of THMs.

2 Material and Methods

2.1 *Site Description and Water Sampling*

For the GIS-based trends analysis of various THMs compounds, the water samples were obtained from the water treatment plants (WTPs) of five different states in India (Fig. 1). Each sample was placed in a 30 mL clean glass vial with a polypropylene cover and a rubber septum that was PTFE-faced. Immediately after the sample collection, 0.010 gm of sodium sulfite was added as a dechlorinating agent for the prevention of further THMs formation and stored in 4 °C till analysis.

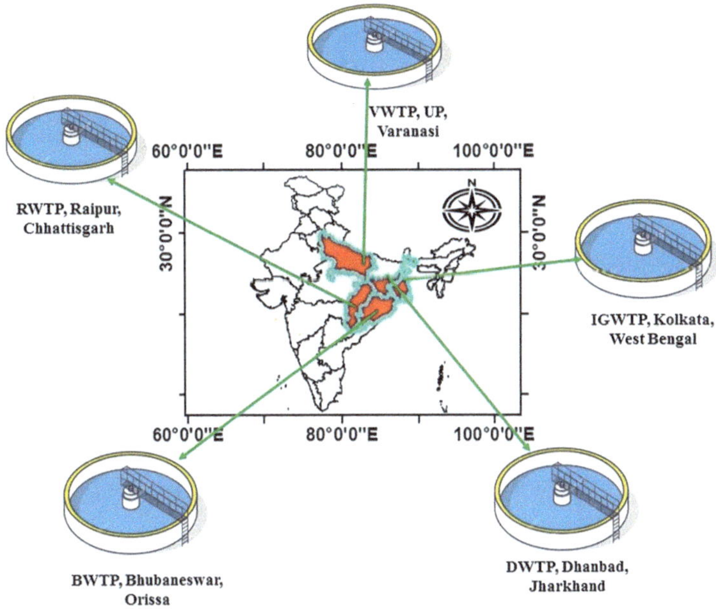

Fig. 1 Study area's location map

2.2 *Analytical Method and Software Used*

The USEPA (551.1) liquid-to-liquid extraction method was used to analyze several THM compounds using a gas chromatograph (GC) (Thermo Fisher, CERES 800 plus) fitted with an electron capture detector (ECD). The condition adopted for GC-ECD conditions is Carrier gas (Nitrogen) flow—60 (ml/min), Oven temperature—40 °C, Detector temperature—250 °C, and Injector temperature—200 °C. The GIS-based spatial distribution of THMs compounds was investigated using 10.3 ArcGIS MAP software for Geospatial Analysis. The radar plot comparative of magnitudes THMs species was performed using Origin Lab 2018 software.

3 Results & Discussion

3.1 *Spatial Distribution of THMs in all Five WTPs*

3.1.1 Spatial Distribution of CF

The regulations of each THMs compound are being proposed by the Bureau of Indian Standards (BIS) 2012. However, the results of the present study suggest that further attention should be paid to the occurrence of chlorinated compounds, especially CF. During the investigation period, the average concentrations of CF for all five WTPs ranged from 319.85 to 380.72 μg/l, which exceeded BIS (2012) standards (200 μg/l) (Fig. 2). The maximum value of CF was observed at VWTP, followed by DWTP > IGWTP, >RWTP, and > BWTP. The variation in the level of CF in these WTPs is attributed to the difference in residual chlorine and pH (Thokchom et al. 2020). During the chlorination process, HOCl is formed by the hydrolysis of chlorine (Cl_2), which is a weak acid that dissociates in water and forms hypochlorite ions (OCl^-) (Eq. 1 and 2). HOCl is dominant under acidic pH (>3.0), whereas in an alkaline condition formation of OCl^- leads. Relatively HOCl is more reactive than OCl^- in water; hence, acidic pH yields more CF (Abdullah et al. 2003). The previous studies also demonstrated that the pH and residual chlorine are the most influencing parameters for the formation of CF (Mahato and Gupta 2022; Park et al. 2019).

$$Cl_2 + H_2O \rightleftharpoons HOCl + Cl - + H^+ \quad (1)$$

$$HOCl \rightleftharpoons OCl - + H^+ \quad (2)$$

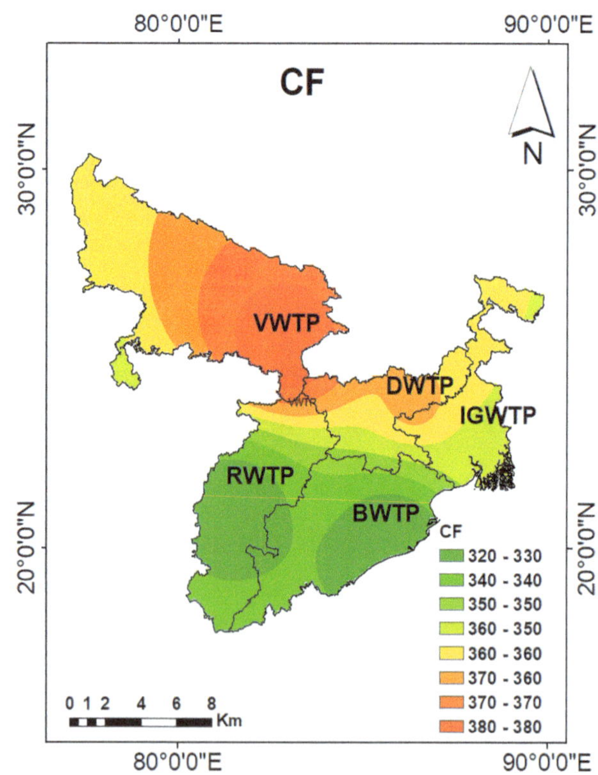

Fig. 2 Spatial distribution map of CF

3.1.2 Spatial Distribution of BDCM

During chlorination, the presence of bromide ions (Br-) causes the formation of hypobromous acid (HOBr), that is approx. 20 times more reactive with THMs precursors than HOCl (Sohn et al. 2006). According to the WHO, the BDCM is the second most toxicological THMs species than CF, which concentrates on human placental trophoblasts, which are responsible for producing a hormone needed throughout pregnancy (WHO 2004). A literature survey revealed that more than 90% of the THMs compounds consist of CF, followed by BDCM (14%) (Chowdhury et al. 2010). In this research, the concentration level of BDCM was monitored (16.90 to 21.71 μg/l) higher after the CF in all five WTPs, which is well supported by the previous reports (Mahato and Gupta 2022; Chowdhury et al. 2010; Mishra et al. 2014; Minashree Kumari 2014). However, their concentrations of BDCM were below the BIS standards (60 μg/l) (BIS. 2012). The spatial distribution map illustrated that RWTP possessed a high level of BDCM, followed by BWTP > IGWTP > VWTP and > VWTPS (Fig. 3). The variation in the level concentration of BDCM may be attributed to the occurrence of bromide ions in the source water of these WTPs (Uyak and Toroz 2007).

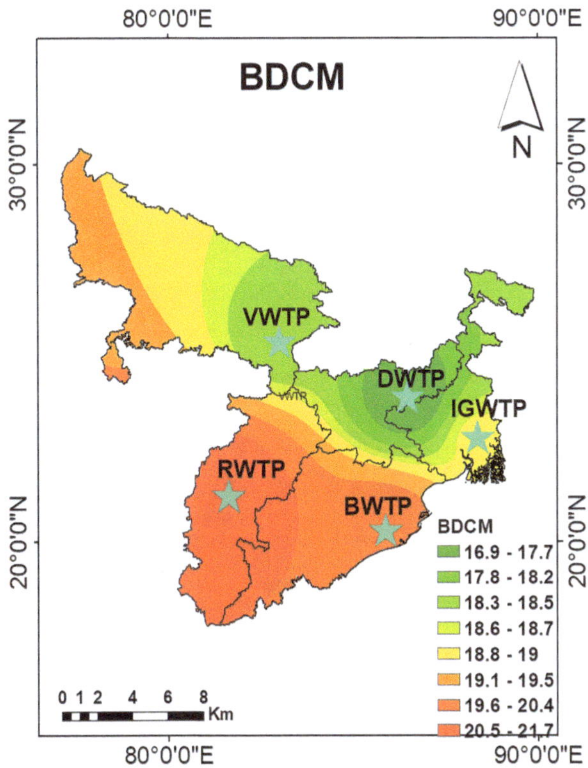

Fig. 3 Spatial distribution of BDCM

3.1.3 Spatial Distribution of DBCM

The formation of brominated THMs in drinking water is thought to be more harmful than chlorinated THMs (Bond et al. 2014). As previously discussed, the occurrence of Br^- in source water rapidly oxidized by free chlorine and formed HOBr, which is more reactive than HOCl. The concentration range of DBCM in all five WTPs varied from 8.45 to 15.52 µg/l. The spatial distribution of DBCM in all five WTPs is depicted in Fig. 4, below the acceptable limit (60 µg/l) of BIS (2012). Moreover, a higher concentration of DBCM was observed in VWTP, followed by RWTP > DWTP > IGWTP and > BWTP. The above findings are consistent with previous reports (Minashree Kumari 2014; Chowdhury et al. 2010).

Overall, the GIS spatial distribution of THMs species indicated that the CF is the principal compound that contributes the highest THMs concentration, followed by BDCM and DBCM in all five selected WTPs. The variation may be due to differences in the availability of raw water sources, organic precursors, a dose of chlorine, operational parameters, geographical location, and, ultimately, the treatment process (Mahato and Gupta 2022; Chowdhury et al. 2010).

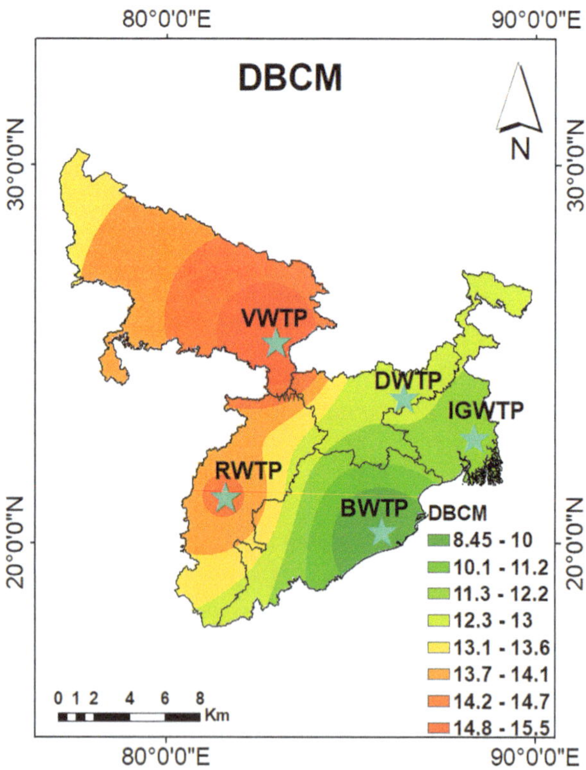

Fig. 4 Spatial distribution of DBCM

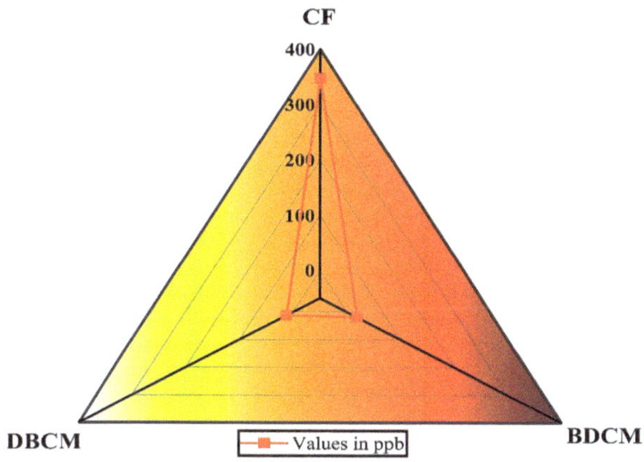

Fig. 5 Comparative of magnitudes of THMs species

3.2 Comparative of Magnitudes THMs Species

A radar plot cluster observation was performed for the identification of the comparative magnitudes of various THMs species (Fig. 5). The radar plot suggested that CF is the most predominant compound by BDCM and DBCM in the drinking water supplies all five selected WTPs. It was also reported in the literature that the occurrence of BDL bromide ions tends to form more chlorinated THMs (Thokchom et al. 2020). The results of the present study are consistent with the observation of sensitivity analysis. Various previous literature also supports the observation of this study (Karim et al. 2011; Danish and Ahmad 2018; Mahato and Gupta 2022).

4 Conclusions

Disinfection by chlorine in drinking water is probably the major source for the formation of THMs compounds due to the presence of NOM. The study investigated the GIS-based spatial distribution of THMs species in the drinking water supplies of five major cities in India. As a whole, the average concentration of CF (350.28 μg/l) was higher than the prescribed guideline value of BIS (200 μg/l), USEPA (80 μg/l), and WHO (200 μg/l). The radar plot comparative of magnitudes analysis also revealed that CF is the most predominant THMs compound, followed by BDCM (19.20 μg/l) and DBCM (12.53 μg/l), and are not exceeded the standard value of various regulatory agencies. The variation in the THMs range value in the chlorinated drinking water supplies of all five selected WTPs is attributed to the differences in the availability of raw water sources, organic precursors, a dose of chlorine, operational parameters, geographical location, and ultimately the treatment process.

The study is useful for the management authorities to reduce the harmful risk of THMs compounds in drinking water supplies.

References

Abdullah MP, Yew CH, bin Ramli MS (2003) Formation, modeling and validation of trihalomethanes (THM) in Malaysian drinking water: a case study in the districts of Tampin, Negeri Sembilan and Sabak Bernam, Selangorm Malaysia. Water Res 37(19):4637–4644

BIS Water–Specification, I.S.D (2012) Bureau of Indian standards. New Delhi, India, pp 1–12

Bond T, Huang J, Graham NJ, Templeton MR (2014) Examining the interrelationship between DOC, bromide and chlorine dose on DBP formation in drinking water—a case study. Sci Total Environ 470:469–479

Chakraborty R (2017) Issues of water in India and the health capability paradigm. Ethics in Science and Environmental Politics 17:41–50

Chowdhury S, Champagne P, McLellan PJ (2010) Investigating effects of bromide ions on trihalomethanes and developing model for predicting bromodichloromethane in drinking water. Water Res 44(7):2349–2359

Clark RM, Goodrich JA, Deininger RA (1986) Drinking water and cancer mortality. Sci Total Environ 53(3):153–172

Danish M, Ahmad T (2018) A review on utilization of wood biomass as a sustainable precursor for activated carbon production and application. Renew Sust Energ Rev 87:1–21

Hasan A, Thacker NP, Bassin J (2010) Trihalomethane formation potential in treated water supplies in urban metro city. Environ Monit Assess 168:489–497

Karim Z, Mumtaz M, Kamal T (2011) Evaluation of trihalomethanes in tap water samples of Karachi City (Pakistan). J Basic Appl Sci 7(1)

Karimi Pasandideh E, Kakavandi B, Nasseri S, Mahvi AH, Nabizadeh R, Esrafili A, Rezaei Kalantary R (2016) Silica-coated magnetite nanoparticles core-shell spheres (Fe 3 O 4@ SiO 2) for natural organic matter removal. J Environ Health Sci Eng 14:1–13

Krzeminski P, Vogelsang C, Meyn T, Köhler SJ, Poutanen H, de Wit HA, Uhl W (2019) Natural organic matter fractions and their removal in full-scale drinking water treatment under cold climate conditions in Nordic capitals. J Environ Manag 241:427–438

Mahato JK, Gupta SK (2020) Guideline development history of Trihalomethanes in drinking water–a review in Indian perspective. Int J Environ Earth Sci

Mahato JK, Gupta SK (2022) Exploring applicability of artificial intelligence and multivariate linear regression model for prediction of trihalomethanes in drinking water. Int J Environ Sci Technol 19(6):5275–5288

Minashree Kumari SG (2014) Factors influencing the formation of Trihalomethanes in drinking water supplies. In: Strategic technologies of complex environmental issues-a sustainable approach, vol 225.

Mishra BK, Gupta SK, Sinha A (2014) Human health risk analysis from disinfection by-products (DBPs) in drinking and bathing water of some Indian cities. J Environ Health Sci Eng 12:1–10

Park KY, Yu YJ, Yun SJ, Kweon JH (2019) Natural organic matter removal from algal-rich water and disinfection by-products formation potential reduction by powdered activated carbon adsorption. J Environ Manag 235:310–318

Rook JJ (1974) Formation of haloforms during chlorination of natural waters. J Soc Water Treat Exam 23:234–243. Find this article online

Sohn J, Amy G, Yoon Y (2006) Bromide ion incorporation into brominated disinfection by-products. Water Air Soil Pollut 174:265–277

Thokchom B, Radhapyari K, Dutta S (2020) Occurrence of trihalomethanes in drinking water of Indian states: a critical review. In: Disinfection by-products in drinking water, pp 83–107

Uyak V, Toroz I (2007) Investigation of bromide ion effects on disinfection by-products formation and speciation in an Istanbul water supply. J Hazard Mater 149(2):445–451
World Health Organization and WHO (2004) Guidelines for drinking-water quality, vol 1. World Health Organization
Zainudin FM, Hasan HA, Abdullah SRS (2018) An overview of the technology used to remove trihalomethane (THM), trihalomethane precursors, and trihalomethane formation potential (THMFP) from water and wastewater. J Ind Eng Chem 57:1–14

Recent Development in Nanotechnology for the Removal of Disinfection By-Products

Bhawana Raj, Rachna, and Umesh Kumar Singh

Abstract Access to inexpensive, clean water sources is one of the UN's Sustainable Development Goals. Rapid world population growth, climatic changes, high energy needs, and municipal and industrial pollution, which endangers freshwater reserves and a sustainable supply, have all contributed to the problem's exponential growth. Removal of disinfection by-products is another significant issue that requires an appropriate solution. When a disinfectant interacts with organic matter that is present in nature, man-made pollutants, bromide, and iodide ions, disinfection by-products (DBPs) are created. The demand for more modern water treatment has been motivated by DBPs, a chemical water treatment problem. New strategies and better materials are required by contemporary water treatment technologies for the management and methods of conservation of water resources. Water contamination is a major cause of the spread of many infectious diseases. The key challenge is to create novel technologies that can provide enough treatment facilities without producing harmful by-products. While many traditional water treatment technologies have been used to prevent the transmission of infectious diseases that are fatal and non-fatal through the water; together with other long-range relevant technology, nanotechnology is one such technique for removing pollutants from water by adsorption or nanoparticles that serve as catalysts for oxidation of chemical or photochemical for the removal of contaminants. Nanotechnology provides the potential of ensuring the long-term availability of clean drinking water and the conservation of water resources by utilizing cutting-edge materials and procedures for conservation, purifying water, and reuse. In this chapter, we examine recent developments in nanotechnologies for water filtration, as well as their prospects, practical difficulties with commercialization, and potential dangers.

B. Raj · Rachna
Department of Environmental Science, Central University of South Bihar, Gaya, Bihar, India

U. K. Singh (✉)
Department of Environmental Science, Central University of South Bihar, Gaya, Bihar, India

Centre of Environmental Studies, University of Allahabad, Prayagraj, Uttar Pradesh, India
e-mail: Umeshsingh@cusb.ac.in

S. Madhav et al. (eds.), *Drinking Water Disinfection By-products*,
https://doi.org/10.1007/978-3-031-49047-7_14

Keywords Nanotechnology · Disinfection by-products · Wastewater treatment · Natural organic matter

1 Introduction

Water is a crucial component of life. But due to the climate change threat, industrialization, and population increase, quality of water and sufficiency have become major issues. Poor quality water has a wide range of negative effects on human welfare in both the economic and social spheres. A significant public health victory of the twentieth century was the purification of drinking water. Water disinfection has been used for a long time and is mostly used to get rid of the microorganisms that cause illnesses that are linked to water. Disinfection is the process of using various chemical and physical techniques to kill or inactivate pathogenic bacteria, cysts, and other germs by 99.9%. A safe, clean, and sufficient supply of water is one of the fundamental building blocks of humanity, sustainability, and life's overall well-being (Shannon et al. 2008). Sterilization is the destruction of pathogenic bacteria and their spores, whereas disinfection is the destruction of pathogenic microorganisms but not their spores (Yoo 2018; Rutala and Weber 2008). Chlorine is the most often used disinfectant for purifying drinking water. Although chlorine benefits us, science has found a link between the DBPs disinfection by-products that are created when ozone, chlorine, chlorine dioxide, or chloramines are used to treat our water supply. These chemical disinfectants combine with organic substances that exist naturally to create a range of compounds that have been related to anomalies in human and lab animal reproduction and development as well as potential carcinogens (Bull et al. 2001; Boorman 1999; Nieuwenhuijsen et al. 2000). In general, all chemical disinfectants produce by-products that are harmful to health (DBPs). These DBPs could have a hydrophobic, hydrophilic, volatile, or non-volatile character. As a result, harmful disinfection by-products (DBPs) can be created. Since chlorine can combine with the organic substances (NOM) present during the disinfection process, this is a problem. By manipulating matter's size and shape, nanotechnology, which is a class of technologies that perform on the scale of the nanometer (i.e., between 1 and 100 nm), can establish materials, devices, and systems with essentially novel properties and functions (Mansoori et al. 2007; Ramsden 2008). One of the most effective options to modernize environmental cleanup methods is through all of this. Because of its potential uses in numerous industries, nanotechnology is gaining momentum on a global basis, including pharmaceutical products (Kiparissides and Kammona 2015; Müller et al. 2015), industries for food items (Duncan 2011; Singh and Upadhyay 2022), energy (Hussein 2015; Liu et al. 2011; Zhang et al. 2011) and treatment of pollution (Brame et al. 2011), offers opportunities for the advancement and transformation of traditional remediation methods that would otherwise be impossible.

2 Disinfectant By-Products

The process of disinfection, which gets rid of microbiological contamination that can lead to infections and pandemic illnesses, is the most crucial one in traditional drinking water treatment systems. Free chlorine, chloramines, ozone, and chlorine dioxide, which are all frequently used chemical disinfectants in the water business, can react with different elements in natural water to create DBPs, many of which are carcinogenic (Mazhar et al. 2020). The use of free chlorine is the classic strategy used most frequently for water and wastewater treatment (Kerwick et al. 2005). Because of its germicidal effects, chlorination has gained widespread acceptance in disinfection; however, the method has fallen behind due to the dangerous DBPs formation like trihalomethanes (THMs), chlorite, halo acetic acids (HAAs), chloropicrin, and chlorophenols (Simpson 2008). The amount of chlorine used, contact hours, pH of the solution, water, or wastewater, temperature, amount of naturally existing organic material present, and bromide ion presence all have an impact on the creation of DBP (Sadiq and Rodriguez 2004; Al-Omair and El-Sharkawy 2007). These hazardous DBPs may have an impact on human health that is carcinogenic, increasing the threat to bladder and also increasing risk of colon cancer (Cantor et al. 1985; Simpson 2008). Aldehydes, followed by ketoacids and carboxylic acids, are the most common non-halogenated organic DBPs produced during ozonation in the presence of organic matter that is naturally occurring. Ozonation creates hypobromous acid when both naturally occurring organic matter and bromide are present, which then causes the creation of brominated organ halogen compounds (e.g., bromoform). There is no direct synthesis of organ halogen DBPs; instead, the most prevalent chlorine dioxide DBPs are chlorite (ClO^{2-}) and chlorate ions. The main chlorine dioxide DBPs, in contrast to the other disinfectants, are produced by the disinfectant's breakdown rather than an interaction with precursors.

3 Types of Nanotechnology Used in the Wastewater Treatment Process

A few procedures are highly useful but largely unsuccessful in reducing DBP development. Nanoparticles are increasingly being used in all facets of life that has increased their potential for use in environmental pollution. Although research is being done on using nanoparticles to reduce environmental pollution, nanoparticle uses will become more significant globally in the future. Many different types of nanoparticles, including nanoparticles of metal oxide, nanoparticles of titanium oxide, and iron of nano-zero valent, are utilized in the disinfection of contaminated water (Sharma et al. 2009; Li et al. 2008; Kurniawan et al. 2012). Some of the effective water treatment methods and equipment made possible by nanotechnology include:

- Photocatalysis
- Nanofiltration
- Nano sorbents

3.1 *Photocatalysis*

The phenomenon of using light to lower a chemical reaction's activation energy or temperature is known as photocatalysis. An appealing alternative for treating water is to use advanced oxidation processes in conjunction with sunshine[61]. Photocatalysis is capable of degrading a wide range of organic and inorganic compounds, including organic acids, estrogens, pesticides, dyes, crude oil, microorganisms (including viruses and chloride-resistant species), and nitrous oxides. Metals like mercury can also be eliminated when paired with filtration or precipitation (Rickerby and Morrison 2007). Semiconductor photocatalysis has received a lot of interest as a potential solution to the world's energy dilemma and to stop environmental degradation (Tong et al. 2012). An appealing approach to water purification is photocatalysis, which uses metal oxide semiconductor nanostructures like oxide of zinc (ZnO), titania (TiO), oxide of tungsten (WO_3), zinc stannate (Zn_2SnO), and others to remove both chemical and biological impurities. Of the reported photocatalysts, TiO_2 is the most auspicious one for the development of the technique since it is the most energetic under UV irradiation, photo stabilize, non-poisonous, and affordable. To improve the sensitivity of light and performance of photocatalytic, TiO_2 photocatalyst should be changed via deposition of metal, doping of ion, dye preparation and clamping of TiO_2 with other materials. This is because the TiO_2 photocatalyst has certain drawbacks. After all, it restricts UV irradiation.

3.1.1 Metal Deposition

Noble metals (Ruthenium, Rhodium, Palladium, Silver, Platinum, Gold) have been used extensively as semiconductors' dopants because their Fermi level is situated at a level of energy underneath the conduction band of semiconductors. In order to facilitate the flow of electrons and prevent charge recombination, the photo-generated electron can be confined by inert metals that have been deposited on the surface. For the photo-electrochemical inactivation of microorganisms, the first inert metal is Pt which pairs with TiO_2. The optical absorption of TiO_2 was found to be extended to the visible light (VL) range by the Pt deposition. Using ceramic tiles coated in Pt/TiO_2, excellent photocatalytic cleaning activity toward E. coli was demonstrated. Another popular technique for photocatalytic water disinfection is to modify the surface of TiO_2 with silver (Ag). To maximize the usage of visible light, it utilizes both the surface plasmon resonance and the electron trapping action of Ag nanoparticles. The most efficient photocatalyst for photocatalytic disinfection of E. coli was Ag-TiO_2, according to the findings of experiments utilizing Pt- and

Ag-loaded TiO_2 loaded with Pt and Ag (use 0.5%–5% Pt or Ag as loading amounts). Investigations on the modification of TiO_2 by metals other than Ag, such as Cu and Zn, were also conducted. It revealed that Zn and Ag had a beneficial impact while Cu harmed the photocatalytic disinfection of bacteria.

3.1.2 Doping of Ion

TiO_2 doping with an iron of non-metals has shown tremendous potential in developing visible-light-driven photocatalysis because the dopants can form impurities that diminish the band gap. The most effective dopant in the field of photocatalytic disinfection is N_2. Rizzo et al. (2014) investigated the suppression of an antibiotic-resistant E. coli strain isolated from a biologically treated urban wastewater effluent using N-doped TiO_2 photocatalysis (Rizzo et al. 2014). When exposed to sunlight, the antibiotic-resistant E. coli strain was more easily inactivated by N-doped TiO_2 than by undoped TiO_2. Another efficient method of improving TiO_2's photocatalytic activity is sulfur (S) doping. In comparison with pure TiO_2, within 60 min, the created S-doped TiO_2 was seen to show powerful visible-light-driven bacterial killing action toward Micrococcus Lyle.

3.1.3 Dye Sensitization

It is well understood that dyes strongly absorb in the VL area. Once excited, the dye's excited states can introduce an electron into TiO_2's conduction band. As a result of O_2 scavenging the supplied electrons, $•O_2$ is eventually created, leading to the synthesis of •OH. At various times, a copper (II) phthalocyanine chloride solution was applied to the TiO_2 layer. The altered TiO_2 films demonstrated bacterial deactivation ability against E. coli under visible light, and the extent to which the dye was covered by the TiO_2 surface was proportional to that activity.

3.2 Nanofiltration

Filtration is one of the most frequent and important steps in the treatment of wastewater in which solid is separated from liquid by using filter membrane. Due to its distinctive charge-based abhorrence property and high rate of percolation, nanofiltration (NF) is an emerging technology, in the field of water treatment. It is one of the techniques based on membrane filtration, along with the sizes and types of particles which can be filtered out. NF is gaining popularity these days since it uses less energy and has low pressure needs (7–30 atm) than reverse osmosis (RO) procedures (20–100 atm) (Cadotte et al. 1988; Drewes et al. 2008). Since the NF membranes have characteristics in between those of RO membranes which are non-porous and those of porous filtration membranes on a small scale, the transfer in nanofiltration primarily happens by using diffusion mechanism; which is also use in the NF membranes and have a size exclusion property (Bowen and Mukhtar 1996). Using NF in the field of treatment of wastewater is comparatively unfamiliar, and many businesses, including those in the textile, pharmaceutical, dairy, petrochemical, and other sectors, are paying close attention to the approach (Bowen et al. 2002; Thanuttamavong et al. 2002). Nearly all organic and inorganic contaminants, including some dangerous bacteria, may be removed from wastewater with NF due to its different mechanism of filtration and the availability of different membranes types (Thanuttamavong et al. 2001; Thanuttamavong et al. 2002). The creation of NF membranes involves the use of the following nanoparticles: -.

3.2.1 Carbon Nanomaterials

Because of their simplicity in production, great mechanical robustness, and strong rejection ability, carbon nanotubes are among the most researched nanomaterials for membrane manufacturing. Carbon nanotube (CNT)-based membranes have been shown to have a high rate of pollutant rejection and good solvent permeability (Dong et al. 2010). Only water can flow through the CNTs' tiny pores (usually 1 nm to 10 nm in diameter), which obstructs chemical and biological pollutants. Some researchers demonstrated bacterial pollutants like E coli and polio virus are effective removed by using carbon nanotubes to eliminate various heavy hydrocarbons. In addition to size-selective exclusion, carbon Nano tubes have revealed a powerful capacity to adsorb various bio-chemical pollutants available in water (Richardson and Plewa 2020; Tamburri et al. 2005).

3.2.2 Metal Oxides

Another affordable option for the creation of NF membranes is metal oxides. In addition, most metal oxides exhibit photocatalytic activity when exposed to light. According to Zhang et al. 2008, a membrane made of titanium dioxide (TiO_2) nanowire mesh is used to filter wastewater to remove humic acid and total organic carbon (TOC) (Zhang et al. 2008). Under UV light irradiation, it can remove approx. 100% of humic acid and more than 90% of TOC by combining the photocatalytic and filtering capabilities of the TiO_2-based membrane. Additionally, the NF membranes demonstrated a notable holding time for organic color in a TiO_2 and gamma-alumina combination (Romanos et al. 2012). It has also been demonstrated that adding silica (SiO_2) nanoparticles to polymeric NF membranes enhances their thermal stability, hydrophilicity, and permeability capabilities (Jin et al. 2012; Singh et al. 2013).

3.2.3 Zeolites

Zeolites are crystalline aluminosilicate minerals that are microporous and have pores that range in size from sub-nanometer to nanometers. Zeolites are extremely ideal for use in RO and NF because of their outstanding stability of thermal and chemical brought on by the dormancy of the aluminosilicate crystals. According to studies (Gevers et al. 2005, 2006), zeolite-stuffed polydimethylsiloxane (PDMS) NF membranes perform exceptionally well in terms of conserving a variety of pollutants of organic nature, having good thermal stability, and minimizing swelling of PDMS network. Additionally, new complexes of metals and saturated/unsaturated hydrocarbons have been separated using zeolites (Turlan et al. 2001; Nikolakis et al. 2001).

3.3 Nano Sorbents

Sorption is the process through which two substances, sorbate and sorbent, interact physically or chemically and become adsorbate and sorbent, respectively. To separate contaminated water, sorbents are frequently utilized in water purification and treatment. The three common processes in the sorption of pollutants in water on the surface of a sorbent are the transport of pollutants from the water to the surface of an adsorbent, adsorption at the surface of an adsorbent, and transit inside the sorbent. Two crucial characteristics of nanoparticles make them particularly effective sorbents. In comparison with bulk particles, they have higher specific surface areas, and they are simple to functionalize with different chemical groups to boost their affinity for the desired pollutants. Additionally, the nanosized pores of Nano sorbents aid in the sorption of pollutants. The following is a description of some of the most popular nanomaterials used as absorbents.

3.3.1 Carbon Nano Sorbents

When it comes to the surface assimilation of several organic and inorganic contaminants in water, carbon nanoparticles have been widely exploited. Activated carbon is the most highly used carbon material out of these nanomaterials because of its high adsorption ability, stability of high thermal, great resistance to abrasion damage, and low cost. Granular activated carbon (GAC) has been utilized for the elimination of several organic pollutants from water (Fairey et al. 2006). According to research by Asenjo et al. (2011), benzene and toluene from industrial effluent have high adsorption capabilities on activated carbon, with 400–500 mgg^{-1} and 700 mgg^{-1}, respectively. Mercury (II), copper (II), cobalt (II), cadmium (II), chromium (III), nickel (II), lead (II), and chromium (VI) heavy metal ions were also discovered to be effective for removal by activated carbon (Kobya et al. 2005; Al-Omair and El-Sharkawy 2007). One-dimensional carbon nanotubes (CNTs) are also garnering much interest as an effective adsorbent material alongside activated carbon because of their huge particular surface area and exceptional thermo-chemical stability. In particular, just operationalizing the surface of the CNTs will allow you to adjust their reactivity (Tamburri et al. 2005; Moore et al. 2003; Hu et al. 2006). Hazardous materials like 1,2-dichlorobenzene, lead, and cadmium can be adsorbed on the surface of the water by using CNTs, investigated by Luan and colleagues (Li et al. 2007). The structure of the CNTs and their surface condition were found to have a significant impact on the adsorption of these pollutants and treating the CNTs with oxidants improved both their adsorption capacity and dispersibility. In a different study using CNTs, Kuo et al. (2008) discovered that the physisorption mechanism is responsible for dye adsorption on the surface of the CNTs. Both CNTs and activated carbon have elevated adsorption rates and capacities. They are also both chemically and thermally stable materials, which makes them excellent for the treatment of water procedures. Because of the small diameters, the CNTs and powdered activated carbon are difficult to completely separate from the water.

3.3.2 Bio Sorbents

Some organic contaminants have been shown to have very low concentrations, often in between picograms or nanograms per liter of water, which makes it difficult for them to be eliminated from the water body (Zhang et al. 2003). It has been discovered that bio sorbents, which are often made from biological or agricultural materials, hold the potential for the effective removal of such pollutants. Bio sorbents provide various benefits over conventional absorbents, including inexpensive, high productivity, less agro and bio sludge, no supplementary needs of nutrients, and the ability to regenerate. Heavy metal ions have been removed from water using bio sorbents (Ngah and Hanafiah 2008). Metal ions adsorbing via the amino acid group's chelation of chitosan, and chitosan-based sorbents have demonstrated encouraging findings with very effective adsorption capability for metal ions (Guibal 2004; Miretzky and Cirelli 2009).

3.3.3 Metal Oxide

Most frequently, manganese (Mn), titanium (Ti), silicon (Si), iron oxide (Fe), and tungsten (W) are utilized as adsorbents (W). Because they are affordable and simple to modify to change their adsorption capability and selectivity, metal oxides are useful as adsorbent materials. Recently, nano sorbents based on iron oxide have been researched to remove different organic contaminants in water (Zhong et al. 2006). Fe-oxide nano sorbents have magnetic properties which allow them to be magnetically isolated from the aqueous phase (Iram et al. 2010; Zhang et al. 2010). Additionally, the nanostructure of tungsten oxide (WO_2) has demonstrated a very high ability for the adsorption of organic dye in water (Jeon and Yong 2010). To successfully remove reactive yellow 84 dyes from wastewater effluents of several textile industries Talleb and colleagues developed a zinc-aluminum layer with double hydroxide nano sorbent (Kesraoui-Ouki et al. 1993) The removal of organic pollutants and heavy metals from wastewater has also shown positive results when non-metallic oxides, such as silica (SiO_2), are used as nano sorbents (El-Safty et al. 2012).

3.3.4 Zeolites

Zeolites are an appealing sorbent for the treatment of water because of their large particular surface area and strong ion exchange capacity. The majority of zeolites can be made economically and are found naturally. Dong et al. (2010) created hexadecyltrimethylammonium (HDTMA)-modified zeolite, which is used to adsorb phenol compounds from water. The HDTMA molecules create a double-layer micelle at the upper part of zeolite, boosting the nano sorbent's ability to adsorb substances. Heavy metal ion adsorption has also been done using zeolites (Ngah and Hanafiah 2008). Chabazite and clinoptilolite, two natural zeolites (Kesraoui-Ouki et al. 1993) are used for the adsorption of lead and cadmium and were examined by Pansini (1996). Using two naturally occurring zeolites that had been produced with NaOH, the authors demonstrated extraordinarily high sorption capabilities for lead (Pb) and cadmium (Cd), with metal removal efficiencies of more than 99%. High levels of porosity in zeolite boost its adsorption capacity, and it has the power to photo-catalytically decrease high-valence metal ions to minimize their toxicity.

3.3.5 Nano Zero Valent Iron (NZVI)

A novel nanomaterial called nonzero valent iron (NZVI) may purge both inorganic and organic contaminants from water. Due to the short lifetime of the highly reactive NZVI, numerous modifications of the material have been investigated, including NZVI with surface changes, NZVI emulsified for improved miscibility, NZVI bimetallic for increased reactivity, and NZVI on carbon support (better distribution). Heavy metals are either immediately adsorbed to the NZVI surface, where they become immobile, as is the case with Zn^{2+} and Cd^{2+}. or they are reduced at the NZVI surface, as is the case with Cu^{2+} and Ag^{2+}.

4 Toxicity of Disinfectant By-Product

Over 700 DBPs have been discovered so far, and of them, approximately 100 have been extensively studied for their traits, life cycles, and potential biological toxicity. Non-steroidal anti-inflammatory medicines, natural organic compounds, and anthropogenic contamination of emerging concern have all served as predecessors for the DBPs. Depending on the source of water and the disinfection technique DBPs varied significantly in their toxicity and had been proven to cause cytotoxicity, carcinogenicity, teratogenicity, and endocrine disruptions (Pansini et al. 1991). Different DBPs have shown different levels of toxicity when compared during the studies. For instance, I-DBPs have been discovered to be less poisonous than Br-DBPs, whereas CI-DBPs have been discovered to be the most toxic of these two. Even though I-DBPs are less hazardous than the other two it still inhibits glyceraldehyde-3-phosphate dehydrogenase (GAPDH) and produces oxidative stress in cells, which results in cytotoxicity and genomic DNA (Richardson and Plewa 2020). In the various structural DBP classes, ido-DBPs are the most hazardous, followed by bromo- and chloro-DBPs, in that order, whereas bromo-DBPs are more toxic than chlorine-DBPs. Interest in nitrogen-containing DBPs (N-DBPs) has been increased in recent years because it has proved to be more toxic than the carbonaceous DBPs (C-DBPs) (Plewa et al. 2008). Although it has been demonstrated that DBPs can contribute to oxidative stress in living organisms' cells and tissues, the relationship between fetal growth restriction and congenital abnormalities caused by DBP exposure during gestation is still not entirely clear, though it is thought to be related to it directly or indirectly (Pals et al. 2011). It is necessary to manage the disinfectant treatment process at different times of the year with a variety of dosages and with varied disinfectants to capture the toxicity of the water since the disinfectant can generate hazardous compounds depending on its precursors, nature, and quantity of doses (Monarca et al. 2000). Quantitative research on in vitro mammalian cells has demonstrated that regulated trihalomethanes (THMs) and halo acetic acids have less cytotoxic and genotoxic effects than do HANs, halo acetamide, and halonitromethanes (HAAs) (Komaki et al. 2014). Chloramines are thought to play a significant role in generating eye and respiratory system irritation. A study conducted on rats showed that Trichloramine inhalation can irritate rats. In another study on rats, 750 ppm of Trihalomethane-chloroform (TCM) and Trihalomethane (THM)-chloroform in the drinking water and 16.17 mg/kg of dose in a month have shown that THM in the drinking water responsible to cause toxicity in the liver, hyperplasia in the urinary bladder and metaplasia in the small intestine of the rat. Four disinfectant agents such as peracetic acid, chlorine, ozone, and UV radiation have shown harmful effects when tested on aquatic organisms (Da Costa et al. 2014). DBP toxicity has reportedly been shown to be exclusive to species like Scenedesmus sp. was the most susceptible to halo acetic acids, but D. magna was significantly affected by halo acetonitrile and trihalomethanes. In addition to selectivity, halogen atoms also contribute to toxicities. Potential reasons for toxicity of DBPs to microalgae include damage to photosystems and oxidative stress (Cui et al. 2021). The U.S. EPA regulates five halo acetic acids, which make up a significant class of drinking water disinfection by-products (DBPs). Iodo- > bromo- > chloroacetic acid is the rank order of the monohalogenated acetic acids (monoHAAs) in decreasing order of toxicity (Pals et al. 2011). 2,2,4-trichloro-5-methoxycyclopenta-4-ene-1,3-dione (TCMCD) in drinking water has been linked to genotoxicity and germplasm damage in human beings. An experiment conducted on Zebrafish with 100 mg/L at 96 h has 100% mortality of embryo development in the Zebrafish (Shen et al. 2010). Chlorinated DBPs have the potential to create epithelial cancers and alter cell proliferation by destroying nucleic acids, posing a risk to human health (Romanos et al. 2012). Fishes are chosen for most of the studies because of their brief embryonic periods and lifecycles. Zebrafish and danio rerio are one of the extensively studied fish species for in vivo toxicity evaluations, with an emphasis on mortality, mobility, and development because of their size, cheap cost, simplicity of handling and care, the quantity of information of embryogenesis mechanisms, daily breeding, and high fecundity. N-nitroso diethylamine (NDEA) bioaccumulated in the Zebrafish in the order of kidney>liver>gill and can cause oxidative damage to fish.

5 Nanotechnology for the Removal of DBP

Nanotechnology is an interdisciplinary technical method that is gaining attraction in water decontamination operations due to the unique qualities of nanoparticles such as enhanced reactivity, higher aspect ratio, and quantum properties. Several nanoparticles are now being employed in the treatment of DBPs. Carbon-based nanoparticles, metal oxide-based nanoparticles, titanium oxide-based nanoparticles, nano Zero valent iron, single component nanomaterial, and metal-nonmetal composites are some examples of these nanoparticles. Since Geim and colleagues extracted single-layer samples of graphene from graphite in 2004, it has sparked global interest due to its remarkable electrical, thermal, and mechanical characteristics (Li et al. 2015). The selective identification of DBP with the magnetic graphene oxide @ gold nanoparticles-molecular (MGO@AuNPs-MIPs) composite with MIP outer layer has been synthesized effectively. Because of MGO@AuNPs' unique mechanical features and extraordinarily wide area, the resultant MGO@AuNPs-MIPs composite has great selectivity and good sensitivity toward DBP molecules (Dong et al. 2010). In a study on graphene flakes, it has showed the adsorption of DBP, namely chlorine oxyanions where anion-π interacts with the π electron-rich graphene flakes and shows a directly proportional relationship between the graphene flakes and the adsorption energy (Li et al. 2006). Various natural and manufactured nanomaterials, including TiO_2, ZnO, and fuller, have demonstrated strong antibacterial capabilities by interfering with the bacterial cell envelope and energy transmission, as well as inhibiting enzyme activity and DNA synthesis (Li et al. 2006). UV/TiO_2 treatment can effectively degrade DBP precursors in less time and thus can decrease disinfection by-product formation potential (DBPfp) more quickly (Gora et al. 2018). Titanium dioxide is a photocatalyst that can remove DBP from water by first adsorbing the pollutant and then degrading it when exposed to UV light (Gora and Andrews 2017)· TiO_2 photocatalysis, in conjunction with the physical removal facilitated by the ceramic membrane filter, offers significant benefits over UV disinfection and minimizes the risk for DBPs formation after the chlorination (Gerrity et al. 2008). Dissolved organic nitrogen (DON) is known for producing carcinogenic nitrogenous disinfectant by-products (N-DBPs) that can be reduced significantly by the sunlight/Cu-TiO_2 system by the process of oxidation. Additionally, high-molecular-weight proteins can be degraded into low-molecular-weight proteins (Li et al. 2015). Numerous nanomaterials, including carbon-based nanomaterials, nano photocatalysts, and nano metallic particles, have demonstrated microbicidal effectiveness. Silver nanoparticles, zero-valent iron (ZVI) nanoparticles, nanostructured photocatalysts, membrane-based nanotechnology, and nano adsorbents are mostly used in marketed nanotechnology-enabled products to handle water and wastewater treatment challenges. Zerovalent iron nanoparticles are more suited for in-situ and ex-situ usage due to their mobility in porous media and reactive endurance. They also have greater reactivity and surface area than ZVI. With the use of zero-valent iron nanoparticles, chlorinated organic compounds with extremely hazardous characteristics due to their limited biodegradability, such as 1,4-chlorinated ethane, TCE, and carbon tetrachloride, can be halogenated (Tseng et al. 2011). Promising results have been obtained when dechlorination and synthetic granular activated carbon and zerovalent iron (GAC-ZVI) composites are used to couple adsorption and degradation of trichloroethylene (TCE) (Tseng et al. 2011). In a study, four disinfectants, namely resorcinol, phloroglucinol, 3-hydroxybenzoic acid, and tannic acid were removed with the help of nanofiltration membranes NF70 and NF270.

6 Future Scope

Nanotechnology has made substantial advances in practically every field of science and technology during the previous two decades. Even though nanotechnology offers significant benefits in the elimination of disinfection by-products, there is a lack of enthusiasm for scaling up these technologies due to a variety of barriers such as harmful residual effects if any, and production issues during scaling up. This challenge, however, can be overcome by using multidisciplinary techniques ranging from chemistry to economics. The use of nanotechnologies to remove DBPs is relatively new, requiring a detailed understanding of the desorption and adsorption processes on a large scale, as well as the release of metal ions from the nanoparticles into the environment and the disposal of nanoparticles after treatment. Also, more and more pilot-scale studies need to apply in the real field as the nanotechnology data can be sometimes site, condition, and matrix-specific, and results from laboratory studies may differ in the actual environment setup. The behavior and mechanism of nanotechnologies will need to be further studied in the future since they may act differently in different water bodies. For instance, TiO_2 nanoparticles are being studied extensively for their potential utility in the elimination of DBPs. Researchers are excited about the practical application of TiO_2 on DBPs to improve water quality based on the experimental results. Nanomaterial-assisted photocatalysis may also include significant technological problems for practical use. Aside from that, the total operating cost of nanotechnology-led water treatment methods will play a significant role in their future adoption. It can be recommended to estimate the costs of wastewater treatment in the actual world and compare them to the expenses of conventional wastewater treatment techniques. The creation of very simple and economical nanotechnologies will remain a research focus.

7 Conclusion

Advanced technologies for water quality management are in high demand and nanotechnology is one of the best solutions for ensuring excellent water quality and removing contaminants. Presently, nanotechnologies are playing an important role in the removal of DBPs and hence, attracting a lot of attention. In the future, graphene-based materials may be able to address water concerns sustainably. In comparison with the conventional methods of removal of DBPs, nanomaterials have special traits including a larger surface area to volume ratios, stronger surface reactivity, and semi conductivity. Nanoparticles such as TiO_2, nAg, MgO, ZnO, and have demonstrated considerable results in the removal of DBPs. The published research offered contradicting findings on the impacts of particle size and optimal dose, which would require more research to better understand, making it impossible to draw judgments about which may deliver the greatest benefits. Nanotechnologies have the potential to enhance all present DBP purification processes and in the development of solar-powered useful water purification devices.

References

Al-Omair MA, El-Sharkawy EA (2007) Removal of heavy metals via adsorption on activated carbon synthesized from solid wastes. Environ Technol 28(4):443–451

Asenjo NG, Álvarez P, Granda M, Blanco C, Santamaría R, Menéndez R (2011) High performance activated carbon for benzene/toluene adsorption from industrial wastewater. J Hazard Mater 192(3):1525–1532

Boorman GA (1999) Drinking water disinfection byproducts: review and approach to toxicity evaluation. Environ Health Perspect 107(suppl 1):207–217

Bowen WR, Mukhtar H (1996) Characterisation and prediction of separation performance of nanofiltration membranes. J Membr Sci 112(2):263–274

Bowen WR, Welfoot JS, Williams PM (2002) Linearized transport model for nanofiltration: development and assessment. AICHE J 48(4):760–773

Brame J, Li Q, Alvarez PJ (2011) Nanotechnology-enabled water treatment and reuse: emerging opportunities and challenges for developing countries. Trends Food Sci Technol 22(11):618–624

Bull RJ, Bull R, Krasner S, Daniel PA (2001) Health effects and occurrence of disinfection by-products. AWWA Research Foundation

Cadotte J, Forester R, Kim M, Petersen R, Stocker T (1988) Nanofiltration membranes broaden the use of membrane separation technology. Desalination 70(1–3):77–88

Cantor KP, Hoover R, Hartge P, Mason TJ, Silverman DT, Levin LI (1985) Drinking water source and risk of bladder cancer: a case-control study. In: Water chlorination: chemistry, environmental impact, and health effects, vol 5, pp 143–149

Cui H, Chen B, Jiang Y, Tao Y, Zhu X, Cai Z (2021) Toxicity of 17 disinfection by-products to different trophic levels of aquatic organisms: ecological risks and mechanisms. Environ Sci Technol 55(15):10534–10541

Da Costa JB, Rodgher S, Daniel LA, Espíndola ELG (2014) Toxicity on aquatic organisms exposed to secondary effluent disinfected with chlorine, peracetic acid, ozone and UV radiation. Ecotoxicology 23:1803–1813

Dong Y, Wu D, Chen X, Lin Y (2010) Adsorption of bisphenol a from water by surfactant-modified zeolite. J Colloid Interface Sci 348(2):585–590

Drewes J, Bellona C, Xu P, Amy GL, Filteau G, Oelker G (2008) Comparing nanofiltration and reverse osmosis for treating recycled water. International Water Association (IWA)

Duncan TV (2011) Applications of nanotechnology in food packaging and food safety: barrier materials, antimicrobials and sensors. J Colloid Interface Sci 363(1):1–24

El-Safty SA, Shahat A, Ismael M (2012) Mesoporous aluminosilica monoliths for the adsorptive removal of small organic pollutants. J Hazard Mater 201:23–32

Fairey JL, Speitel GE, Katz LE (2006) Impact of natural organic matter on monochloramine reduction by granular activated carbon: the role of porosity and electrostatic surface properties. Environ Sci Technol 40(13):4268–4273

Gerrity D, Ryu H, Crittenden J, Abbaszadegan M (2008) Photocatalytic inactivation of viruses using titanium dioxide nanoparticles and low-pressure UV light. J Environ Sci Health A 43(11):1261–1270

Gevers LE, Vankelecom IF, Jacobs PA (2005) Zeolite filled polydimethylsiloxane (PDMS) as an improved membrane for solvent-resistant nanofiltration (SRNF). Chem Commun 19:2500–2502

Gevers LE, Vankelecom IF, Jacobs PA (2006) Solvent-resistant nanofiltration with filled polydimethylsiloxane (PDMS) membranes. J Membr Sci 278(1–2):199–204

Gora S, Liang R, Zhou YN, Andrews S (2018) Settleable engineered titanium dioxide nanomaterials for the removal of natural organic matter from drinking water. Chem Eng J 334:638–649

Gora SL, Andrews SA (2017) Adsorption of natural organic matter and disinfection byproduct precursors from surface water onto TiO2 nanoparticles: pH effects, isotherm modelling and implications for using TiO2 for drinking water treatment. Chemosphere 174:363–370

Guibal E (2004) Interactions of metal ions with chitosan-based sorbents: a review. Sep Purif Technol 38(1):43–74

Hu X, Wang T, Qu X, Dong S (2006) In situ synthesis and characterization of multiwalled carbon nanotube/au nanoparticle composite materials. J Phys Chem B 110(2):853–857

Hussein AK (2015) Applications of nanotechnology in renewable energies—a comprehensive overview and understanding. Renew Sust Energ Rev 42:460–476

Iram M, Guo C, Guan Y, Ishfaq A, Liu H (2010) Adsorption and magnetic removal of neutral red dye from aqueous solution using Fe3O4 hollow nanospheres. J Hazard Mater 181(1–3):1039–1050

Jeon S, Yong K (2010) Morphology-controlled synthesis of highly adsorptive tungsten oxide nanostructures and their application to water treatment. J Mater Chem 20(45):10146–10151

Jin L, Shi W, Yu S, Yi X, Sun N, Ma C, Liu Y (2012) Preparation and characterization of a novel PA-SiO2 nanofiltration membrane for raw water treatment. Desalination 298:34–41

Kerwick MI, Reddy SM, Chamberlain AHL, Holt DM (2005) Electrochemical disinfection, an environmentally acceptable method of drinking water disinfection? Electrochim Acta 50(25–26):5270–5277

Kesraoui-Ouki S, Cheeseman C, Perry R (1993) Effects of conditioning and treatment of chabazite and clinoptilolite prior to lead and cadmium removal. Environ Sci Technol 27(6):1108–1116

Kiparissides C, Kammona O (2015) Nanotechnology advances in diagnostics, drug delivery, and regenerative medicine. In: The nano-micro interface: bridging the micro and nano worlds, pp 311–340

Kobya M, Demirbas E, Senturk E, Ince M (2005) Adsorption of heavy metal ions from aqueous solutions by activated carbon prepared from apricot stone. Bioresour Technol 96(13):1518–1521

Komaki Y, Mariñas BJ, Plewa MJ (2014) Toxicity of drinking water disinfection byproducts: cell cycle alterations induced by the monohaloacetonitriles. Environ Sci Technol 48(19):11662–11669

Kuo CY, Wu CH, Wu JY (2008) Adsorption of direct dyes from aqueous solutions by carbon nanotubes: determination of equilibrium, kinetics and thermodynamics parameters. J Colloid Interface Sci 327(2):308–315

Kurniawan TA, Sillanpää ME, Sillanpää M (2012) Nanoadsorbents for remediation of aquatic environment: local and practical solutions for global water pollution problems. Crit Rev Environ Sci Technol 42(12):1233–1295

Li Q, Mahendra S, Lyon DY, Brunet L, Liga MV, Li D, Alvarez PJ (2008) Antimicrobial nanomaterials for water disinfection and microbial control: potential applications and implications. Water Res 42(18):4591–4602

Li X, Wang X, Li L, Duan H, Luo C (2015) Electrochemical sensor based on magnetic graphene oxide@ gold nanoparticles-molecular imprinted polymers for determination of dibutyl phthalate. Talanta 131:354–360

Li XQ, Elliott DW, Zhang WX (2006) Zero-valent iron nanoparticles for abatement of environmental pollutants: materials and engineering aspects. Crit Rev Solid State Mater Sci 31(4):111–122

Li YH, Zhao YM, Hu WB, Ahmad I, Zhu YQ, Peng XJ, Luan ZK (2007, March) Carbon nanotubes-the promising adsorbent in wastewater treatment. In: Journal of physics: conference series (Vol. 61, No. 1, p. 698). IOP Publishing

Liu Z, Bai H, Xu S, Sun DD (2011) Hierarchical CuO/ZnO "corn-like" architecture for photocatalytic hydrogen generation. Int J Hydrog Energy 36(21):13473–13480

Mansoori GA, Mohazzabi P, McCormack P, Jabbari S (2007) Nanotechnology in cancer prevention, detection and treatment: bright future lies ahead. World Rev Sci Technol Sustain Develop 4(2–3):226–257

Mazhar MA, Khan NA, Ahmed S, Khan AH, Hussain A, Rahisuddin, Changani F, Yousefi M, Ahmadi S, Vambol V (2020) Chlorination disinfection by-products in municipal drinking water–a review. J Clean Prod 273:123159

Miretzky P, Cirelli AF (2009) Hg (II) removal from water by chitosan and chitosan derivatives: a review. J Hazard Mater 167(1–3):10–23

Monarca S, Feretti D, Collivignarelli C, Guzzella L, Zerbini I, Bertanza G, Pedrazzani R (2000) The influence of different disinfectants on mutagenicity and toxicity of urban wastewater. Water Res 34(17):4261–4269

Moore VC, Strano MS, Haroz EH, Hauge RH, Smalley RE, Schmidt J, Talmon Y (2003) Individually suspended single-walled carbon nanotubes in various surfactants. Nano Lett 3(10):1379–1382

Müller B, Zumbuehl A, Walter MA, Pfohl T, Cattin PC, Huwyler J, Hieber SE (2015) Translational medicine: nanoscience and nanotechnology to improve patient care. In: The nano-micro interface: bridging the micro and nano worlds, pp 289–310

Ngah WW, Hanafiah MM (2008) Removal of heavy metal ions from wastewater by chemically modified plant wastes as adsorbents: a review. Bioresour Technol 99(10):3935–3948

Nieuwenhuijsen MJ, Toledano MB, Eaton NE, Fawell J, Elliott P (2000) Chlorination disinfection byproducts in water and their association with adverse reproductive outcomes: a review. Occup Environ Med 57(2):73–85

Nikolakis V, Xomeritakis G, Abibi A, Dickson M, Tsapatsis M, Vlachos DG (2001) Growth of a faujasite-type zeolite membrane and its application in the separation of saturated/unsaturated hydrocarbon mixtures. J Membr Sci 184(2):209–219

Pals JA, Ang JK, Wagner ED, Plewa MJ (2011) Biological mechanism for the toxicity of haloacetic acid drinking water disinfection byproducts. Environ Sci Technol 45(13):5791–5797

Pansini M (1996) Natural zeolites as cation exchangers for environmental protection. Mineralium deposita 31:563–575

Pansini M, Colella C, De Gennaro M (1991) Chromium removal from water by ion exchange using zeolite. Desalination 83(1–3):145–157

Plewa MJ, Wagner ED, Muellner MG, Hsu KM, Richardson SD (2008) Comparative mammalian cell toxicity of N-DBPs and C-DBPs
Ramsden J (2008) Essentials of nanotechnology. BookBoon
Richardson SD, Plewa MJ (2020) To regulate or not to regulate? What to do with more toxic disinfection by-products? J Environ Chem Eng 8(4):103939
Rickerby D, Morrison M (2007) Report from the workshop on nanotechnologies for environmental remediation
Rizzo L, Sannino D, Vaiano V, Sacco O, Scarpa A, Pietrogiacomi D (2014) Effect of solar simulated N-doped TiO2 photocatalysis on the inactivation and antibiotic resistance of an E. Coli strain in biologically treated urban wastewater. Appl Catal B Environ 144:369–378
Romanos GE, Athanasekou CP, Katsaros FK, Kanellopoulos NK, Dionysiou DD, Likodimos V, Falaras P (2012) Double-side active TiO2-modified nanofiltration membranes in continuous flow photocatalytic reactors for effective water purification. J Hazard Mater 211:304–316
Rutala WA, Weber DJ (2008) Guideline for disinfection and sterilization in healthcare facilities
Sadiq R, Rodriguez MJ (2004) Disinfection by-products (DBPs) in drinking water and predictive models for their occurrence: a review. Sci Total Environ 321(1–3):21–46
Shannon MA, Bohn PW, Elimelech M, Georgiadis JG, Marinas BJ, Mayes AM (2008) Science and technology for water purification in the coming decades. Nature 452(7185):301–310
Sharma YC, Srivastava V, Singh VK, Kaul SN, Weng CH (2009) Nano-adsorbents for the removal of metallic pollutants from water and wastewater. Environ Technol 30(6):583–609
Shen X, Shao Z, Xian Q, Zou H, Gao S, Zhang J (2010) Detection method and toxicity study of a new disinfection by-product, 2, 2, 4-trichloro-5-methoxycyclopenta-4-ene-1, 3-dione (TCMCD), in chlorinated drinking water. Water Res 44(3):974–980
Simpson DR (2008) Biofilm processes in biologically active carbon water purification. Water Res 42(12):2839–2848
Singh AK, Pandey RP, Jasti A, Shahi VK (2013) Self-assembled silica nanocrystal-based antibiofouling nanofilter membranes. RSC Adv 3(2):458–467
Singh V, Upadhyay S (2022) Nanotechnological revolution: carbon-based nanomaterials detecting pollutants. In: Nanotechnology for environmental pollution decontamination. Apple Academic Press, pp 101–134
Tamburri E, Orlanducci S, Terranova ML, Valentini F, Palleschi G, Curulli A, Brunetti F, Passeri D, Alippi A, Rossi M (2005) Modulation of electrical properties in single-walled carbon nanotube/conducting polymer composites. Carbon 43(6):1213–1221
Thanuttamavong M, Oh JI, Yamamoto K, Urase T (2001) Comparison between rejection characteristics of natural organic matter and inorganic salts in ultra low pressure nanofiltration for drinking water production. Water Sci Technol Water Supply 1(5–6):77–90
Thanuttamavong M, Yamamoto K, Oh JI, Choo KH, Choi SJ (2002) Rejection characteristics of organic and inorganic pollutants by ultra low-pressure nanofiltration of surface water for drinking water treatment. Desalination 145(1–3):257–264
Tong H, Ouyang S, Bi Y, Umezawa N, Oshikiri M, Ye J (2012) Nano-photocatalytic materials: possibilities and challenges. Adv Mater 24(2):229–251
Tseng HH, Su JG, Liang C (2011) Synthesis of granular activated carbon/zero valent iron composites for simultaneous adsorption/dechlorination of trichloroethylene. J Hazard Mater 192(2):500–506
Turlan D, Urriolabeitia EP, Navarro R, Royo C, Menendez M, Santamaria J (2001) Separation of Pd complexes from a homogeneous solution using zeolite membranes. Chem Commun 24:2608–2609
Yoo JH (2018) Review of disinfection and sterilization–back to the basics. Infect Chemother 50(2):101–109
Zhang L, Song X, Liu X, Yang L, Pan F, Lv J (2011) Studies on the removal of tetracycline by multi-walled carbon nanotubes. Chem Eng J 178:26–33

Zhang S, Niu H, Hu Z, Cai Y, Shi Y (2010) Preparation of carbon coated Fe3O4 nanoparticles and their application for solid-phase extraction of polycyclic aromatic hydrocarbons from environmental water samples. J Chromatogr A 1217(29):4757–4764
Zhang X, Du AJ, Lee P, Sun DD, Leckie JO (2008) TiO2 nanowire membrane for concurrent filtration and photocatalytic oxidation of humic acid in water. J Membr Sci 313(1–2):44–51
Zhang ZL, Hong HS, Zhou JL, Huang J, Yu G (2003) Fate and assessment of persistent organic pollutants in water and sediment from Minjiang River estuary, Southeast China. Chemosphere 52(9):1423–1430
Zhong LS, Hu JS, Liang HP, Cao AM, Song WG, Wan LJ (2006) Self-assembled 3D flowerlike iron oxide nanostructures and their application in water treatment. Adv Mater 18(18):2426–2431

SPRINGER NATURE

GPSR Compliance

The European Union's (EU) General Product Safety Regulation (GPSR) is a set of rules that requires consumer products to be safe and our obligations to ensure this.

If you have any concerns about our products, you can contact us on ProductSafety@springernature.com

In case Publisher is established outside the EU, the EU authorized representative is:

Springer Nature Customer Service Center GmbH
Europaplatz 3
69115 Heidelberg, Germany

Kolibri 360 GmbH
Ferdinand-Jühlke-Straße 7
99095 Erfurt, Deutschland
produktsicherheit@kolibri360.de